CISM COURSES AND LECTURES

The series presents lecture notes, monographs, edited works and proceedings in the field of Mechanics, Engineering, Computer Science and Applied Mathematics.
Purpose of the series in to make known in the international scientific and technical community results obtained in some of the activities organized by CISM, the International Centre for Mechanical Sciences.

INTERNATIONAL CENTRE FOR MECHANICAL SCIENCES

COURSES AND LECTURES - No. 352

DIAGNOSTICS OF ROTATING MACHINES IN POWER PLANTS

Proceedings of the CISM/IFToMM Symposium
October 27-29, 1993
Udine, Italy

EDITED BY

G. DIANA
TECHNICAL UNIVERSITY OF MILAN

SPRINGER-VERLAG WIEN GMBH

Le spese di stampa di questo volume sono in parte coperte da
contributi del Consiglio Nazionale delle Ricerche.

This volume contains 256 illustrations

Originally published by Springer Verlag Wien-New York in 1994

In order to make this volume available as economically and as rapidly as possible the authors' typescripts have been reproduced in their original forms. This method unfortunately has its typographical limitations but it is hoped that they in no way distract the reader.

ISBN 978-3-211-82613-3 ISBN 978-3-7091-2706-3 (eBook)
DOI 10.1007/978-3-7091-2706-3

CONTENTS

IDENTIFICATION TECHNIQUES IN ROTORDYNAMICS

R. Nordmann
University of Kaiserslautern, Kaiserslautern, Germany

In order to find out the condition of a rotating machine during operation usually vibrations (output signals) are measured at specified machine locations. **Indications** can be determined from these signals, which allow a **diagnosis of possible failures** (fault parameters) based on the support of expert knowledge and/or good rotordynamic models. However, the success of failure diagnostic would probably be much better, if in addition to the output signals more information would be available about rotor excitation forces. This leads to the idea of using **Identification Techniques** with measured input and output signals. Frequency response functions can than be calculated and an extraction of physical and/or modal parameters, expressing the actual system condition, is possible. This contribution describes the basic relations of **Identification Techniques in rotordynamics**, including rotordynamic models, measurement and excitation methods and parameter estimation procedures. Some applications of parameter identification are presented together with the corresponding test procedures and test apparatus.

INTRODUCTION -DIAGNOSTICS BY MEANS OF MODELS

Rotating machinery systems like turbines, pumps, compressors are used in different areas of modern technique. Users of such machines expect that they run safe and reliable and have good efficiency and high availability. In order to fulfill these requirements mechanical problems have to be considered very careful. Particularly the mechanical shaft-vibrations in connection with strains and stresses of rotating and nonrotating machine parts are of great importance for a machines life time. Bending vibrations for example are caused by unbalance, process forces or selfexcitation-mechanisms like oil-film bearings, seals, aerodynamic forces, internal friction etc.

These undesired vibrations should be investigated in an early stage of the machine design. Well developed computer simulation tools are available today, in order to predict the machines dynamic behaviour and to avoid trouble after commissioning and start up. These tools are mostly based on the Finite Element Method and allow to include all important components (shaft, bearings, seals, pedestals foundations) with their corresponding mechanical effects. By simulation the following informations about the rotordynamic characteristics can be found (Fig. 1), e.g. for the case of bending vibrations:

- **Natural Frequencies, damping, mode shapes, stability**
- **Unbalance Vibrations**
- **Transient Vibrations**

Such results are not only important for the machine design. They can also be very helpful for an evaluation of the machines dynamic behaviour and correspondingly the machines mechanical condition during operation. Therefore it seems to be obvious to work also with a rotordynamic model as a **diagnostic model** in order to find out possible **machine malfunctions** with respect to location and specification.

The idea of using rotordynamic "**diagnostic models**" is based on the fact, that machine failures like unbalance, worn out clearances, shaft cracks etc. have a large influence on the rotordynamic behaviour. That means on the other hand that measured vibration signals of typical machine parts (bearing pedestals, shafts etc.) can give hints about failures by comparison with reference data and **identification** of possible **fault parameters** (Fig. 1).

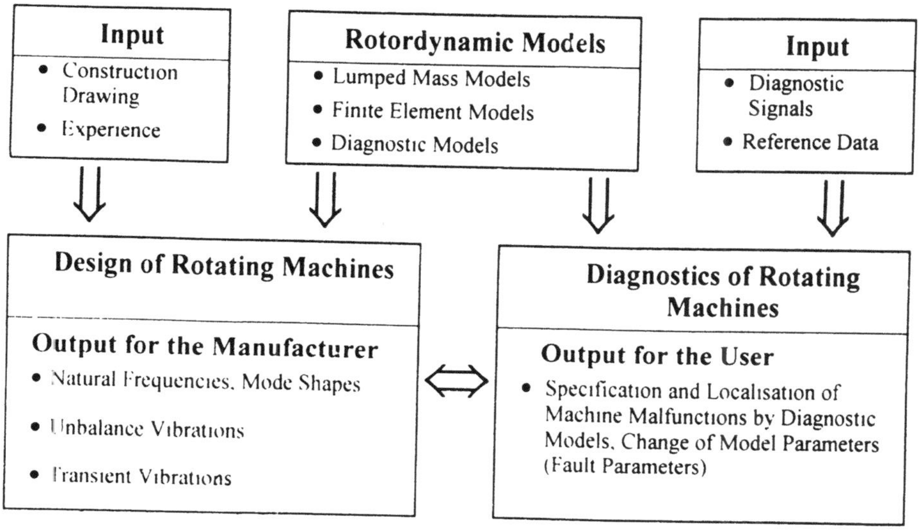

Fig. 1: Models for the Design and Diagnostics of Rotating Machines

The use of vibration signals in the procedure of **failure diagnostics** requires to define **indications**, that can be processed from the signals but can also be determined from rotordynamic models. Such **indications** are for example **time signals** and **orbits**, **frequency spectral data**, **run up** or **run down curves** and **transfer functions** (see Fig. 2).

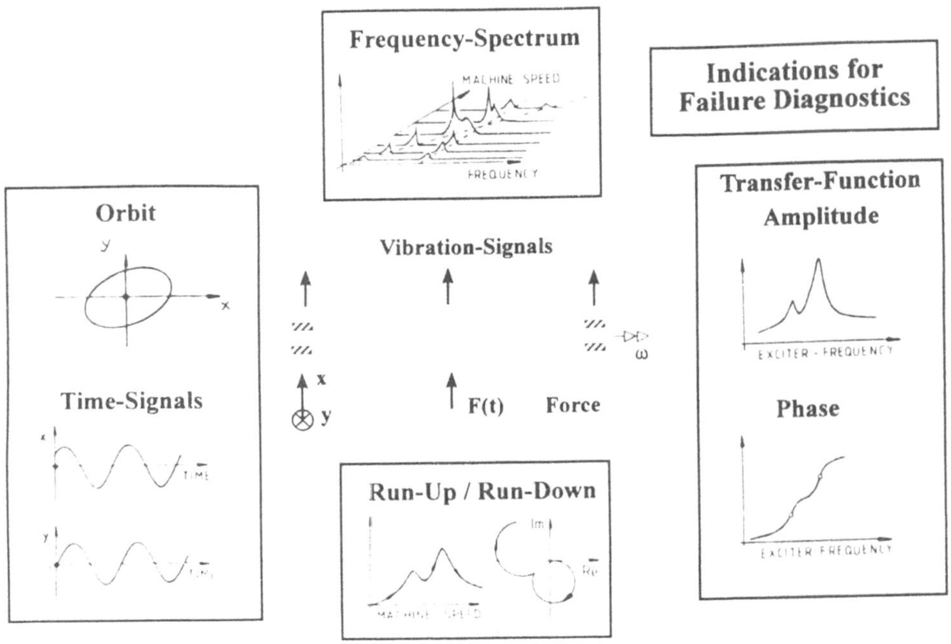

Fig. 2: Indications for Failure Diagnostics in Rotating Machinery

With todays modern tools of **sensors** and **signal processing techniques** they can easily be found. However, the indications have to be interpreted in comparison with reference data. For this very important task of failure diagnostics the knowledge of experts in rotordynamics is needed. Figure 3 points out for a simple example how an expert would work in this procedure. By considering the **indications** or **failure symptoms** (orbits and frequency spectra) explanations are given for the observed behaviour (failure diagnostics), based on the knowledge of the rotordynamic behaviour in case of typical effects (Unbalance, rubbing, self-excited vibrations etc.).

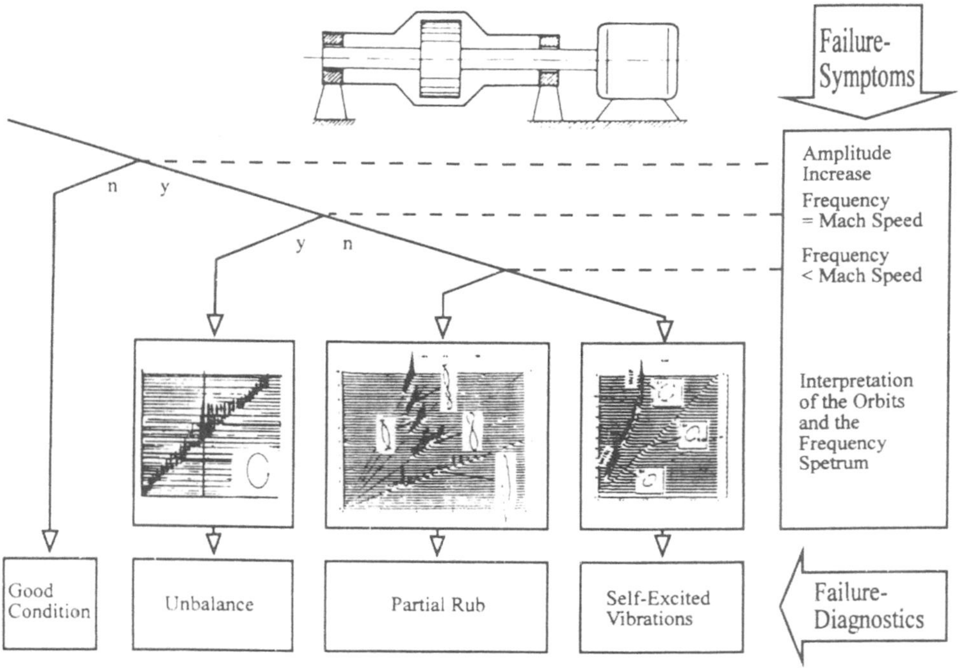

Fig.3: Failure Symptoms and Failure Diagnostics

Rotordynamic models, either used for the simulation in the **design process** or in **diagnosis**, are based on input-output relations, as pointed out in Figure 4. In a rotating machinery system input quantities are usually forces like unbalance forces, process forces, stochastic forces or also test forces Due to these time dependent disturbances the system reacts with vibrations (output signals). This may be vibrations of the shaft, the bearings or other components. It is known, that the output quantities are a function of the input, but they do also depend on the dynamic characteristics (mass, damping, stiffness) of the rotating machinery system with its special operating conditions.

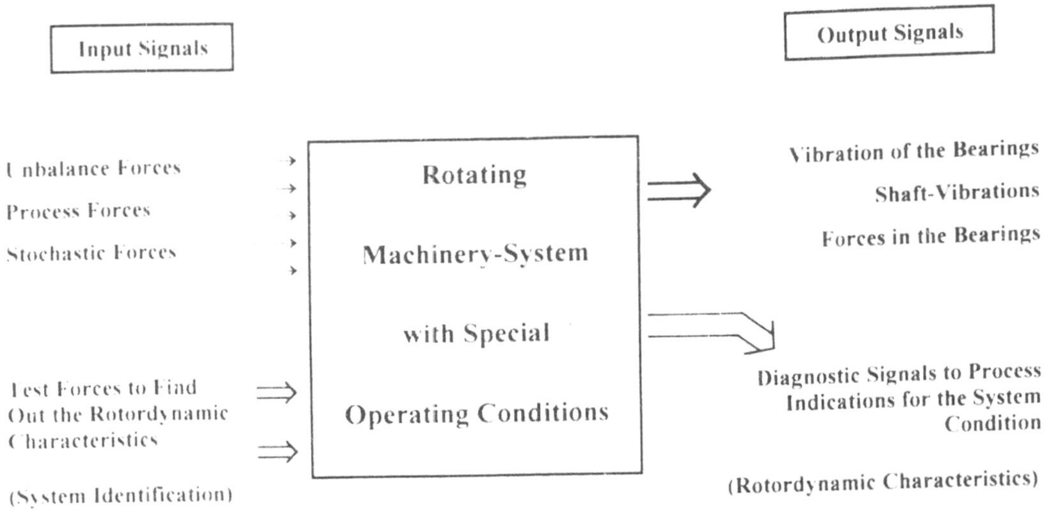

Fig 4. Rotating Machinery System with Input- and Output-Signals

Using a rotordynamic model in the design process means, that output signals are calculated, when the model characteristics and the input data as well are known or assumed.

In the **diagnostic** procedure usually a part of the overall output signals are measured by sensors and defined as **diagnostic signals**. After processing these signals well defined **indications** are obtained. The diagnostic method evaluates deviations of the indications by means of diagnostic models in order to give a message about possible **machinery malfunctions** and **fault parameters** (Figure 5). One of the problems in this evaluation is, to find out whether the changes of output signals are caused by changing input signals (e.g. unbalance) or by variation of some system parameters (mass, damping, system), which characterize the dynamic behaviour and the machine condition.

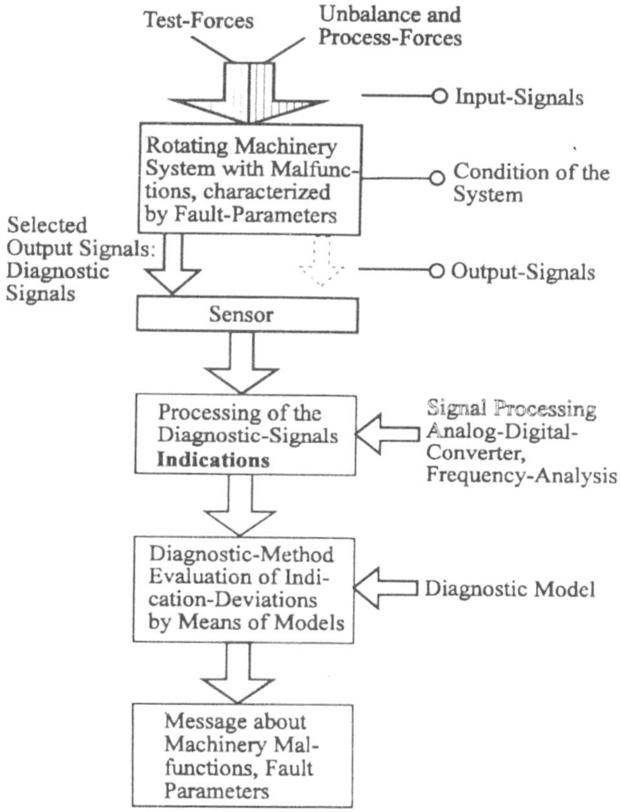

Fig.5: Failure Diagnostic Procedure with Diagnostic Model

To avoid these uncertainties it would be helpful for diagnostics, to investigate the two influences (input, system dynamic characteristics) on the output signals seperately. One possibility to find out system fault parameters in the sense of dynamic characteristics would be, to use **Identification Techniques in Rotating Machinery**. This techniques use measured input and output signals in order to identify the dynamic characteristics.

IDENTIFICATION OF ROTORDYNAMIC CHARACTERISTICS

Dynamic Characteristics of Rotors

For many rotating machines the dynamic behaviour can be described by means of linear models. In these models the relations between input- and output quantities are given in terms of differential equations, expressing the dynamic equilibrium of inertia-, damping-, stiffness and external forces. The **physical parameters** mass, damping and stiffness assembled in the matrices **M**, **D**, **K**, are representative for the **dynamic behaviour.** (Fig. 6). An equivalent possibility to characterize the dynamics of a system is to use **modal parameters**: natural frequencies ω_j, damping values α_j and mode shapes Φ_j. Input-output relations in the frequency domain are known as **complex frequency response functions**. They can be subdivided in frequency dependent compliances $\overline{H}_{k\ell}(\Omega)$ (flexibilities) and stiffnesses $\overline{K}_{k\ell}(\Omega)$ and represent another form for the rotordynamic behaviour.

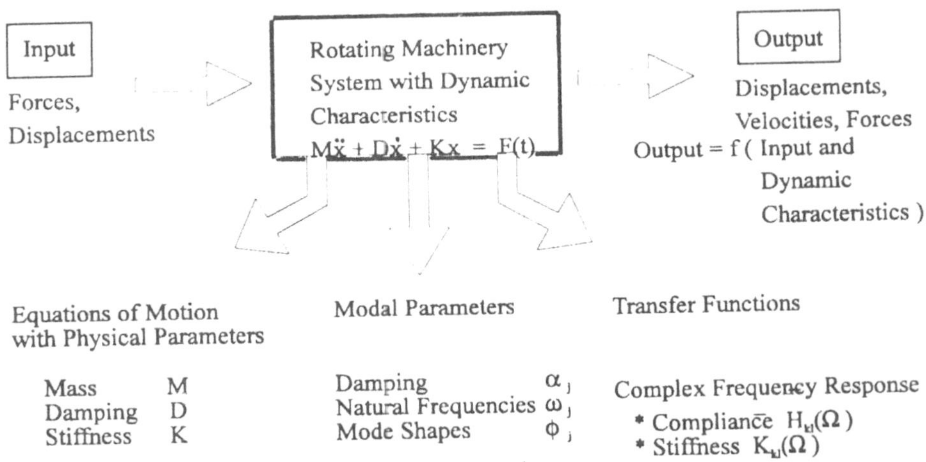

Fig.6: Dynamic Characteristics of a Rotordynamic System

To know the physical parameters, the modal parameters or the frequency response functions is an important assumption to evaluate the condition of a dynamic system. By observing these quantities an early detection of system failures is possible. To perform this an identification of the dynamic characteristics during operation is neccessary.

Identification in this sense means to excite a rotating machine artificially during operation and to pick up the corresponding system response (Fig. 7). From the measured input and output signals the dynamic characteristics can be calculated by means of well known input/output relationships in the time or frequency domain.

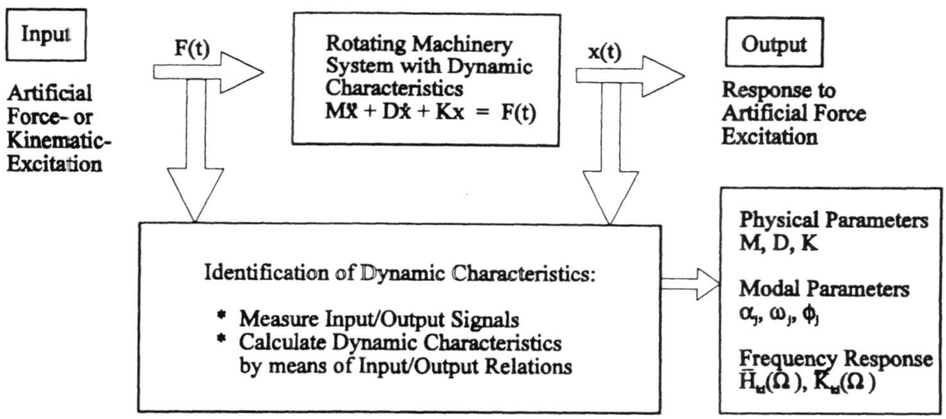

Fig.7: Identification of Dynamic Characteristic of a Rotating Machinery System

Identification procedures to determine the physical or modal parameters of a rotordynamic system consist of different tasks (Fig. 8). At first a **model structure** for the investigated system has to be defined, e.g. a linear model described by differential equations. With assumed system parameters input-output functions, e.g. **impulse response** functions (time domain) or **frequency response** functions (frequency domain) can be calculated. Such functions can also be measured based on the **measurements of input/output** signals with a following signal processing. Finally in the **parameter estimation** task, the selected functions of the model are fitted to the corresponding measured functions by variation of the physical or modal parameters in the model. This iterative procedure of improving the system parameters is interrupted if the correlation of model and measurement is good enough. In the following some basic considerations are discussed for the three subtasks of identification.

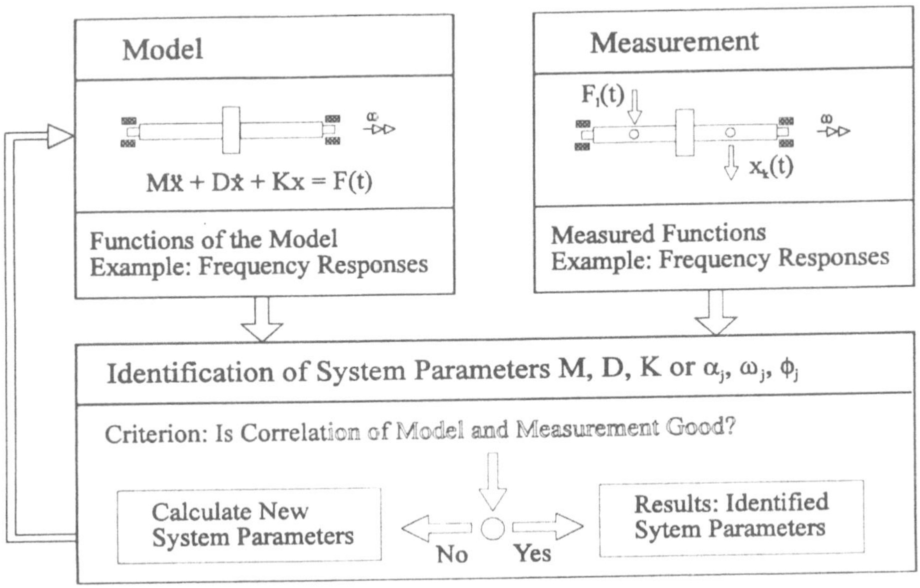

Fig.8: Identification Procedure for the Parameters of the System

Rotordynamic Model

A very important assumption for a successful identification of system parameters is the selection of the correct model. A model is characterized by its structure and its parameters. In many applications a priori knowledge is present for the structure of an actual system. In this case identification concentrates on the determination of the unknown parameters (Parameter Identification).

Different questions have to be checked before getting the best suited model for a special application:

- Is the system **linear** or **nonlinear**;
- Are the **parameters** dependent on **time** or **constant** (shaft with unequal moments of inertia, cracked shaft)?
- Do the system **matrices M, D, K** depend on the **rotational speed** (gyroscopic effects, oil film coefficients)?
- Are the system **matrices M, D, K symmetric**, **skewsymmetric** or **nonsymmetric** (gyroscopic effects, oil film bearings, seals)?
- Is the number of **degrees of freedom** high or low?

In many cases of rotating machinery systems the model structure has the following form

$$\mathbf{M}\ddot{\mathbf{x}} + \mathbf{D}(\omega)\dot{\mathbf{x}} + \mathbf{K}(\omega)\mathbf{x} = \mathbf{F}(t) \tag{1}$$

which represents a time invariant linear model with nonsymmetric matrices **D** and **K**, depending on the rotational speed ω.

After definition of the model the best suited input/output relations for the identification have to be found. Possible functions are impulse response functions and convolution integrals in the time domain or complex stiffness and compliance functions in the frequency domain.

As an example the complex dynamic stiffness matrix $\overline{\mathbf{K}}(\Omega)$ can directly be expressed in terms of the system matrices

$$\begin{aligned} \left(\mathbf{K} - \Omega^2\mathbf{M} + j\Omega\mathbf{D}\right)\hat{\mathbf{x}}(\Omega) &= \hat{\mathbf{F}}(\Omega) \\ \overline{\mathbf{K}}(\Omega) &= \mathbf{K} - \Omega^2\mathbf{M} + j\Omega\mathbf{D} \end{aligned} \tag{2}$$

where Ω is the frequency of excitation and $\hat{\mathbf{x}}, \hat{\mathbf{F}}$ are complex amplitudes. It can be shown, that a matrix element $\overline{K}_{k\ell}$ can simply be presented by the elements $K_{k\ell}, D_{k\ell}, M_{k\ell}$ (see Figure 9).

Harmonic Input: $\mathbf{F}(t) = \hat{\mathbf{F}} e^{j\Omega t}$

Harmonic Output: $\mathbf{x}(t) = \hat{\mathbf{x}} e^{j\Omega t}$

$$\left[\mathbf{K} - \Omega^2\mathbf{M} + j\Omega\mathbf{D}\right] \cdot \hat{\mathbf{x}}(\Omega) = \hat{\mathbf{F}}(\Omega)$$

$$\boxed{\overline{\mathbf{K}}(\Omega) = \mathbf{K} - \Omega^2\mathbf{M} + j\Omega\mathbf{D}}$$

For each Matrix Element:

$$\overline{K}_{k\ell}(\Omega) = \underbrace{K_{k\ell} - \Omega^2 M_{k\ell}} + j\Omega \underbrace{D_{k\ell}}$$

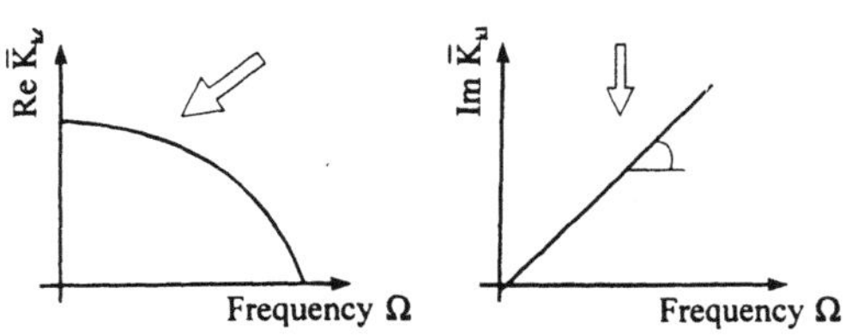

Fig. 9: Real and Imaginary Part of Complex Dynamic Stiffness

The inverse function of $\overline{\mathbf{K}}(\Omega)$ is the complex dynamic compliance

$$\left[\mathbf{K} - \Omega^2\mathbf{M} + j\Omega\mathbf{D}\right]^{-1} \hat{\mathbf{F}}(\Omega) = \hat{\mathbf{x}}(\Omega)$$
$$\overline{\mathbf{H}}(\Omega) = \left[\mathbf{K} - \Omega^2\mathbf{M} + j\Omega\mathbf{D}\right]^{-1} \tag{3}$$

The relations between $\overline{\mathbf{H}}(\Omega)$ and **K**, **D**, **M** are now nonlinear. However, compared to $\overline{\mathbf{K}}(\Omega)$ the compliance functions $\overline{\mathbf{H}}(\Omega)$ can be measured more easily.

In **Modal Analysis** usually $H_{k\ell}(\Omega)$-functions are used. They depend in the following way on the modal parameters

$$H_{k\ell}(\Omega) = \frac{\hat{x}_k(\Omega)}{\hat{\bar{F}}_\ell(\Omega)} = \sum_i \frac{\varphi_{ki}\psi_{\ell i}}{j\Omega - \lambda_i} \tag{4}$$

where $\lambda_i = \alpha_i + j\omega_i$ are the complex eigenvalues and φ_{ki}, $\psi_{\ell i}$ complex right hand and left hand eigenvector components.

Measurement Techniques

The objectives of the measurement task are to excite a rotordynamic system artificially by force or kinematic excitation, to measure input and output signals and to process functions that are used later on for the parameter estimation. In case of linear systems frequency domain functions are often used (Figure 10).

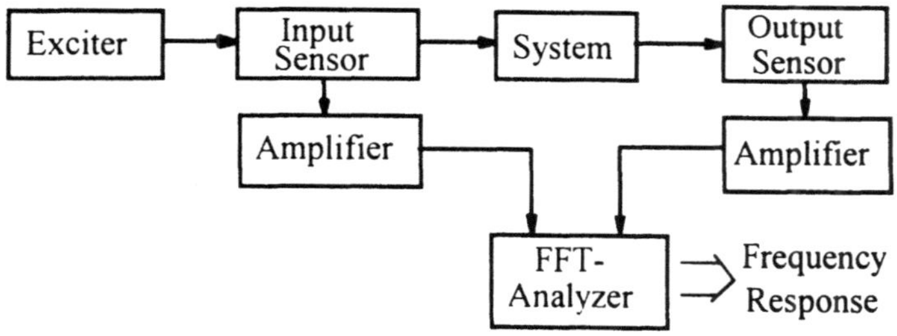

Fig. 10: Test Configuration to measure Frequency Response Functions

Figure 11 points out the main differences between measuring compliance functions (force excitation) or stiffness functions (kinematic excitation). The compliance $H_{k\ell}(\Omega)$ is defined as output $\hat{x}_k(\Omega)$ divided by the force of excitation $\hat{F}_k(\Omega)$, where all other forces are considered to be zero.

On the other hand the relation $\overline{K}_{k\ell}(\Omega) = \dfrac{\hat{F}_k}{\hat{x}_\ell(\Omega)}$ is found from the forces $\hat{F}_k$.that have to be applied when the system is kinematicly excited by $\hat{x}_k(\Omega)$ only and all other displacements are set equal to zero.

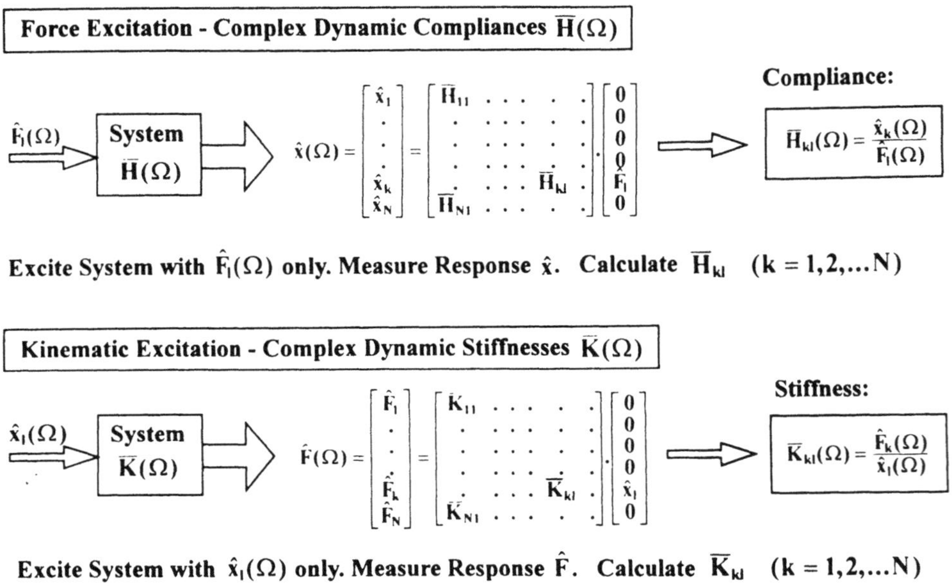

Fig. 11: Measurements of Dynamic Compliances and Stiffnesses

One of the main problems to apply identification techniques in rotordynamics is the excitation of a rotating structure during operation. It is neither easy to have access to the rotor nor to measure forces, especially when the signal to noise ratio is not good.

Different equipments for excitation have been developed and used for identification either at real machines or at small test rigs. Some of them are schematically shown in Figure

12. They include the **preload of a shaft** with a following snap back (Morton), the **hammer impact method** (Nordmann, Schöllhorn, Lund, Tonnesen), **unbalance excitation** via a second shaft with different running speed (Muszynska, Bently), shaking the **shaft via a rider** (Roger, Ewins) and the use of active **magnetic bearings** (Neumer, Matros, Nordmann). The excitation signal depends strongly on the possibilities of the exciter type. Figure 13 points out the main characteristics, advantages and disadvantages of some typical signals like sinusoidal, impulse, random and force preload. It is also shown which of the equipments can be used in order to realize special signal types. One of the advantages of the **magnetic bearing** exciter is, that different signal types can be generated and that a **contactless force excitation** is possible.

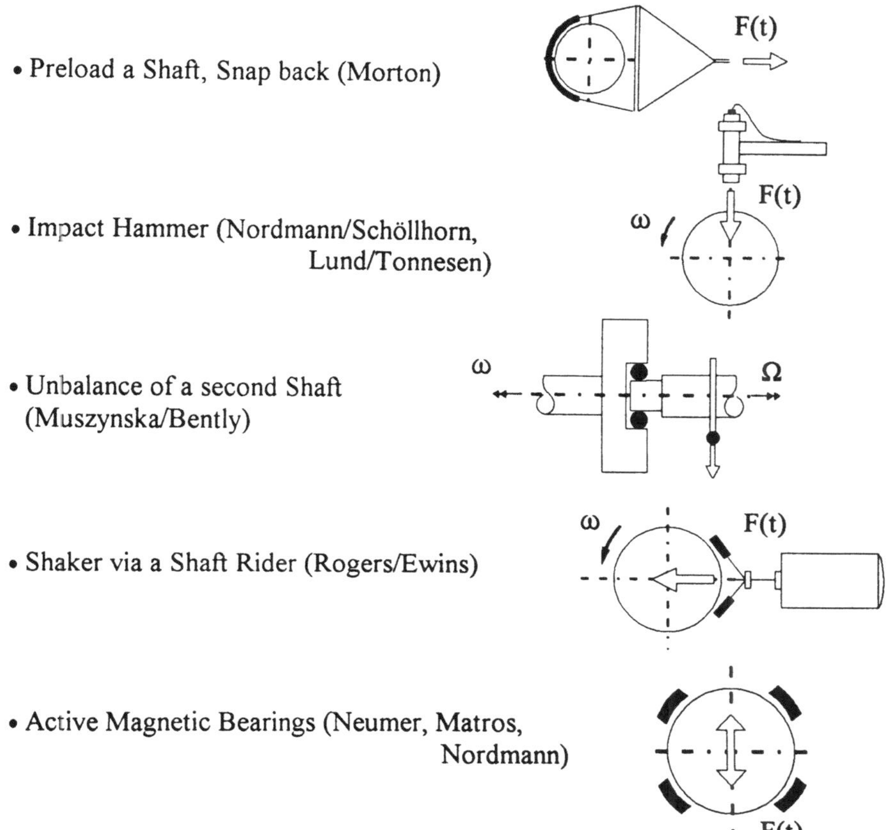

Fig. 12: Excitation Equipments for Rotating Machinery

Sinusoidal (Unbalance, Shaker, Magnetic Bearings, Hydraulic Exciter)

High Energy in one Frequency
Signal to Noise Ratio is high
Effort to excite many Frequencies
Suited to find Nonlinearities
Very often used: Swept sine

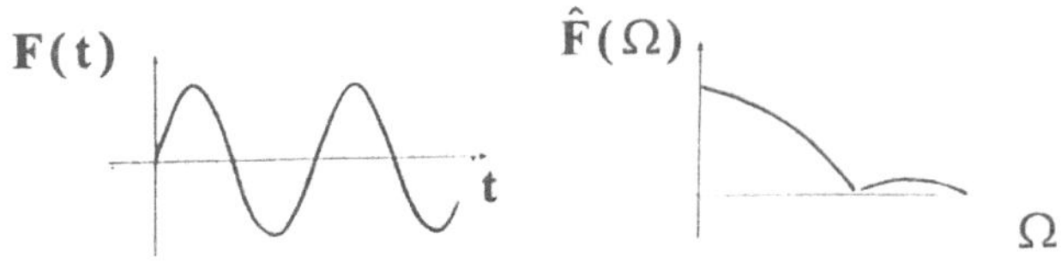

Impulse (Impact Hammer)

Low Energy in one Frequency
Signal to Noise Ratio is Low (Accuracy)
Fast Procedure (Broadband, Easy to move)
May Exceed Linear Range

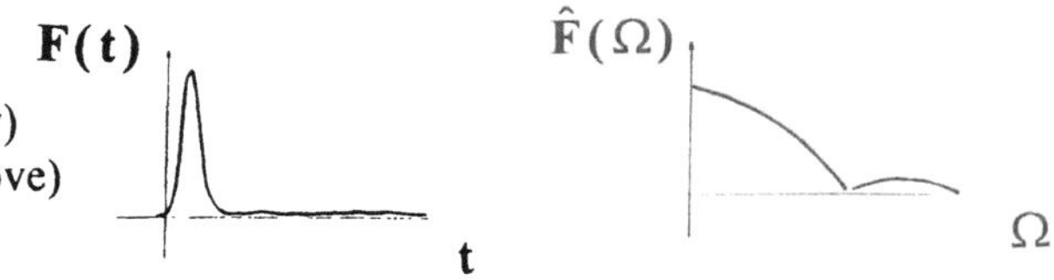

Random (Shaker, Magnetic Bearings, Hydraulic Exciter)

Frequency Spectrum constant
Signal to Noise Ratio is good
Relative fast Procedure (Broadband)
Exciter Equipments necessary
Difficult to move

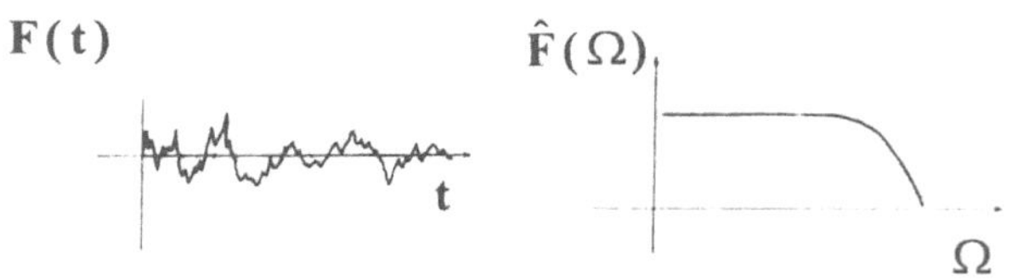

Force-Preload (Snap back)

High Energy in Low Frequency Range
Signal to Noise Ratio good only for low Freq
Broadband Excitation
Special Equipment required
Difficult to move

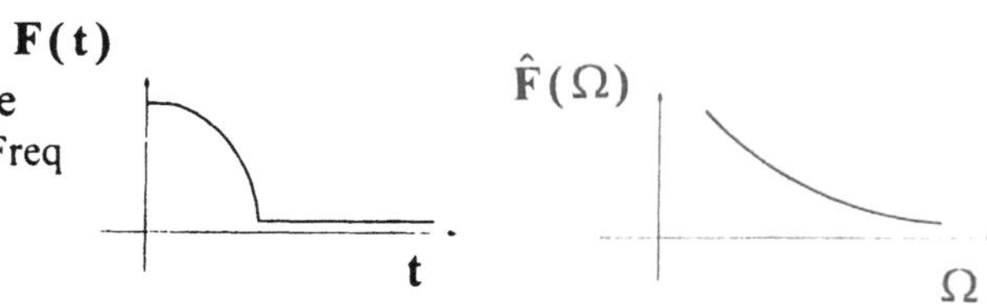

Fig. 13: Typical Signals for Excitation

Parameter Estimation

When input/output functions of the system have been measured in the frequency or time domain, corresponding model functions are fitted to the measured ones in order to estimate the unknown physical or modal parameters. In general **parameter estimation** procedures like **least squares** method, **instrumental variables**, **maximum likelihood** can be used for this task (Figure 14). When applying one of these methods it is necessary to consider the different possibilities with respect to the measured data (time or frequency domain), the signal processing (online or offline), the available numerical algorithms (direct, linear, iterative, nonlinear) and the type of error (input, output or equation error).

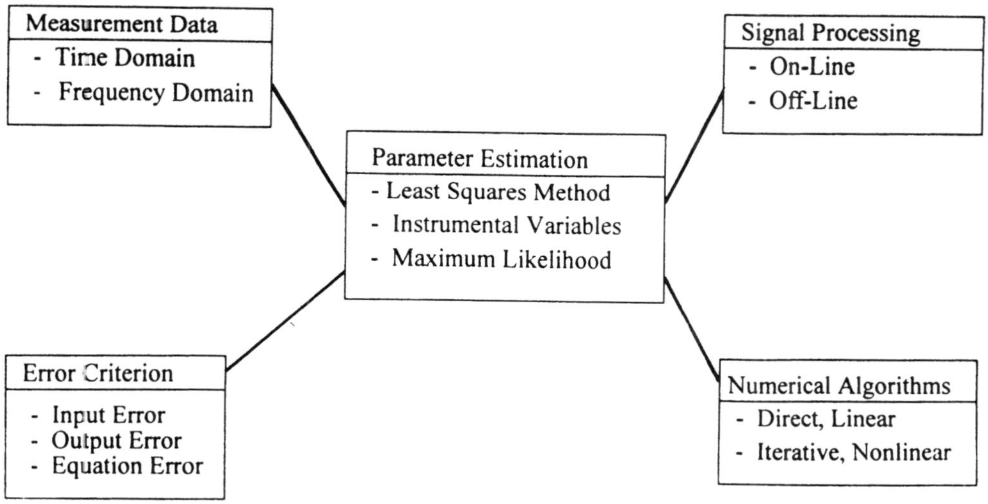

Fig. 14: Parameter Estimation Procedures

EXAMPLES - IDENTIFICATION TECHNIQUES IN ROTORDYNAMICS

With a few exceptions most of the known identification techniques in rotordynamics were used at laboratorium test rigs. The following cases are only some typical examples and do not include all known applications that were presented in the literature.

Morton /1/ was one of the first, who identified dynamic parameters of a full operating turbomachine by means of a **broadband excitation** technique (Figure 15). In this case the rotating **shaft is preloaded** with a static force via an additional **foil bearing**. Due to a breaking link the preload is suddenly taken off and the shaft is excited by a **force jump**, which is measured with a strain gange. The method was successfully used to determine rotordynamic coefficients of oil film bearings from measured frequency response functions.

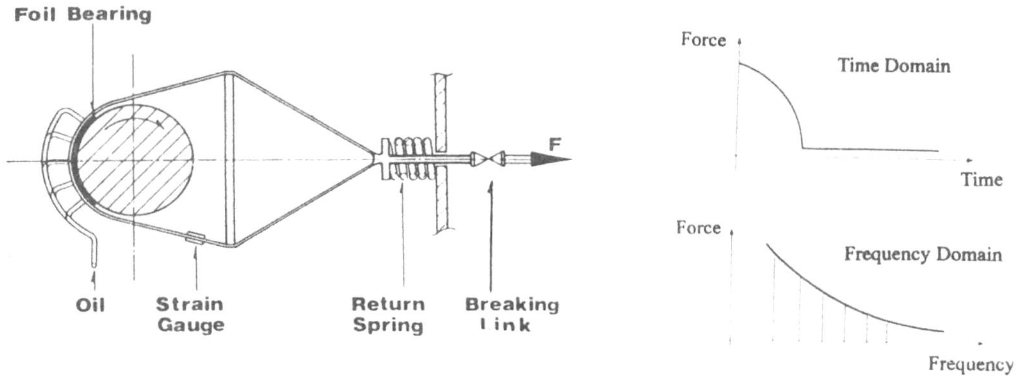

Fig. 15: Identification Technique by Morton

The **impulse technique** method using an impact hammer was tested by **Nordmann/Schöllhorn** /2/ and **Lund/Tonnesen** (3) as well. The operating shaft is excited by a hammer impact which is also a **broadband excitation**. Force and displacement response are measured with suited sensors in the time domain and transferred to the frequency domain by means of FFT, leading to complex **compliance functions** (Figure 16). These functions contain the modal parameters of the rotating systems (see Equ. 4), that can be determined by parameter estimation. Figure 16 shows some results in terms of natural frequencies, damping and mode shapes of a **Laval-Rotor**, supported in **oil film bearings** /2/. It can be seen that the system may become unstable, with increasing running speed.

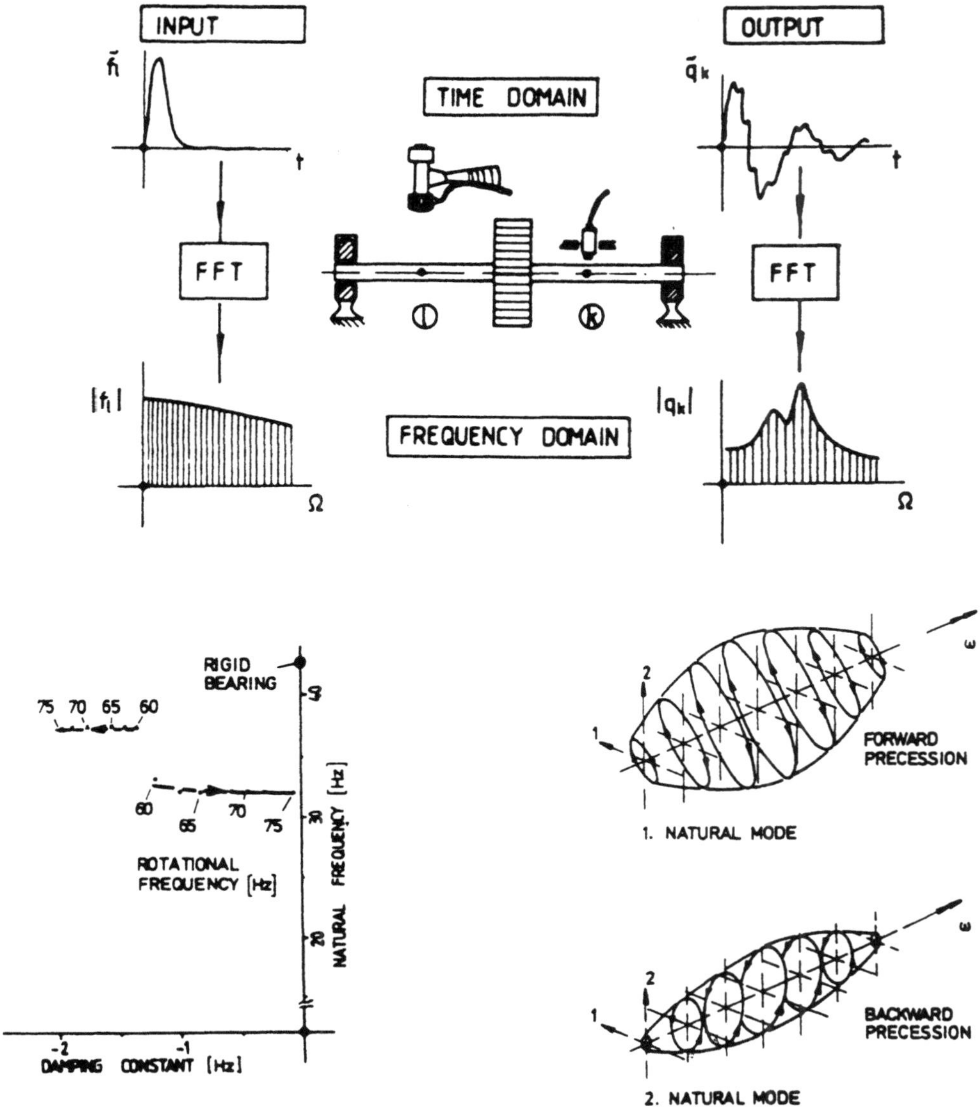

Fig. 16: Identification Technique by Nordmann/Schöllhorn

One of the disadvantages of the broadband excitation is the fact, that the energy is distributed over a frequency range. This may lead to poor results, if the signal to noise ratio is not high enough.

In opposite to the previous procedures **Muszynska/Bently** /4/ excite a rotor system by a sinusoidal force, coming from the **unbalance excitation** of a **second shaft** running with a different rotational speed compared to the investigated rotor (nonsynchronous excitation). The unbalance force is transferred in a connecting point between the two rotors (Fig. 17). The authors applied the method at test rigs in order to identify **modal parameters** and **fluid dynamic coefficients** (seals, bearings) as well. Both compliance and stiffness frequency functions were used in these tests. The procedure offers the advantage of having the complete excitation energy in one frequency line. However, the procedure is time consuming due to many test runs for various frequencies of excitation.

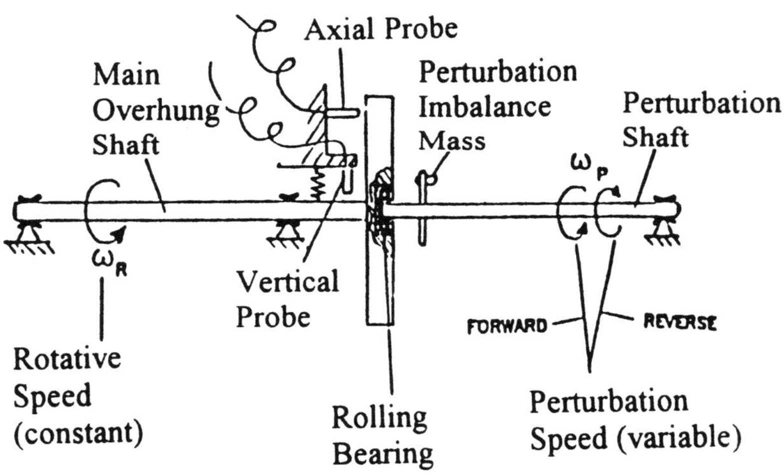

Fig. 17: Identification Technique by Muszynska/Bently

In the test rig of **Kanki/Kawakami /5/ seal dynamic coefficients** were identified by an excitation of the test rig casing with two **hydraulic actuator** systems working in two perpendicular directions (Figure 18). Again, from the measured quantities sinusoidal force and displacement frequency response functions are calculated and the dynamic coefficients are estimated.

Besides force preload, hammer impact, unbalance, electrodynamic and hydraulic shaker systems **magnetic bearings** are possible **modern actuators**, that can be used in identification procedures for rotating machinery.

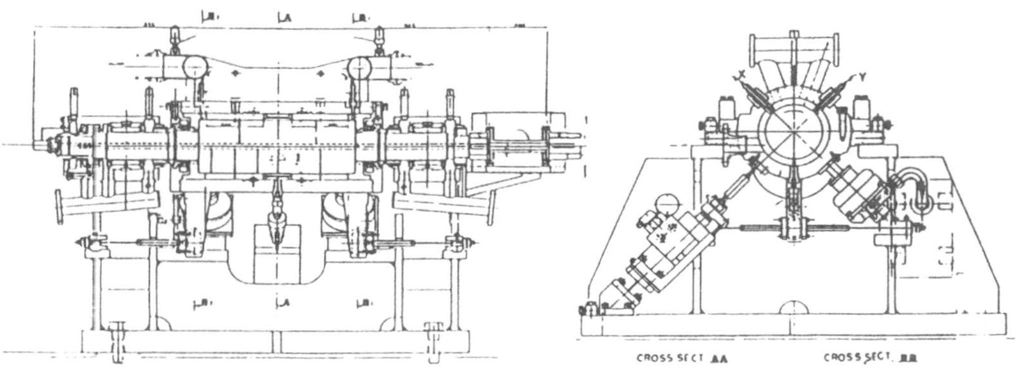

Fig. 18: Identification Technique by Kanki/Kawakami

Nordmann/Matros/Neumer /6/ use this technique for the identification of motion dependent **fluid/structure** interaction forces of centrifugal pump components like seals, impellers, balance pistons and bearings. The **component tester** is shown in Figure 19. It consists mainly of a rigid shaft with an overhung component like impeller or seal, the casing, two radial magnetic bearings and an axial bearing. The magnetic bearings have a double function. They support the shaft but do also work as **actuator elements** in order to excite the shaft kinematicly: **displacement excitation**. Different shapes like radial and tilting motions of the shaft can be realized and it is also possible to select different types of signals (sinusoidal, random etc.).

Duo to the displacement excitation dynamic stiffness functions can be determined, which allow a direct identification of the dynamic forces or dynamic coefficients. Figure 20 presents as an example the real and imaginary part of a dynamic stiffness frequency function (see also Figure 9).

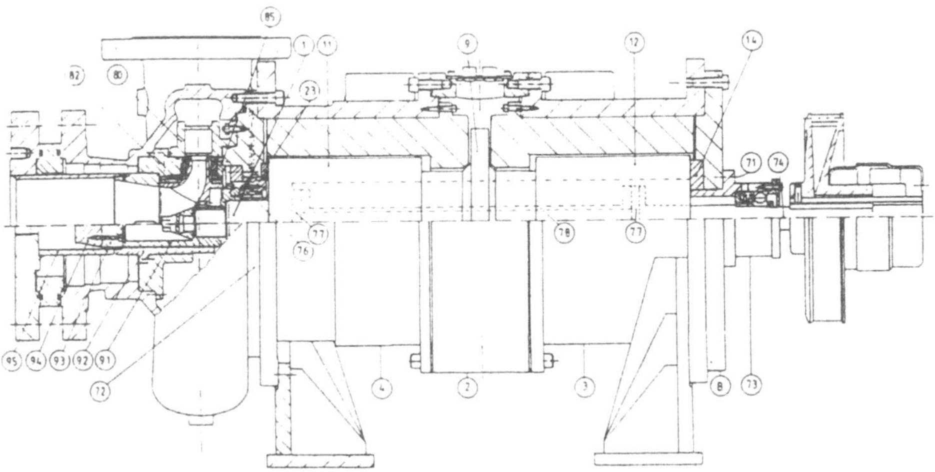

Fig. 19: Component tester with magnetic bearings

The magnetic bearing seems to be a very interesting actuator tool for future identification tasks. However, there are still problems to solve, e.g. the accuracy of force measurements. New test rigs are in use in order to improve the identification techniques based on active magnetic bearings.

$$Re = -(K - M\Omega^2) \qquad Im = -\Omega D$$

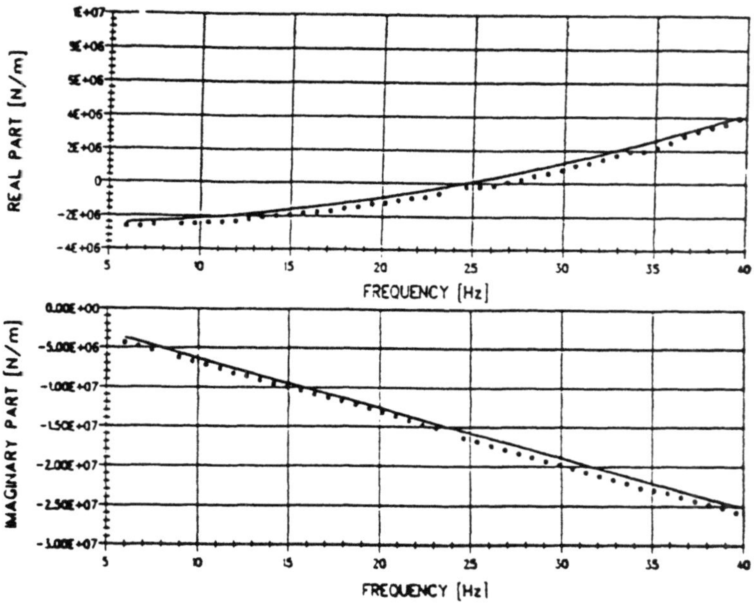

• : Measured — : Fitted

Fig. 20: Real and Imaginary Part of a Dynamic Stiffness Frequency Response

REFERENCES

/1/ Morton, P.G.: "The Derivation of Bearing Characteristics by Means of Transient Excitation applied directly to a Rotating Shaft", IUTAM-Symposium, Dynamics of Rotors, Lyngby Denmark, Springer Verlag 1975, pp. 350-370.

/2/ Nordmann, R., Schöllhorn, K.: "Identification of Modal Parameters of an Elastic Rotor with Oil Film Bearings", Journal of Vibrations, Acoustics, Stress, and Reliability in Design, 1984, Vol. 106.

/3/ Tonnesen, J., Lund, J.: "Experimental Techniques for Rotordynamic Analysis, CISM Course No. 297, Rotordynamics 2, Ed. by N.F. Rieger, Springer-Verlag, 1988.

/4/ Muszynska, A., Bently, D.: "Modal Testing and Parameter Identification of Shaft/Fluid Lubricated Bearing System. IV Int. Modal Analysis Conference, Union College, Schenectady, NY, Los Angeles, CA, February 1986.

/5/ Kanki, H., Kawakami, T.: "Experimental Study on the Dynamic Characteristics of Pump Annular Seals", Vibr. in Rotating Machinery IMechE Conf., York, 1984.

/6/ Nordmann, R., Matros, M., Neumer, T.: "Parameter Identification in Rotating Machinery by Means of Active Magnetic Bearings", IFTOMM, 4th Int. Conference on Rotor Dynamics, Chicago, Illinois, 1994.

VIBRATION DIAGNOSTIC EXPERT SYSTEM FOR STEAM TURBINES

H. Kanki, C. Yasuda and S. Umemura
Takasago R. & D. Center,
Mitsubishi Heavy Industries, Takasago City, Japan

R. Itoh and C. Miyamoto
Control System Design Section,
Takasago Machinery Works,
Mitsubishi Heavy Industries, Takasago City, Japan

T. Kawaguchi
Kansai Electric Power Company, Inc., Osaka Japan

ABSTRACT

A vibration diagnostic system has been developed for the main steam turbine of a 375MW fossil power plant. The diagnostic system continuously monitors the shaft vibration of the main steam turbine as well as various process data. Also, the system uses an AI expert system to diagnose the cause of vibration owing to the turbine failure or malfunctions as they occur or tend to occur.

The system configuration, functions and applications are presented.

1. INTRODUCTION

In recent years, nuclear power stations in Japan have had an aggregate capacity corresponding to 18% of the nation's entire electric power generation and are their role as base load plants is being planned. On the other hand, fossil power plants are playing the role of mainly absorbing fluctuations in electricity demand. Hence latest fossil power plants are now operated in the Daily-Start-Stop (DSS) mode or in the Weekly-Start-Stop (WSS) mode. These operation modes frequently introduce thermally and mechanically severe transient conditions to turbines and often cause turbine stops owing to unexpected troubles. In turbine generator plants operated in DSS or WSS mode, therefore, it is necessary to improve reliability of machinery systems and control systems, and at the same time, to reinforce monitoring of these systems.

To improve reliability of turbine generator plants, Mitsubishi Heavy Industries, Ltd. (MHI) has developed and has been using a rotating machinery failure diagnostic system (the Machinery Health Monitoring System) [1], which diagnoses the health of turbine generator plants by monitoring vibration data continuously. To respond to the above-mentioned new demand for fossil power plants in recent years, therefore, MHI has developed a vibration diagnosis expert system for steam turbines for the purpose of upgrading the diagnostic functions of the conventional Machinery Health Monitoring System and of supplementing flexibility of the conventional system.

This new system is an expert system which detects turbine troubles in an early stage by continuously monitoring shaft vibration of steam turbines as well as various process data. In addition, it automatically diagnoses the cause of detected trouble, on the basis of the result of vibration data analysis and of a diagnostic knowledge base prepared by turbine experts.

2. MALFUNCTIONS AND VIBRATION OF STEAM TURBINES

The most important matter in diagnosing malfunctions or failures is to find the relation between causes and symptoms of malfunctions or failures. When malfunctions or failures occur with the rotor system of a turbine generator, the symptoms are usually detected as vibration phenomena; therefore, it is necessary to have a sufficient knowledge of the relationship between types of malfunctions and conditions of resultant vibration.

There are various forms of vibration problems which occur with rotor systems of steam turbine generators; therefore, they can be classified into forced vibration and self-excited vibration in principle [2]. Vibration problems are summarized in Table 1. Forced vibration is caused by external forces inherent in rotor systems of steam turbine generators, and is not abnormal essentially in itself. However, when an external force becomes larger than usual, for some reason, or a frequency of an external force coincides or approaches very closely to a natural frequency of various parts of the steam turbine generator rotor system, which are cases of so-called resonance, the vibration gets abnormally large. Examples of forced vibration are an unbalance vibration of a turbine rotor owing to bending in the rotor and a resonance between the critical speed and operating revolution speed of a turbine rotor.

Table 1 Classification of vibration problems

Item \ Classification	Forced vibration	Self-excited vibration
Kind	· Unbalance vibration · Vibration owing to hydraulic pulsations · Higher harmonic vibration	· Oil whip · Steam whirl · Whirl by internal friction
Judgement	Essentially normal but abnormal when the level of vibration is high	Basically abnormal and countermeasures indispensable
Countermeasure	Since external forces are too large or there is resonance in many cases, a countermeasure is taken in either case	· Improvement on bearing seals which causes the problem · Increase in external damping

On the other hand, self-excited vibration occurs independently of the external force inherently existing in rotor systems of steam turbine generators mentioned above and should not occur under normal conditions. An example is oil whip which occurs by oil films of bearings. The frequency of self-excited vibration is entirely unrelated to frequencies of external forces and coincides with the natural frequency of the mechanical system. This is an important characteristic of self-excited vibration. As explained in the above, the generation mechanisms of forced vibration and self-excited vibration fundamentally differ between each other with their great difference in vibration characteristics as well.

From the viewpoint of failure diagnosis, therefore, it is necessary to start out by finding to which type a vibration problem belongs. For this purpose, the frequency of vibration generated is the key for judgement. Namely, the frequency of a forced vibration coincides with that of the external force, and the frequency of a self-excited vibration coincides with natural frequency of the mechanical system.

For rotor systems of steam turbine generators, the relationship between various vibration problems and phenomena are summarized in Table 2 [3]. As it is clear from the table, vibration phenomena generated from various failures are different, and it is a distinctive point that frequencies are markedly different in particular.

Table 2 (1) Relationship between various turbine failures and vibration phenomena

Cause of vibration	Frequency of vibration	Characteristics and phenomena of vibration
1 Unbalance of rotor*	The frequency of vibration coincides with the frequency of rotor revolution	
(1) Poor balancing		o Vibration corresponding 1 1 with rotation
(2) Thermal rotor bending		o The amplitude of vibration changes with thermal load It is relatively hard for it to be affected with small rotors
(3) Rotor bending owing to contact between stationary part and rotating part		o There are cases where vibration suddenly increases because of rubbing
(4) Wear and corrosion of rotor		o Vibration gradually increases with time
(5) Deposit or falling-off of foreign materials		o Vibration gradually increases as foreign materials deposit increasingly, and suddenly increase when deposited materials fall off
(6) Deformation or damage of rotor		o Vibration gradually increases as deformation progresses and suddenly increases when damaged
(7) Loosing n various parts		o There are cases where vibration is caused by loosing at fitting parts because of heat
(8) Unbalance of coupled rotor system		o Even if a rotor is balanced by itself, unbalance vibration may occur when rotors are coupled
2 Poor alignment*	Generally the same as above Lowest order of natural frequency of rotor system	
(1) Poor centering		o As in the case of poor alignment, if centering of both faces is incorrect, bearing loads are liable to become uneven and the sensitivity to unbalance vibration gets high
(2) Misalignment	In special cases, there is a large vibration component that at times has a frequency of two times the revolution frequency	o When centering is extremely poor, bearings may float or get an uneven contact and oil whip or sub-harmonic vibration may sometimes occur.
(3) Thermally poor alignment		o Bearing pedestals or casings may be thermally expanded and incorrect centering may occur
(4) Sinking foundation		o Vibration gradually increases with years
3 Mis-coupling*	Mainly with a frequency of vibration matching with the revolution speed	
(1) Poor acuracy of coupling		
(2) Uneven contact of tightening bolts		
(3) Poor tooth contact of gear couplings	There are cases where peculiar vibration phenomena occur together	o There are cases where peculiar vibration phenomena occur relating to the natural frequency of rotor system
(4) Poor lubrication of gear couplings	There are cases where peculiar vibration phenomena occur together	o There are cases where resonance vibration of rotor system is unsteadily induced and seizing phenomena occur at gear tooth surfaces
(5) Vibration caused by hydraulic coupling	There are cases where peculiar vibration phenomena occur together	
4 Wear of bearings* Excessive metal gap Excessive back clearance of bearing metals Poor lubrication		
5 Critical speed of rotor system*		
(1) Critical speed	Same frequency with the revolution speed	o Vibration suddenly increases in the vicinity of shaft critical speed; however, it returns to the normal level when the shatt speed passes the range of critical speed.
(2) Second order critical speed	Frequency of two times the revolution speed	o Vibration increases with the revolution speed in the vicinity of 1/2 of the first order critical speed. This vibration is mainly caused by asymmetry in rotor rigidity in many cases.
6 Oil whip or oil whirl**	Frequency below 1/2 of the revolution speed (natural frequency of rotor system)	o This type of vibration often occurs in high speed rotating machines using journal bearings, and is a self-excited vibration based on oil films of bearings The characteristics of this vibration are as follows: (a) Frequency is equal to the critical speed of rotor systems. (b) Vibrations of this type sometimes occur at the revolution speed, exceeding two times of the first order critical speed of rotor systems. (c) The vibration frequency is nearly the same irrespective of the revolution speed of rotors. (d) The points where vibration occurs and disappears deviate when the revolution speed is increasing and when decreasing. (e) The vibration occurs and disappears suddenly. (f) Once the vibration occurs, it does not decrease even if the revolution speed is increased. (g) The whirl direction of vibration is the same with the direction of rotation. Furthermore, there are cases where rotors whirl at the revolution speed higher than the first order critical speed and also lower than double of the first order critical speed. In such cases, the frequency of vibration is approximately 1/2 of the revolution speed and the rotor itself does not bend much. The former and the latter are occasionally distinguished by being called "oil whip" and "oil whirl," respectively.

Notes*. Forced vibration ** Self-excited vibration

configuration of a general expert system is shown in Fig.1. This is a diagnosing system using a knowledge base which is a set of knowledge for diagnosis collected from human experts. The knowledge base includes not only simple numerical data but also non-numerical knowledge. There have been several methods of expression proposed for knowledge bases, but the commonly used IF-THEN method is adopted for the vibration diagnostic expert system for steam turbines developed this time. This is because FMECA (Failure Modes Effects and Criticality Analysis) [4] has been adopted to infer causes of troubles in steam turbine generator rotor systems and because fault trees have been well developed for trouble analysis. In other words, fault trees can easily be converted into the rules of the IF-THEN form, and the diagnostic knowledge base can be easily constructed on the basis of the conventional FMECA by adopting the IF-THEN method.

4. SYSTEM CONFIGURATION

The configuration of the vibration diagnostic expert system for steam turbines is explained in this chapter.

4.1 Hardware configuration

Fig.2 shows an external view of the apparatus for the steam turbine vibration diagnostic expert system. A CRT for diagnosing is on the left side and that for monitoring is on the right side in the figure.

Fig.3 shows the hardware configuration of the system which is basically composed of a data monitoring sub-system and a diagnosis sub-system. The data monitoring sub-system is a data acquisition and monitoring system to continuously monitor vibration and various process data in rotor systems of steam turbine generator. The sub-system is provided with vector filters and a spectrum analyzer for vibration data analysis and with an A/D convertor for data acquisition. The data monitoring sub-system is basically the same with the conventional rotating machinery failure diagnostic system and a 16 bit mini-computer system.

On the other hand, the diagnosis sub-system is the expert system section to diagnose abnormal vibration and is a 32 bit mini-computer system. The data monitoring sub-system is connected to the diagnosis sub-system according to the standard network of IEEE-802.3 for data communication.

Fig.2 Overview of apparatus for the steam turbine vibration diagnostic expert system

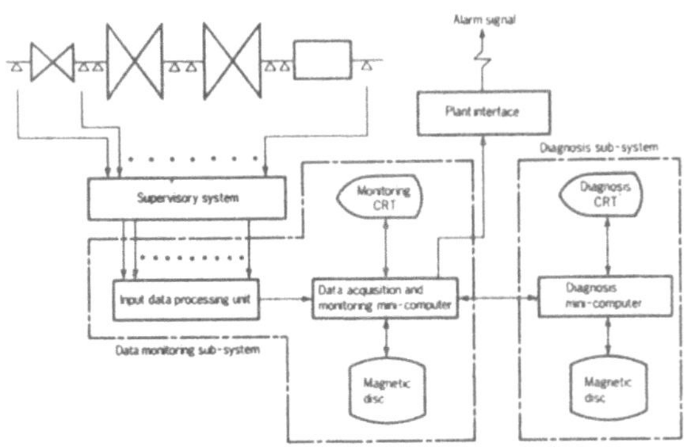

Fig.3 Hardware configuration of diagnostic system

Table 2 (2) Relationship between various turbine failures and vibration phenomena

Cause of vibration	Frequency of vibration	Characteristics and phenomena of vibration
7 Whirl owing to rub between rotating parts and stationary parts** (1) Friction	Vibration with a comparatively low frequency	o These are the cases of "friction whirl" which occur with poorly lubricated bearings having excessively loose shaft guides or excessively large clearances. If the rotor surface and the bearing surface are in contact and whirl together without sliding, the whirling speed becomes equal to the revolution speed x (shaft radius/shaft clearance) This means that the rotor whirls at a considerably high speed
8 Sub-harmonic resonance owing to mechanical gaps or nonlinearity*	Whirling at a frequency corresponding to 1/2, 1/3, 1/4, the revolution speed	o If mechanical gaps or non-linear elements exist in a system, sub-harmonic resonance phenomena occur and the system sometimes vibrates severely This type of vibration has the following characteristics (a) Sub-harmonic vibrations only occur in a limited range of frequency and force of forced vibration (b) With symmetric non-linear systems, an odd number of sub-harmonic vibrations such as on the order of 1/3, 1/5, occur (c) With asymmetric non-linear systems, an even number of sub-harmonic vibrations such as on the order of 1/2, 1/4, occur (d) As in the case of non-linear first order vibration, jumping or disappearing phenomena occur in the amplitude at the upper and lower limit frequencies (e) The range where sub-harmonic vibrations occur depends largely on the magnitude of forces, size of mechanical gaps, and damping factors (f) The range where sub-harmonic vibrations on the order of 1/N occur is on the vicinity of Q=N This type of vibration is similar phenomenally to the oil whip phenomena described in 6 above However, it is difficult to distinguish these vibrations just by frequency analysis using ordinary vibration analyzers and it is necessary to record the vibration wave patterns on oscillographs and to surely confirm the wave patterns Nevertheless, sub-harmonic vibrations occur very rarely as compared with the oil whip phenomena
9 Improper foundation* (1) Incorrect installation level (2) Improper tightening of foundation bolts (3) Poor grouting (4) Poor foundation stiffness	Generally with a frequency equal to the revolution speed	o If the foundation stiffness is low, comparatively large vibrations occur even without having any particular vibration failure, and, in many cases, large problems result With rotating fluid machinery in particular, fluid flow is smooth in the vicinity of the maximum efficiency point ; however, many flow turbulences exist in the low load range and act as random exciting forces causing the machinery to vibrate in many cases.

3. INTRODUCTION OF EXPERT SYSTEM

In the conventional rotating machinery failure diagnostic system, attention is paid mainly to the frequency components of the generated vibration described in the previous chapter. And, the diagnoses are carried out automatically by a frequency weighted point method using cause and symptom matrices. This method, however, has the following weak points:

(1) The method is effective for small scale problems; however, it becomes difficult to infer causes when the matter for diagnosis is complicated and of a large scale.

(2) The method deals with numerical data only and cannot deal with various data which are difficult to express numerically but necessary for vibration experts for diagnosis. For this reason, the method is useful to classify causes of vibration, but it cannot diagnose deeply enough to determine the primary causes.

(3) The diagnostic system itself becomes inflexible by using this method, therefore, it is not easy to newly add or change diagnostic algorithms.

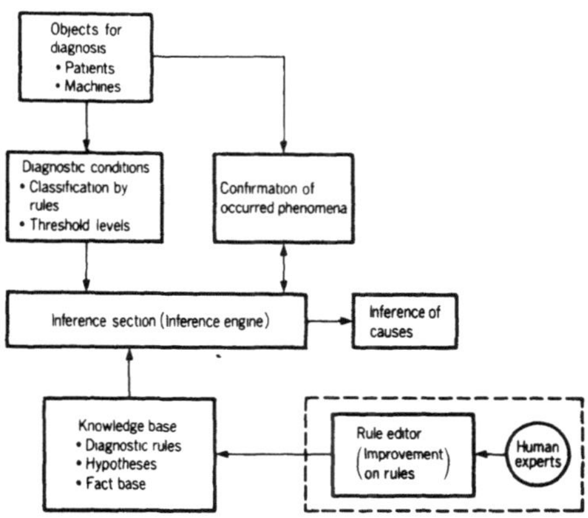

Fig.1 Configuration of general expert system

To solve the above problems, it is possible to introduce expert systems into the diagnostic system. Expert systems are an application of knowledge engineering that make it possible to utilize knowledge and diagnostic inference processes used for diagnosis by experts. The

4.2 Software configuration

Fig.4 shows the software configuration of the system. The data monitoring sub-system is operated under the control of a real-time operating system and consists of a control program, a data acquisition program, a monitoring program, an alert processing program, a man-machine interface program and a communication control program. The data acquisition program is a program to acquire vibration data and process data received from turbine supervisory systems, and record them in the data base. The monitoring program checks acquired vibration data and analyzed results of these data and judges if there are any faulty conditions. The alert processing program executes processing such as alarming and diagnosis sub-system starting when the monitoring system detects any abnormal conditions. The man-machine interface program graphically displays various compiled data and analyzed results of these data. The communication control program controls data transmission with the diagnosis sub-system through Ethernet according to the standard of IEEE-802.3.

On the other hand, the software of the diagnosis sub-system consists of a diagnosis control program, a diagnosis program, various simulation programs, and a communication control program. Each of the above-mentioned programs is executed under the control of UNIX operating system. The diagnosis control program is a kind of monitor program to control all other programs. The diagnosis program is the expert system itself, which executes diagnosis automatically according to the knowledge base prepared by the turbine experts. The simulation programs simulate unbalance vibrations and also calculate optimum balance weights for field balancing. The result of this simulation is utilized to diagnose causes of turbine vibration troubles as a model base. The communication control program performs data transmission with the data monitoring sub-system through Ethernet according to the standard of IEEE-802.3.

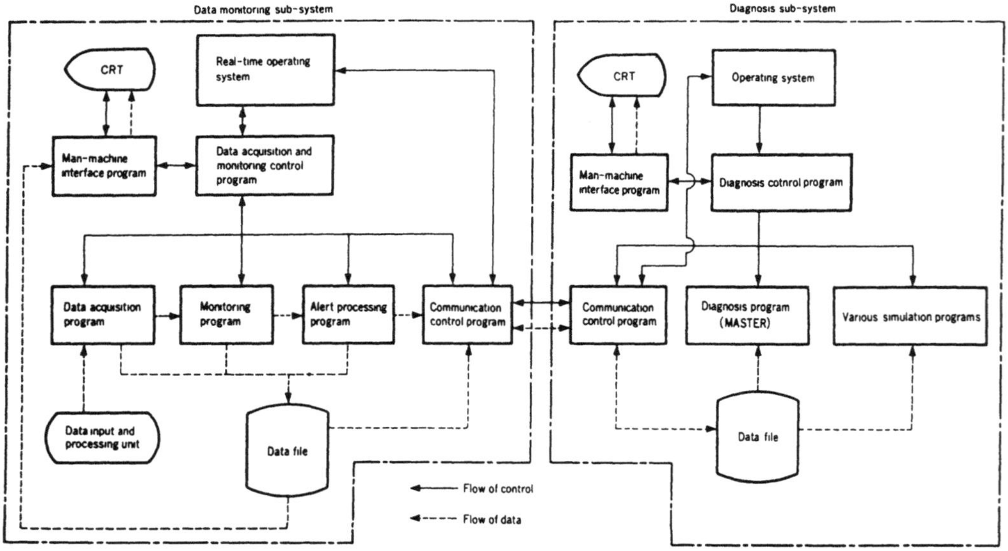

Fig.4 Software configuration of diagnostic system

5. DIAGNOSIS PROGRAM

This chapter explains the diagnosis program which constitutes the brain of the steam turbine vibration diagnosis expert system.

5.1 Configuration of diagnosis program

Fig.5 shows the configuration of the diagnosis program. As shown in the figure, the configuration is of a general expert system which executes diagnosis on the basis of a knowledge base as a set of various diagnosis rules collected from turbine experts. However, the special feature of this program is that it contains a model base, shown in the rectangular space surrounded by broken lines in the figure, in addition to the ordinary knowledge base. In other words, it is made possible to diagnose the cause and location of troubles by comparing vibration response data simulated by the above-mentioned unbalance vibration simulator, with the fact base of measured data. Diagnostic data, obtained from a mathematical model subject to diagnosis such as simulator, is called a model base. Utilizing such a model base, diagnostic accuracy can be improved, as compared with the case of diagnosing only by knowledge bases collected from human experts.

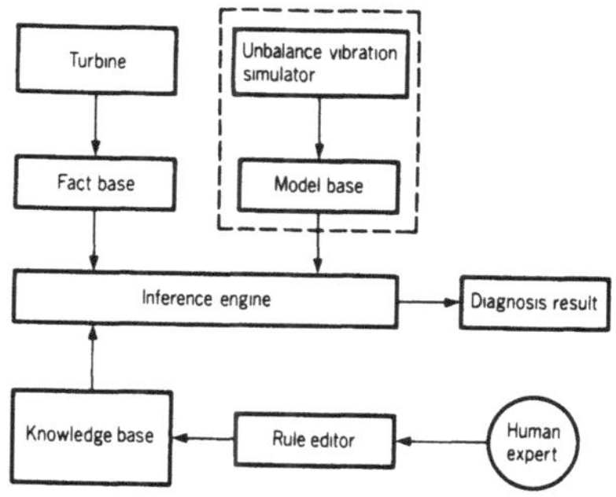

Fig.5 Configuration of diagnosis program

When encountered with vibration troubles of turbines, turbine experts diagnose, in general, using not only knowledge obtained through their own experience but also results of various simulation calculations. Utilization of the model base described above follows this process.

5.2 Inference mechanism

The inference mechanism of this system uses MASTER (MHI Assistant System for Tracing Expert Resolution) developed by MHI [5].

The MASTER can handle both the rule-type knowledge base as knowledge expression function in IF-THEN form, and the frame-type knowledge base which enables structural expressions. Furthermore, as an inference process, it is possible to have a forward or backward chaining inference for the rule-type knowledge base and to retrieve information and inference by frame handling for the frame-type knowledge base.

For the steam turbine vibration diagnostic expert system, a production system adopted from among those inference functions of MASTER as described above. It has the largest past experience in forward chaining inference which uses the rule-type knowledge as diagnostic expert system. Furthermore, the certainty factor, *CF* value, is introduced into MASTER to express the fuzziness of knowledge. Specifically, it is the certainty factor defined according to EMYCIN [6] that expresses the certainty of a matter occurring in consideration using a real number between +1.0 and –1.0. The number +1.0 means an absolute certainty of occurrence and the number –1.0 means an absolute certainty of non-occurrence.

Furthermore, certainty factors are defined not only for each phenomenon, but also for each rule. In executing each rule, the certainty factor according to the definition of EMYCIN is calculated.

5.3 Description of diagnosis rule

As explained in the above, the steam turbine vibration diagnostic expert system uses a production system according to forward chaining inference. Hence, the diagnosis rules are first of all expressed by decision trees. Fig.6 shows an example of decision tree expressions for diagnosis rules.

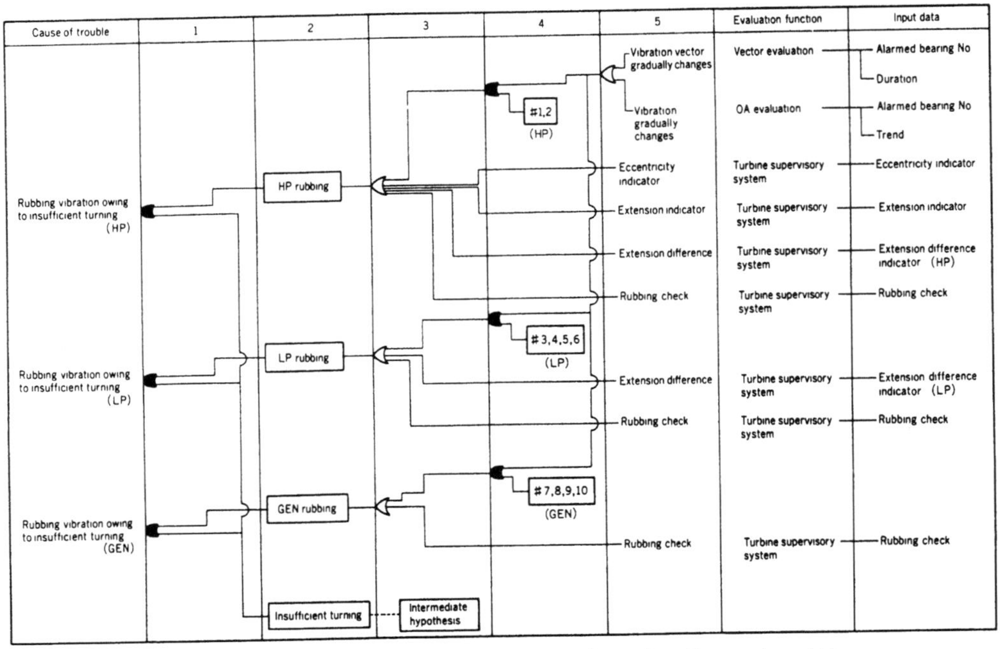

Fig.6 Example of decision tree expressions for diagnosis rules

As shown in Fig.6, each rule is expressed basically with AND or OR. The relationship between AND and OR, the basic forms of each rule, and the IF-THEN rule are shown in Fig.7. It clearly shows trigger phenomena which start execution of the rules. This is the point different from ordinary decision trees and fault trees. Namely, Fig.7(a) means that this rule is executed only when phenomenon A is proved (CF_a>0). In addition, Fig.7(a) means that the certainty factor CF_c of the conclusion C is calculated from the certainty factor CF_a of phenomenon A, the certainty factor CF_b of phenomenon B and the certainty factor CF_r of the rule according to the EMYCIN method.

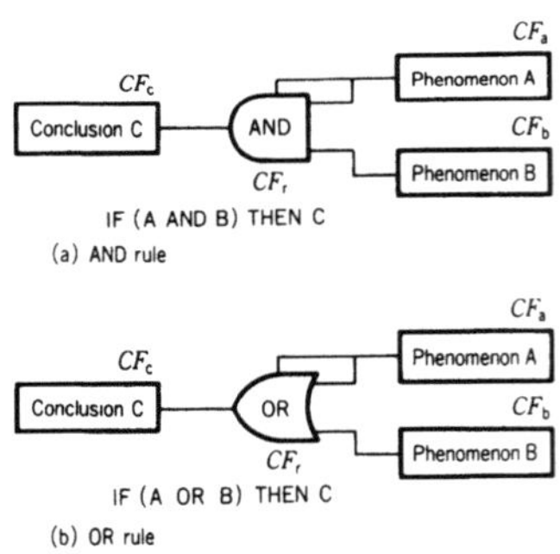

Fig.7 Relationship between decision trees and IF-THEN rules

5.4 Evaluation function

In Fig.7, the proof of the first stage phenomena downstream from most of the decision trees (at the right side in the figure) is executed by evaluation functions. The 14 functions shown in Table 3 are provided as evaluation functions for the steam turbine vibration diagnostic expert system. Each evaluation function proves first stage phenomena by calculating and outputting certainty factors for the first stage phenomena using observed data in the fact bases.

The 12th evaluation function in Table 3 is to perform simulation for unbalanced vibration response mentioned previously in Paragraph 5.1. This function evaluates phenomena by comparing trends of measured unbalanced vibration and trends of theoretically predicted vibration.

Furthermore, evaluation functions for interactive input are provided for such data that are not measured on-line, if necessary.

In calculating certainty factors using evaluation functions, certainty factors are determined by membership functions considering that observed data and results of analyzing these data are a fuzzy set. Taking the example of a phenomenon that "vibration level is high," the certainty factor is evaluated by the measured value of vibration level by using the membership function as shown in Fig.8. In the figure, the standard value is the average value of turbine vibration level under normal operation, and the certainty factor becomes 0.0 at this point. A level of 100μm is the alarming value for the plant. Similar membership functions are prepared for various vibration analysis results and process data; however, membership functions are determined on the basis of Stuart's 3σ theorem by statistically processing data under normal operation of turbines and using the average values and standard deviations of these data, in principle.

Table 3 List of evaluation functions for diagnosis

No	Name of function	Content
1	Evaluation function for vibration OA level.	Evaluation of values and trends of vibration OA level.
2	Evaluation function for vibration spectrum	Evaluation of each banded frequency component of vibration spectrum.
3	Evaluation function for differences between vibration transducers	Evaluation of variances in indicated values of vibration transducers
4	Evaluation function for vibration vectors	Evaluation of values and trends of revolution speed component vibration
5	Evaluation function for higher harmonic components	Evaluation of values and trends of higher harmonic component vibration vectors.
6	Evaluation function for relationship between vibration vectors and process data	Evaluation of relationship between various process data and vibration vectors
7	Evaluation functions for deviations in Nyquist Plot	Evaluation of magnitude and direction of deviations in Nyquist Plot
8	Evaluation function for vibration at low revolution speed	Evaluation of vibration at the turbine start-up and at the low revolution speed
9	Evaluating function for first order critical speed	Evaluation of frequency and amplitude of vibration at first order critical speed when increasing the revolution speed.
10	Evaluation function for natural frequencies	Evaluation of natural frequencies of each part.
11	Evaluation function for turbine supervisory system	Evaluation of indicated values of various turbine supervisory system
12	Evaluation function for influence coefficient vectors	Evaluation of comparisons between measured vectors and influence coefficient vectors.
13	Evaluation function for balance weight mounting condition	Evaluation of comparisons between balance weight mounting positions and influence coefficient vectors.
14	Interactive function.	Processing of various interactive inputs data.

6. CASE HISTORY

The steam turbine vibration diagnostic expert system has already been installed for the steam turbine generator units of a 375MW fossil power plant actually in service. In this chapter, some examples are presented how the system in service automatically detected abnormal conditions and diagnosed failures.

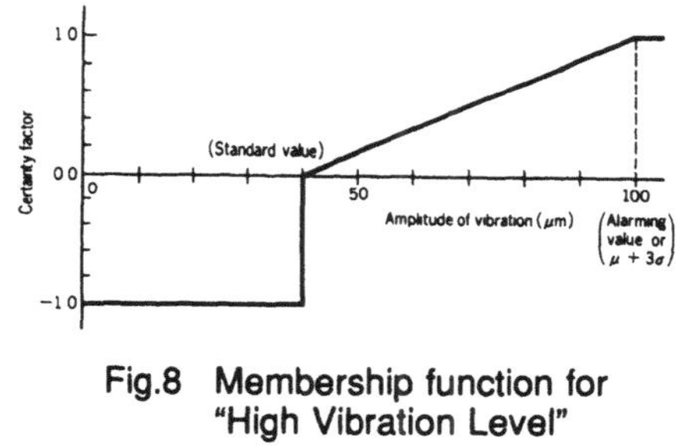

Fig.8 Membership function for "High Vibration Level"

(1) Diagnosis example 1

Result of diagnosis: Shaft bending owing to insufficient turning

The diagnostic system detected an excessive vibration during starting operation of the turbine and issued an alarm. In this case, the diagnosis result was "shaft bending owing to insufficient turning" and the system indicated the countermeasure that "continue warming-up if the vibration is within the permissible level. If the vibration is exceeding the permissible level, stop the turbine and restart the turbine after performing turning sufficiently."

Fig.9 shows the diagnosis result displayed on the diagnosis CRT. Together with the diagnosis result, countermeasures and the decision tree diagram used for diagnosis are shown on the screen. Fig.10 shows the trend in shaft vibration spectrum during starting operation and indicates that changes are mainly in the revolution component showing the characteristic of shaft bending.

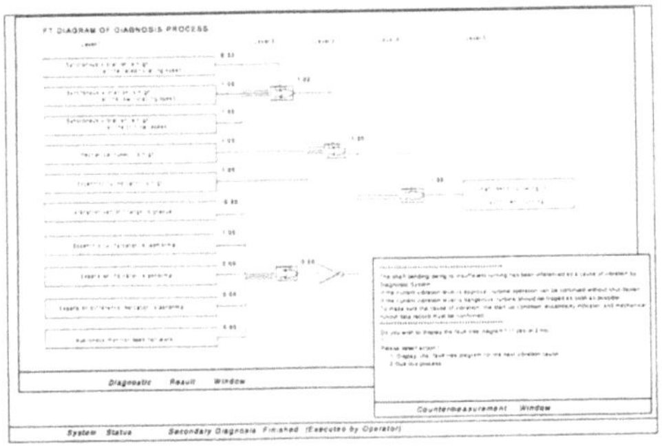

Fig.9 CRT display of diagnosis result – Example 1

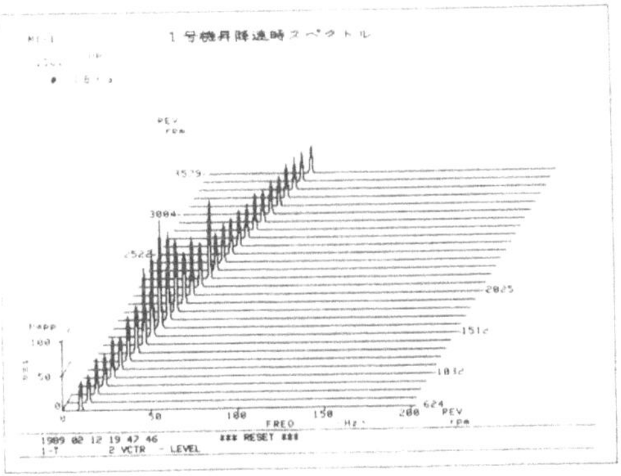

Fig.10 Three-dimensional spectrum display

(2) Diagnosis example 2

Result of diagnosis: Miscontact of vibration transducer shoe
Abnormal rubbing in generator

The diagnostic system detected changes in vibration during rated power operation (100% load) of the turbine and issued an alarm. In this case, the diagnosis results were "miscontact of vibration transducer shoe" and "abnormal rubbing in generator." The certainty factor was +1.0 for both phenomena. Fig.11 shows the diagnosis result displayed on the diagnosis CRT and Fig.12 shows the trend of shaft vibration vectors under the alarm condition. In the diagnosis result shown in Fig.11, the above-mentioned two causes are indicated together. Fig.12 shows that the shaft vibration vector of the No.7 bearing is making a closed loop with a cycle of about 20 minutes. Vibration of the No.7 bearing is the cause of alarm and the vibration vector is indicating the typical characteristic of rubbing vibration. Fig.13 shows vibration spectrum at each bearing under the alarm condition. For the No.7 bearing, low frequency components in the vicinity of 5Hz are distinctive showing the characteristic of miscontact of vibration transducer shoes.

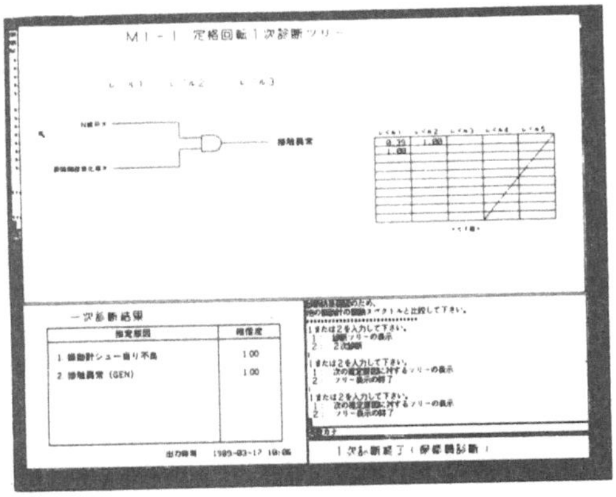

Fig.11 CRT display of diagnosis result – Example 2

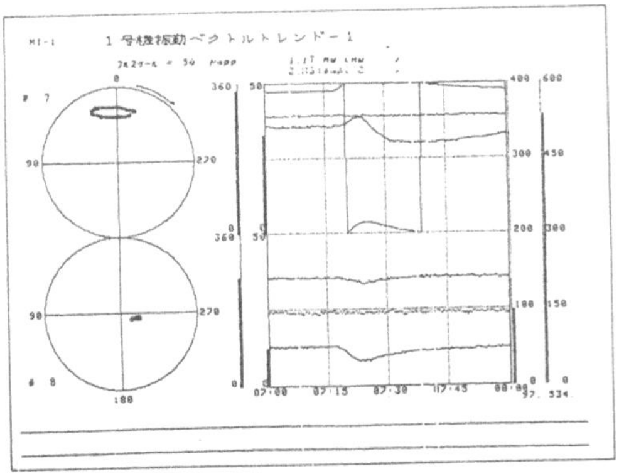

Fig.12 Display example of vibration vectors

In view of the above diagnosis result, the seal ring part of the generator was overhauled for inspection. Slight rubbing marks were found at the sealing part and the seal ring was adjusted and reassembled. In addition, slight uneven rubbing marks were found in inspecting sliding parts of the vibration transducer for No.7 bearing.

The diagnosis results described above agreed with views of turbine vibration experts, and at the same time, the appropriateness of the diagnosis results was actually confirmed by overhaul inspections etc..

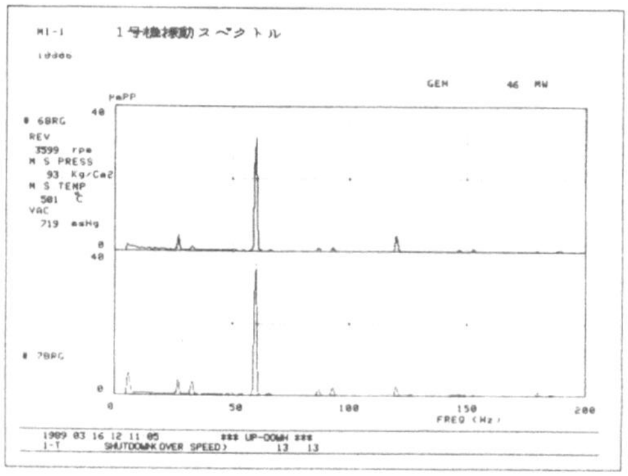

Fig.13 Shaft vibration spectrum

7. CONCLUSION

For the purpose of upgrading the diagnostic functions of the rotating machinery failure diagnostic system (Machinery Health Monitoring System), by means of the conventional causal matrix method, and of supplementing flexibility of the system, a steam turbine vibration diagnostic expert system was developed applying knowledge engineering to the automatic diagnostic functions of the conventional system. The new system is an expert system provided with a knowledge base constructed from diagnosis knowledge of turbine experts as well as with a model base using unbalance vibration response simulators. The new system was applied to an actual steam turbine generator unit and its performance has been confirmed.

The vibration diagnostic expert system for steam turbines was developed and applied for practical use through joint efforts with the Kansai Electric Power Co., Inc., to whom the authors wish to express their sincere gratitude.

REFERENCES

1. H. Yamaguchi, M. Higuchi, C. Yasuda, T. Sasaki, H. Masada and R. Ito: Diagnostic System Based on Vibration Measurement for Rotating Machines, Mitsubishi Juko Giho Vol.24 No.5 (1987).
2. C. Yasuda: Fault Diagnosis and Predictive Maintenance of Steam Turbine, Turbo Machinery Vol.16 No.7 (1988).
3. K. Shiraki and H. Kanki: Vibration Problem in Machinery Industry, Study of Machinery Vol.29 No.7 – Vol.30 No.1 (1977).
4. Collacott, A. R.: Mechanical Diagnosis and Condition Monitoring, Chapman & Hall, 1977.
5. K. Noguchi, R. Murata, T. Yoshikawa, H. Kobayashi, and K. Yoshida: Tool for Building Expert System – MASTER, Mitsubishi Juko Giho Vol.24 No.5 (1987).
6. Van Melle, W. A.: A Domain Independent System that Aids in Constructing Knowledge Based Consultation Programs, Ph. D. Dissertation, Stanford Univ., STANCS-80-820 (1980).

PARAMETERS FOR VIBRATION ANALYSIS ON ROTATING MACHINERY

J.M. Gilstrap
Bently Nevada Corporation, Minden, NE, USA

ABSTRACT

For the most common rotating machinery problems, vibration analysis is usually a straightforward procedure. The primary vibration source is often estimated by using only vibration amplitude and frequency data. However, with the high cost of today's machinery and the even larger investment in plant processes, an "estimate" of the machinery problem is not good enough. In order to increase the accuracy of vibration analysis decision making, other measurement parameters must be used. This paper will present some of the lesser used techniques by way of machinery case histories.

INTRODUCTION

In industry today, much emphasis is placed on **productivity**. In all areas of operation, increased productivity is necessary to remain competitive in the world marketplace. Machines must be more productive; they must remain on-line, with fewer problems, for longer periods of time. Scheduled machinery outages must be not only infrequent, but when they do occur they must be of short duration. If a machine must be shutdown to fix a problem, the source of the problem must be well known, so that it can be fixed quickly, and the machine can be brought back on-line as soon as possible.

This means that increased productivity is necessary also for **people**. Machinery specialists must become more efficient and accurate in their task of fault analysis. When they recommend that a machine be shutdown, they must know with little doubt exactly what the problem is, and what must be done to the machine to bring it back to an acceptable operating condition.

For the past several decades, machinery specialists have used vibration frequency spectrum data as their basic tool for rotating machinery analysis. "Everyone knows" that if the largest vibration component is at shaft rotative frequency (1X), then the problem is rotor imbalance. "Everyone knows" that misalignment produces vibration at twice shaft rotative frequency (2X). However, there are many cases where 1X vibration is **not** due to rotor imbalance. And other parameters will indicate shaft misalignment **long before** the 2X vibration frequency occurs (if it occurs at all).

Spectrum analysis is not accurate enough to be the only criterion upon which to base a critical machine shutdown decision. Other parameters must be evaluated in order to increase the accuracy of the machinery specialist's decision. The following case histories demonstrate the use of some of these other parameters.

SHAFT AVERAGE RADIAL POSITION

When XY proximity probes are installed for radial vibration monitoring, vibration relative to the bearing is **not** the only parameter that should be monitored. In addition to the transducers' dynamic ("ac") signal output that represents shaft vibration, there is a "dc" signal output that represents the average position of the shaft relative to the bearing clearance.

When a shaft begins to rotate in a fluid film bearing, it moves away from the bearing wall and "rides" on the lubricating film. With some unidirectional applied radial force (preload), the shaft remains close to the bearing wall and there forms the normal fluid wedge and minimum film thickness region. By monitoring the dc output of proximity transducers, one can determine if a shaft has attained its "normal" radial position in the bearing or not.

First the XY proximity transducer dc voltages are measured with the machine stopped (ideally in a "warm" condition, for example, immediately after a shutdown). Then when the machine is at full speed and load, and thermal equilibrium is obtained, the dc voltages are again measured, using the initial "at rest" readings as a reference. The difference between the initial readings and the current readings represents the operating average shaft radial position in the bearing.

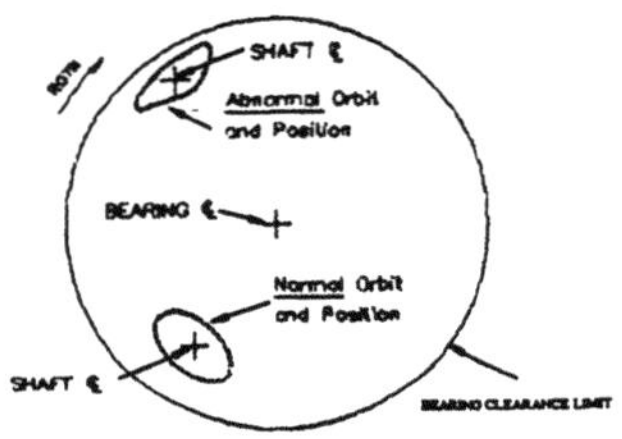

Figure 1.

"Normal" and "Abnormal" Radial Shaft Positions

Figure 1 shows the "normal" average shaft radial position region for a clockwise rotating system. The shaft should be in the lower left-hand quadrant of the bearing. If XY proximity transducer signals indicate a shaft position in the upper left-hand quadrant (or indeed, any location other than the "normal" one), this would be considered "abnormal" and would justify further investigation.

ATTITUDE (POSITION) ANGLE

Figure 2 shows a "normal" average shaft radial position for a counterclockwise rotating system. The "normal" position of a heavy horizontal rotor is in the lower right-hand quadrant of the bearing. The included angle between a line in the radial load force direction and a line connecting the bearing center and the average shaft position center, is called the attitude angle. The bearing or machine manufacturer should define the range of "normal" attitude angles for particular bearing designs. For example, attitude angles are in the range of 30 to 45 degrees for typical cylindrical bearings. Angles are much less for pivot shoe or tilting pad bearings, typically 5 to 15 degrees.

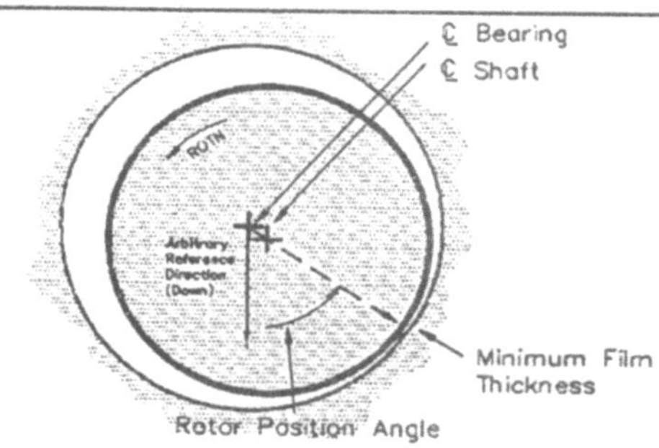

Figure 2.
Attitude Angle with Gravity Preload

For horizontal rotor systems, usually the predominant applied force is radially downward due to gravity (as in Figure 2). But the applied load may be in any other radial direction, especially for fluid handling machines with a single direction discharge, and vertical machines that have no radial load due to gravity. For single discharge volute machines, for example, the radial preload comes from the direction of the discharge (high pressure) piping (Figure 3).

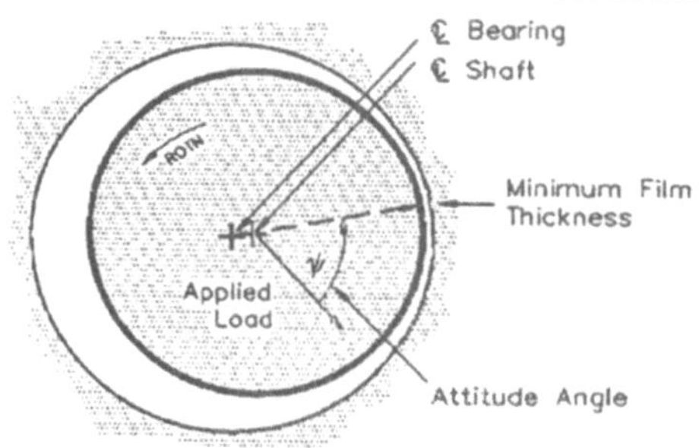

Figure 3.
Attitude Angle with any Direction Preload

ECCENTRICITY RATIO

Eccentricity ratio is the ratio of the distance between the bearing geometric center and the average shaft position center to the radial bearing clearance. This is (under normal conditions) a value between zero and one. See Figure 4. Measuring the eccentricity ratio for a given condition provides an indication of the degree of stability (or lack of stability) of the rotor/bearing system. "Normal" eccentricity ratios are high, values close to one. Generally, values from 0.7 to 0.9 indicate a "normal" condition. Values of less than 0.7 indicate a less stable condition (with a value close to zero being, by definition, a condition of bearing instability)

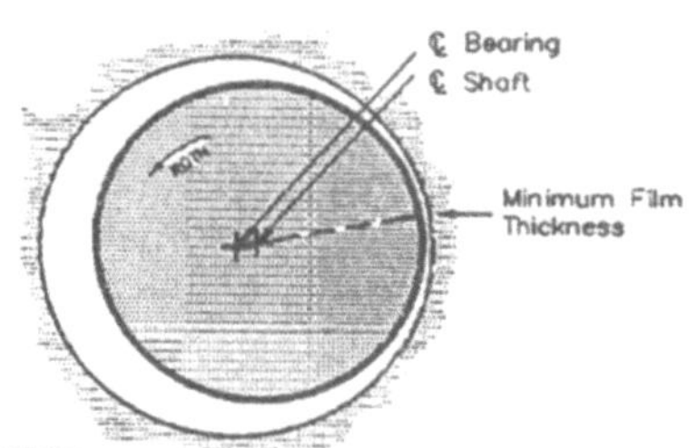

Eccentricity Ratio:

$$\varepsilon = \frac{\text{Shaft to Bearing ℄ Distance}}{\text{Radial Clearance}} \qquad \text{Radial Clearance} = \frac{\text{Diametral Clearance}}{2}$$

Figure 4. Eccentricity Ratio

A SIMPLE CASE OF MISLEADING SPECTRUM ANALYSIS

Consider the data shown in Figure 5. This machine appears to be operating "normally." The vibration amplitude is relatively low as compared to the bearing clearance, the shaft average position seems normal (lower left-hand quadrant for this clockwise rotating shaft, the orbit shape is "normal" (basically elliptical due to the gravity preload on this horizontal machine), and the spectra indicate only 1X vibration, usually associated with rotor imbalance.

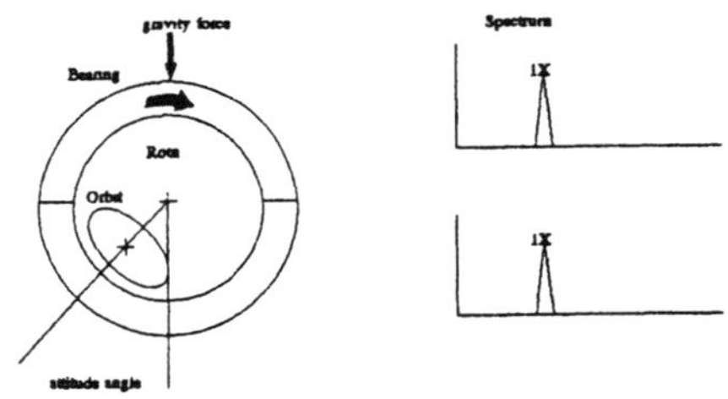

Figure 5.
"Normal" Machine Condition

Now refer to Figure 6. The spectrum data is identical to that in Figure 5. But the state of the machine is quite different from that in Figure 5. The position of the shaft in the bearing is not normal for a horizontal rotor, and the orientation of the elliptical orbit indicates a radial load force from the "nine o'clock" position, perhaps from some misalignment, thermal distortion or other source. In this case, using spectrum analysis alone would lead one to conclude that the condition of this machine is normal, when it is not normal at all.

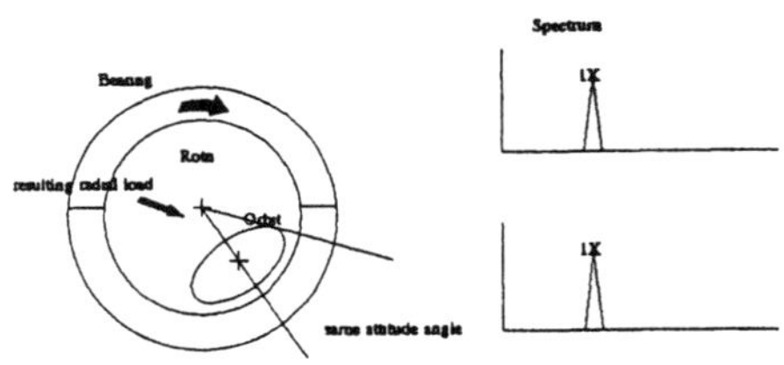

Figure 6.
"Not Normal" Machine Condition with "Normal" Spectra

Figure 7 shows yet another example. Here again, the spectrum data is the same as in Figures 5 and 6. But here again, the state of the machine is different. The orbit is circular (usually not normal for a machine with a gravity preload) and the shaft position is at the bottom center of the bearing (again, not normal). In this case, some additional radial load is acting vertically and to the right, against the natural gravity preload, with just enough force to move the shaft slightly more away from the bearing wall and to change the orbit from the "normal" elliptical shape to a circular shape. This condition is a "precursor" of the fluid-induced instability. Here again, using spectrum analysis alone would lead to the wrong conclusion.

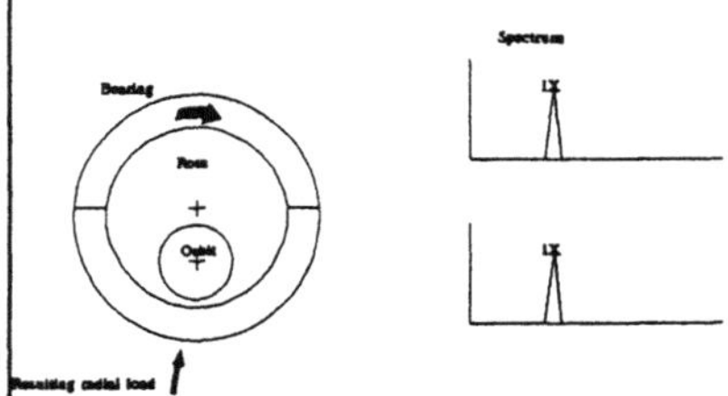

Figure 7.
Another "Not Normal" Machine Condition;"Normal" Spectrums

TRENDING AND MONITORING SHAFT AVERAGE RADIAL POSITION

The benefits of monitoring and trending shaft position during the operation of all critical machinery are shown by a case history. A condensing steam turbine was experiencing static charge on its rotor. This situation is not uncommon, where water droplets moving across the last stages of the turbine create such a charge.

The charge would increase until it was of sufficient magnitude to arc across the point of minimum clearance between the rotor and the stationary machine parts. In this case, the minimum distance was at a radial bearing (the minimum film area at the bottom of the bearing. The static discharge began to erode the bearing babbitt (white metal), gradually increasing the radial bearing clearance. This caused the shaft to move down into the enlarging bearing clearance.

For many days, the situation went unnoticed. The misalignment caused by the shaft position change was not yet large enough to cause any significant change in overall vibration amplitude. Fortunately, the monitoring systems at this facility provided a trend of radial probe gap dc voltages. For several days, readings indicated that at one bearing the shaft was moving downward (Figure 8).

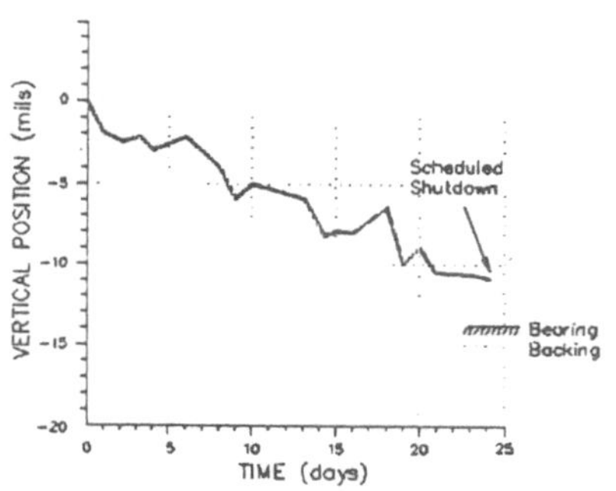

Figure 8. Trend Plot of Shaft Radial Position (Vertical)

The machinery specialist reviewed that data and correctly determined what was happening. However, operations personnel were unconvinced of the problem. After all, there had been no increase in vibration amplitude, the parameter they considered best for determining the health of the machine. Here was the classical conflict between Operations and Maintenance. Operations is paid to keep the plant operating, producing product that represents revenue for the company. Maintenance is paid to keep machinery in good condition, but in this case, evidence was not strong enough to indicate that the machine had a serious problem.

Finally after another week showed the same trend, the machinery specialist concluded that if the analysis was correct, then only a small amount of bearing material was left. In a few more days, there would be contact between the shaft and the bearing backing. The risk of an unnecessary shutdown was lost revenue; but the risk of continued operation was a serious machinery failure, which would represent much more lost revenue due to the lengthy outage that would result.

This time the evidence was believed and the machine was shut down. Upon inspection, it was found that the remaining bearing thickness was only about three mils (75 micrometers). Thus, the trending of shaft radial position saved the machine.

This and other similar incidents over the years has caused Bently Nevada to incorporate probe dc gap alarms into our standard monitoring systems.

It is now common practice for many users to monitor and trend shaft average radial position during machine startup. Once machinery is at full speed and load, and thermal equilibrium is reached, each radial bearing is checked for the proper radial shaft position. Some systems, for example Bently Nevada's Transient Data Manager (TDM2) or ADRE, provide a record of shaft position changes during the machine startup. Figure 9 shows an example of this type of data. As can be seen, for this bearing location the shaft position appears to be too high in the clearance. In fact, this particular machine was determined to have significant vertical misalignment.

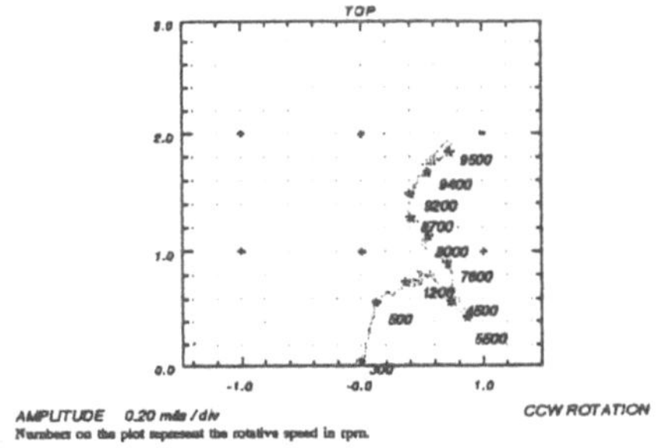

Figure 9.
Shaft Average Radial Position during Machine Startup

LOW AMPLITUDE DOES NOT NECESSARILY MEAN A HEALTHY MACHINE

After the startup of a gas compressor at a chemical facility, vibration amplitude readings were considered "normal" for all bearings on the machine. The plant did not have an on-line computerized monitoring/trending system, nor did the monitors have probe gap dc voltage alarms. However, it was standard practice to periodically measure and document probe dc voltages by hand, especially after the startup of a machine.

In this case, the probe gap data revealed a shaft position change much larger than normal on one bearing. But here again, there was a difference of opinion between Operations and Maintenance. Because the vibration amplitudes for both transducers at this bearing were very low, less than 20 micrometers. It was difficult to convince the people whose job it was to keep the plant operating, to instead shut it down for a bearing inspection.

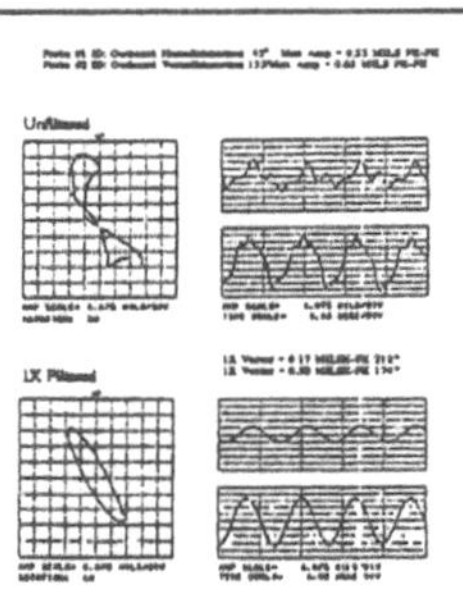

Figure 10.
"High Preload" Orbit on a Gas Compressor

Fortunately, the machinery specialist used additional data to convince Operations Management that there was a problem. The shaft orbit at this bearing location (Figure 10) was not "normal" at all. The shaft appeared to have a strong radial load, producing an orbit similar to the classical "figure 8." In addition, a careful calculation of the assumed shaft position change indicated that the shaft must be **outside** of the original bearing clearance, that is, the bearing must be damaged (Figure 11).

The orientation of the orbit in Figure 10 would normally indicate a strong radial load from the one o'clock direction, or about 90 degrees from the direction indicated by the shaft position change. It was then assumed that the unusual orbit shape must be the result of contact between shaft and bearing at the location indicated by the shaft position change (about 10 o'clock as seen in Figure 11). Disassembly and inspection of the bearing in question confirmed this analysis. There was loss of bearing material on two pads of this tilt pad bearing.

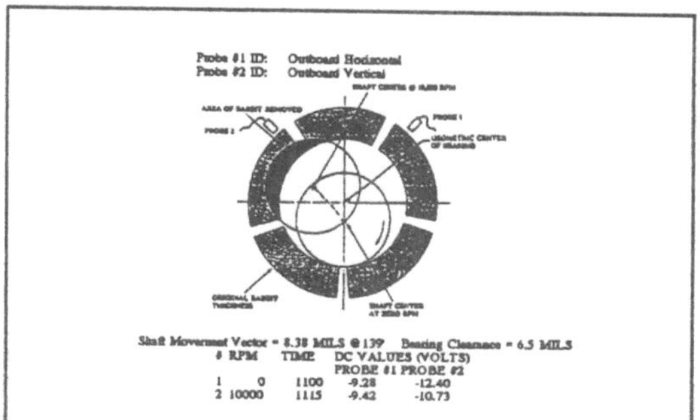

Figure 11.
Shaft Position Change Indicates a Damaged Bearing

DECREASING VIBRATION AMPLITUDE CAN INDICATE A MACHINE PROBLEM

It is generally true that an increase in vibration amplitude represents a machine problem that is getting worse. However, the converse may not be true. Decreasing amplitude does not necessarily mean the problem is getting better. An example is illustrated by a case history. On a vertical pump, 2X vibration amplitude increased over a period of seven days. Then the amplitude decreased over the next nine days (Figure 12). Plant operators correctly assumed that a machine problem was getting worse for the first seven days. But then they incorrectly assumed that for the next nine days the problem was going away.

Fortunately, a machinery specialist was called to look at the data. The specialist saw that the initial amplitude increase was accompanied by a change in phase angle. But when the amplitude began to return to the previous level, the phase value did not return to its previous level. Instead the phase angle continued to change in the same direction. The plot in Figure 12 looks like a classical

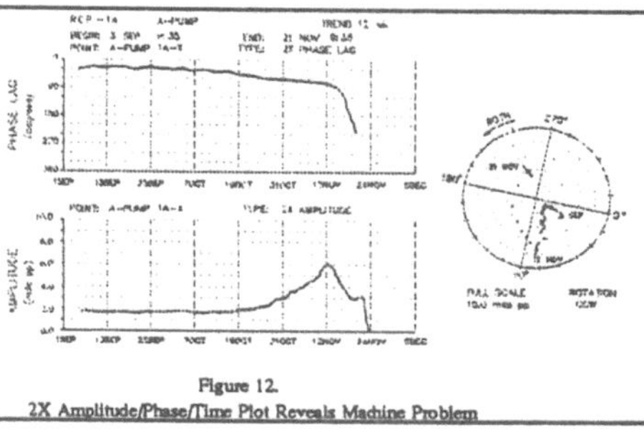

Figure 12.
2X Amplitude/Phase/Time Plot Reveals Machine Problem

transient machine response...an amplitude increase and then decrease, with an associated change in phase angle, as a function of shaft rotative speed...as the rotor passes through its balance resonance. However, in this instance the rotor speed was constant.

Therefore, if the rotor speed was not moving through the region of the resonant frequency, then the resonant frequency must have moved through the rotor speed region. This was the correct assumption. A shaft crack was causing the stiffness of the rotor system to decrease, thereby decreasing the frequency of the resonance.

VIBRATION PHASE ANGLE AS A TREND PARAMETER

Virtually all vibration monitoring systems allow the user to trend amplitude as a function of time. The previous case history is an excellent example of the importance to trend other parameters in addition to vibration amplitude. Vibration phase angle is one of these very important parameters. While vibration at shaft rotative frequency (1X) is usually associated with rotor imbalance, there are many other machine malfunctions that also can produce vibration at this frequency. These include, but are not limited to, loose bearings, excessive bearing clearance (bearing wear), shaft bow (static, as from a bent shaft), shaft resonance excitation, and as shown above, shaft cracks. Even misalignment can cause a change in 1X vibration amplitude **and phase**.

Trending vibration phase angle is a good way to sort through the possible causes of 1X vibration. For example, if resonance excitation is the problem, phase will change with a change in shaft rotative speed. Even rotor imbalance does not necessarily cause a change in the **amplitude** of 1X component. But if the balance state does change, there will be either a change in 1X amplitude, **or 1X phase**, or both.

A good on-line computerized monitoring system will provide the capability to trend not only vibration amplitudes and frequencies, but phase angles as well, for at least the 1X and 2X vibration components. Actually, amplitude and phase values can be monitored together as a true **vector** quantity. Individual software alarm setpoints are set for both amplitude and phase. This technique allows some variation in amplitude or phase (or both) during "normal" machine operation.. It is not uncommon for some vector changes to occur, for example, as a function of machine process conditions. This method of vector monitoring is known also as **Acceptance Region** monitoring (Figure 13). The Bently

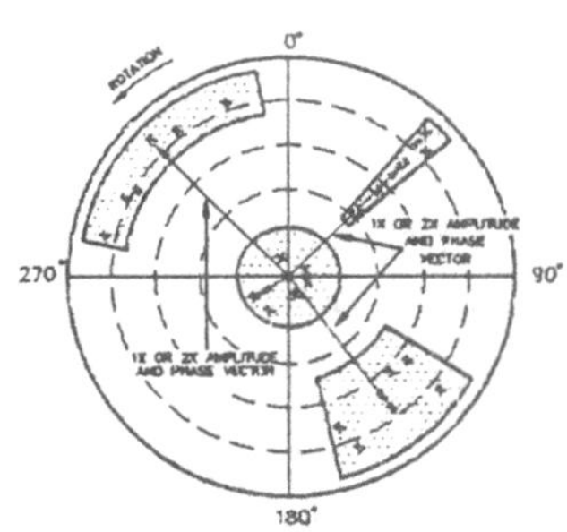

Figure 13. Acceptance Region Monitoring (hardware or software)

Nevada Vector Monitor even provides alarm setpoints (with relay contacts) for phase, in addition to amplitude setpoints for 1X and 2X components.

Phase angle measurement is only possible with a once-per-shaft-revolution reference signal, as provided by a Keyphasor® transducer system. It is highly recommended to install a Keyphasor transducer on all critical **drivers** (turbines, motors, etc.) and on both sides of a gearbox (at least one Keyphasor transducer should be provided for each shaft rotative speed in the machine train).

ENHANCED SPECTRUM ANALYSIS

A recent development by Bently Nevada, called **Full Spectrum**, is an enhancement to traditional frequency spectrum analysis. One disadvantage of the traditional spectrum plot is that it provides no information at all regarding the **direction of motion** of individual frequency components. Many common machine problems produce forward precession of the shaft in the bearing. That is, the direction of the shaft orbit (direction of vibration) is the same as the direction of shaft rotation. However, several malfunctions, including misalignment, rubs, housing resonances, and shaft cracks, can produce **reverse precession** in some circumstances. Under reverse precession, the path of the shaft centerline (orbit) is in the direction opposite to the direction of shaft rotation. The concept is this: determining the precession direction of individual vibration frequency components will provide additional information for classifying particular machine malfunctions.

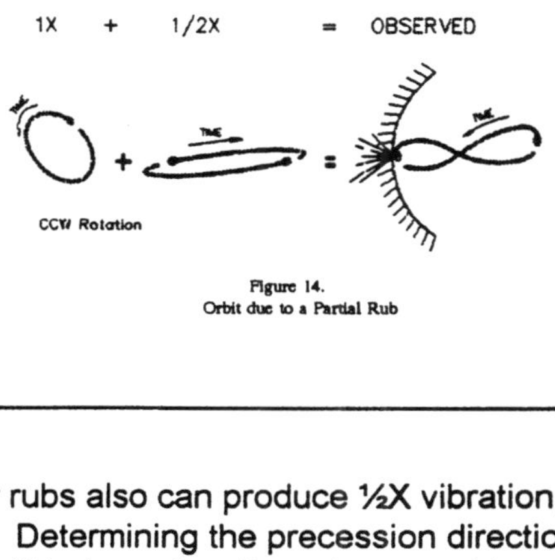

Figure 14.
Orbit due to a Partial Rub

For example, two common machinery problems each can produce vibration at or near ½X. When a fluid-induced instability produces vibration at (or slightly less than) ½X, the direction of precession at that frequency is **always forward**. Rotor rubs also can produce ½X vibration, but the precession direction is usually **reverse**. Determining the precession direction of particular vibration frequency components will help to classify various machinery malfunctions.

The (unfiltered) orbit shape on the right side of Figure 14 results from a partial rotor rub. There are two frequency components in this vibration signal, one at 1X and one at ½X. The 1X component is caused by imbalance. (Indeed it is often excessive imbalance that causes rubs in the first place.) It can be seen that the 1X filtered orbit (on the left of Figure 15) is forward. (In this case, it is assumed that the Keyphasor signal produces a blank space followed by a bright spot on the orbit, with respect to time.) Vibration due to pure imbalance **always** produces forward orbits. The center orbit in Figure 14 shows the motion due to the rub. This is an orbit filtered at ½X, and can be seen to be in reverse direction.

Another way to view this data is to use Bently Nevada's new **Full Spectrum™**. We have developed a signal conditioning technique that differentiates between frequency components that are forward and those that are reverse. Figure 15a shows a conventional frequency spectrum. It can be seen that there is a component at 1X and one at or near ½X. By observation of this data alone, it is not possible to determine if the ½X component is the result of a rub or an instability.

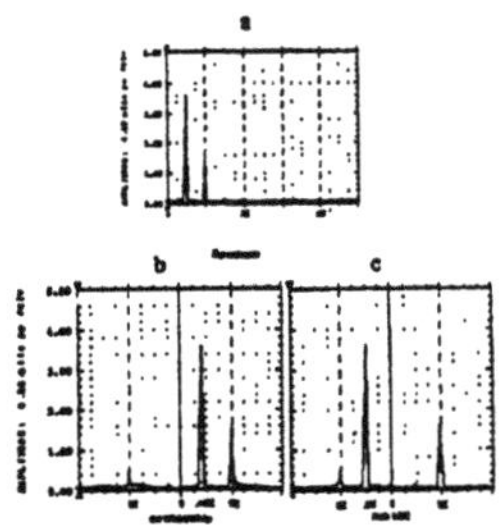

Figure 15a-c.
Plus and Minus Spectrum Presentations

Figure 15b shows the Plus and Minus Spectrum that would result if the problem were a fluid-induced instability. It can be seen that the ½X component is on the positive side of the spectrum presentation. Thus this component's direction is forward, and is most likely due to instability. Figure 15c shows the Plus and Minus Spectrum that would result if the problem were a rub. It can be seen that this time the ½X component is on the negative side of the spectrum presentation. Thus this component's direction is reverse, and is most likely due to a rub.

This same technique also can be used to determine the ellipticity of an orbit at a particular frequency. By definition, a circular orbit can be described by a single rotating vector. An elliptical orbit is described by two vectors rotating in opposite directions. Therefore, a full spectrum for an elliptical orbit will consist of one component on the positive side of the spectrum and another component, at the same frequency, on the negative side. Thus, a Full Spectrum can be used to determine the ellipticity of a filtered orbit.

CONCLUSIONS

It could be possible for a machinery specialist to perform correctly on the job the majority of the time using only the traditional methods of vibration amplitude and frequency analysis. However, in today's competitive marketplace, with a high emphasis placed on productivity, it is not good enough to be right only "a majority of the time." One must attempt to be correct almost **all** of the time. While **perfect accuracy** in decision making may not be attainable, one's chances of success are greatly improved if many different data sources are used, and the comparison of all the data leads to the same conclusion.

Parameters for Vibration Analysis on Rotating Machinery **BENTLY NEVADA CORPORATION**

In addition to amplitude and frequency analysis, many other parameters can be used to confirm one's initial conclusions. These include, but are not limited to, the measurement of average radial shaft position in the bearing, vibration phase angle, the shape of orbits, the direction of shaft precession, and Plus and Minus Spectrum. For diagnostics on critical rotating machinery, where a wrong decision can lead to great expense due to either an unnecessary plant shutdown or continued operation with resulting machinery damage, we strongly recommend that **all available parameters** be used to increase the accuracy of the decision making process.

THE PRACTICAL APPLICATION OF ON-LINE SYSTEMS TO THE DIAGNOSIS OF MACHINE PROBLEMS

P. Bastow and R. Fenner
Schlumberger Technologies, Farnborough, Great Britain

ABSTRACT

The diagnosis of problems on rotating machinery, particularly large turbines, has been significantly advanced in recent years by the use of systems monitoring and analysing vibrations on line. The specification of many of these systems has been developed however from the requirements of the vibration analyst. Consequently the emphasis has been placed on the speed and capacity at which they can gather their data, rather than the presentation of information to the Station engineers in simple formats relevant to the operation of their machine.

This paper describes the features of an on-line turbine monitoring system developed for the UK Power Industry which is now in operation in five UK Power Stations. The communication of information to the operational and maintenance staff is a key feature of the system and has enabled them to fully participate in routinely identifying and resolving turbine problems. Several examples are used to illustrate how the data acquired have been used in this process.

1.0 INTRODUCTION

The 1990's is forecasted to be a decade of change in the electrical power generation industry. The publicly owned utilities of Western Europe are being privatised; the privately owned companies face stiffer competition and therefore need to become more competitive. This is leading to reductions in the number of people directly employed by the generating companies and a greater emphasis placed on contracted-out services. This is particularly true of the technical support departments, where expertise can be consulted as required.

Several factors combine to require the large turboalternators in our power stations to operate for longer periods and at higher efficiencies than originally planned. These include:

a) the economic factor; with the high capital cost of new plant existing stations are having their operational lives extended

b) the environmental factor; the ecological objections to both fossil fuel and nuclear stations demands more economical use of existing sites instead of building new ones.

More flexibility is also expected of the generating plant, particularly the turbogenerators. These machines are now expected to operate within different regimes to the original specification including:

a) Multiple run-ups during a twenty-four hour period and a minimum time to achieve synchronisation to the supply

b) Frequent changes of load size and type

c) Upgraded MW and MVAR operating levels

In view of the above, it is important that vibration and other data are collected from the turbine during all of its operating modes. This data, viewed with respect to time, will gradually contribute to a sound understanding of the condition of the machine, and increase the operators ability to diagnose and understand potential problems and apply corrections before damage is caused to the plant or revenue is lost by having to reduce output power.

In order to achieve higher availability, shorter outages and implement a predictive maintenance strategy, an increased awareness of the condition of all the associated plant is required. This in turn requires a sophisticated monitoring system, capable of regularly and frequently measuring a number of points on all of the machines which are critical to the production output. The data produced by this system must be presented in a variety of formats, suitable for interpretation by various grades of plant personnel, including the control room operators and engineering staff.

The use of vibration analysis is emerging as a valuable technique to assist in the diagnosis of faults or adverse operating conditions within rotating machinery while the machine is still running. However, vibration data is rarely sufficient in isolation; bearing temperatures and oil pressure, power output, eccentricity, steam temperature and many other parameters are both useful and necessary in determining the exact nature of the problem, it's severity, and deciding on the most appropriate course of action.

2.0 SYSTEM USERS

The main users of a turbogenerator monitoring system can be divided into two general categories:

1. Control room staff

2. Engineering staff

The control room is responsible for the day-to-day operation of the generating plant, including output to the distribution system, plant availability and overall efficiency.

Engineering, which may be located on the power station site, at the utilities central technical centre or contracted out to a consulting company, are responsible for the overall condition of the machinery and for providing the necessary expertise to diagnose and overcome operational problems as they occur.

These two user groups require different displays and data presentation, but a considerable amount of interaction. For example, it is usual for the engineering group to configure the operators display set together with the default screens for each machine state.

2.1 Control Room Operators

The control room requires a system which will automatically display vibration parameters which are relevant to the current operating state of the plant, highlighting potential problem areas. The system will detect machine state changes and modify the displays accordingly. The displays must be informative, but able to be interpreted at a glance. Small changes in the operating behaviour of the machine can often be smoothly and quickly corrected by the station staff before they become a significant event. Characteristically these displays will show:

a) Run-down data for one or more bearings [1]. This is typically displayed as a Bode diagram of vibration amplitude and phase against machine speed with a reference trace included for comparison purposes. These parameters are required to be displayed in real time.

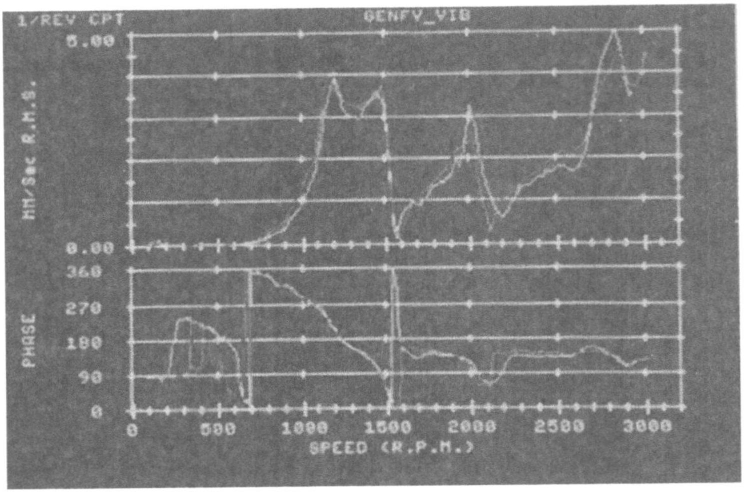

Figure 1, Bode diagram of run-down, with reference trace.

b) On-load data shown as a bar chart, [2], with each of the bars displayed below the appropriate bearing on a mimic diagram representing the particular plant item. The bars of the chart change colour when a preset alert level is exceeded, shown in this figure as a horizontal line. Vibration is normally displayed for all bearings, but in this example eccentricity is also shown for the HP and IP stages.

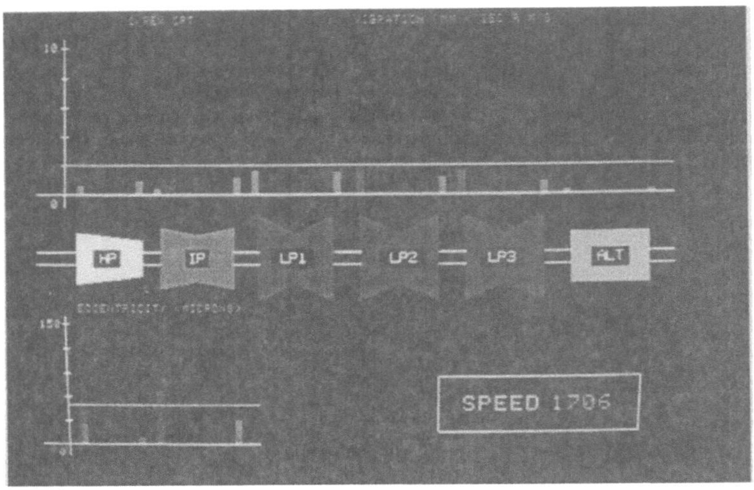

Figure 2, Default Screen for On-load Operation

c) Real time run-up data to show the current machine condition with respect to reference data. This display can assist the operator to accelerate the machine through it's critical speeds. The vibration amplitude can be displayed against speed [3] or in polar form [4].

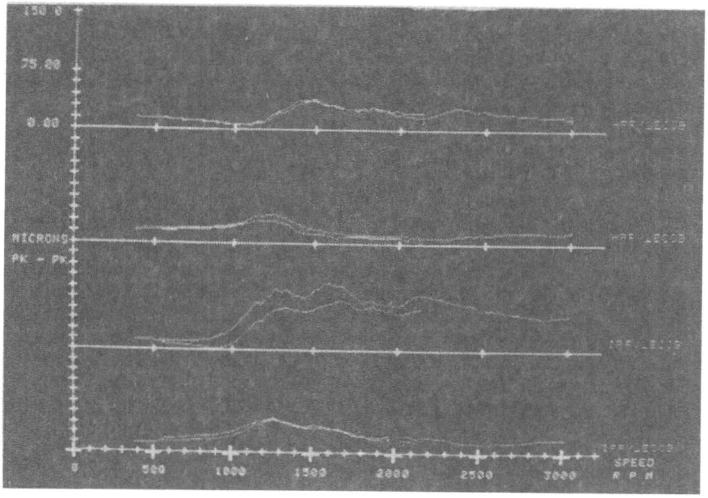

Figure 3, Run-up displaying eccentricity against speed for four bearings, with reference trace.

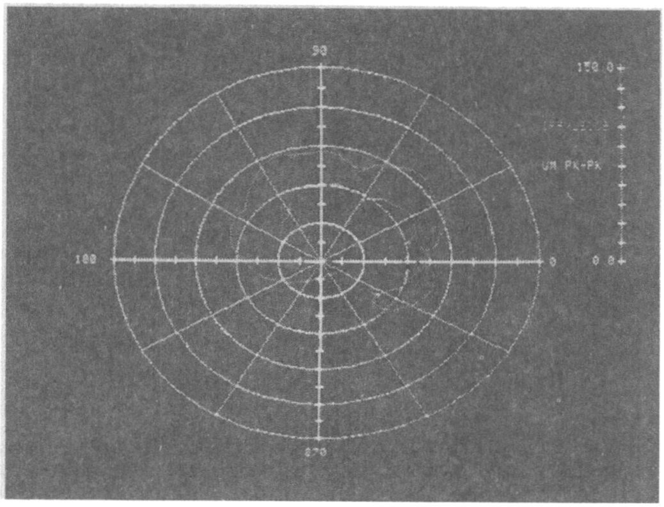

Figure 4, Run-up displayed in polar form, with reference trace

The condition monitoring system must be designed and integrated in such a way that it is an essential component of a tripartite group consisting of the control room operator, the technical specialist and the data measured by the system. Operator intervention is possible via a touch screen with a restricted menu structure. This allows selection of alternative displays.

For a condition monitoring system to be truly effective, the control room staff must be alerted to any abnormal conditions as they occur. This alert will be given before the plant supervisory system detects an alarm condition and will operate on a wider range of parameters, for example, vector alerts on vibration parameters. The control room will then decide whether specialist assistance is required.

2.2 The Vibration Specialist

Diagnosing the cause of vibration within rotating machines requires, among other things, the flexibility to display and compare data from numerous sources. It has been suggested that of the problems detected by unusual vibration in rotating machines, 90% are attributable to thermal irregularities. The ability to combine a number of different parameters on a single display against time allows events to be closely correlated with each other. The multiplot shown in [5] can display any selected parameters, dynamic or static, against a selected time base. The vertical axis is scaled as 0 to 100% of the scale values printed at the side of the display.

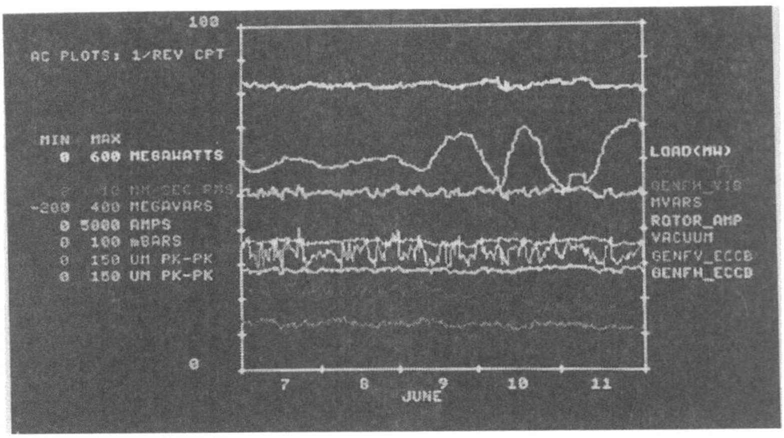

Figure 5, Example of a multiplot.

On detecting an alert condition, the monitoring system will store high resolution data for a period of time prior to the event in a separate buffer. Additionally, spectral data may have been stored according to pre-configured instructions. This data, together with the normal database of the system, will provide the specialist with an insight to the machine condition and any unusual events leading up to the alert state.

The specialist combines this data from the monitoring system, his knowledge and experience of the machinery and both the operational and intuitive knowledge of the control room operator to diagnose the nature and location of the problem. This diagnosis is performed in association with the operator; this retains his interest in the condition and assures his active participation in the correction.

As a result of this, the control room will then be given a modified operating procedure for the plant, and use the monitoring system to regulate this.

3.0 The On-Line Vibration Monitoring System

The on-line vibration monitoring system was originally developed to monitor turbogenerator units during warm starts.

The system used in the following case studies consists of an HP3497 scanner, Schlumberger 1250/1251 Multichannel Frequency Response Analysers for harmonic analysis of the vibration signals and an HP A600 computer for control of the monitoring processes, display generation and storage of data. Software has been developed for plotting both real time and historical data in a variety of different formats. Subsequent to these particular studies the complete monitoring system has been updated and significantly expanded in capability.

At each of the twelve main turbogenerator bearings two eddy current probes are mounted in the bearing, one in the vertical and one in the horizontal plane, measuring the relative shaft-to-bearing motion. Two Sensonics VEL velocity transducers are mounted on each main bearing pedestal similarly in the vertical and horizontal plane. The system validates the output from each transducer prior to entry into the database.

When the unit is on load, sixty ac signals from the pedestal vibration transducers and shaft probes are harmonically analysed to give the amplitude and phase values for the 1x, 2x and 3x components on each scan. This information is stored together with the values of 92 dc (static) signals from the turbovisory system. One complete scan of the turbogenerator unit takes approximately 5 minutes and is initiated every 11 minutes, there being 128 scans per day in steady state operation. The system automatically detects the operating state of the turbogenerator, using four separate criteria to validate the selection.

4.0 CASE STUDIES

4.1 Light Rubbing of the HP Rotor

The overall rms level of HP eccentricity as displayed on the turbovisory and chart recorders began to vary in an erratic manner, whilst the machine was on steady load. This happened for short periods, between which the eccentricity returned to normal. The level would gradually increase by maybe 100% to a value beyond 100 micrometers peak to peak, remaining there for an hour before decreasing. Other abnormal behaviour included an increase in overall level followed by a decrease then an increase again, following a cyclic pattern, the mean period being constant at anything between half an hour and two hours. Several cycles would be completed before stabilisation.

After several attempts by Operations staff to influence this by opening and closing governor valves, it was apparent that it was not possible to control events in a satisfactory manner. A period of intensive monitoring of the shaft probes during one excursion, however, clarified the situation. The once per revolution amplitude and phase from the probes was recorded at approximately six minute intervals for over two hours. The results can be seen in [6]. Associated with each amplitude change was a significant phase change. Movements at either end of the HP and IP rotors occurred in an anti phase manner and by far the largest amplitude variation was observed at the HP rear bearing, a location not monitored by the station turbovisory equipment. This "slow roll" of the shaft vibration vector is known to be indicative of a shaft rub. Thus from this data and from data acquired during other excursions it was predicted that the HP shaft was rubbing somewhere very close to the HP rear bearing.

During the next suitable outage the pedestal cover was removed and the pedestal oil wiper was found to be full of carbon, caused by rubbing on the shaft. When the machine was brought back into service the eccentricity problem had disappeared. Over twelve months later the problem reappeared and was immediately diagnosed from the on load shaft data and carbon removed from the oil wiper at the next outage. [7] shows the "projected" modes of the HP and IP shafts (the measured components when projected along the HP front vertical vector) at the start and halfway through the excursion in [6]. It is clearly evident from these diagrams where the problem lay.

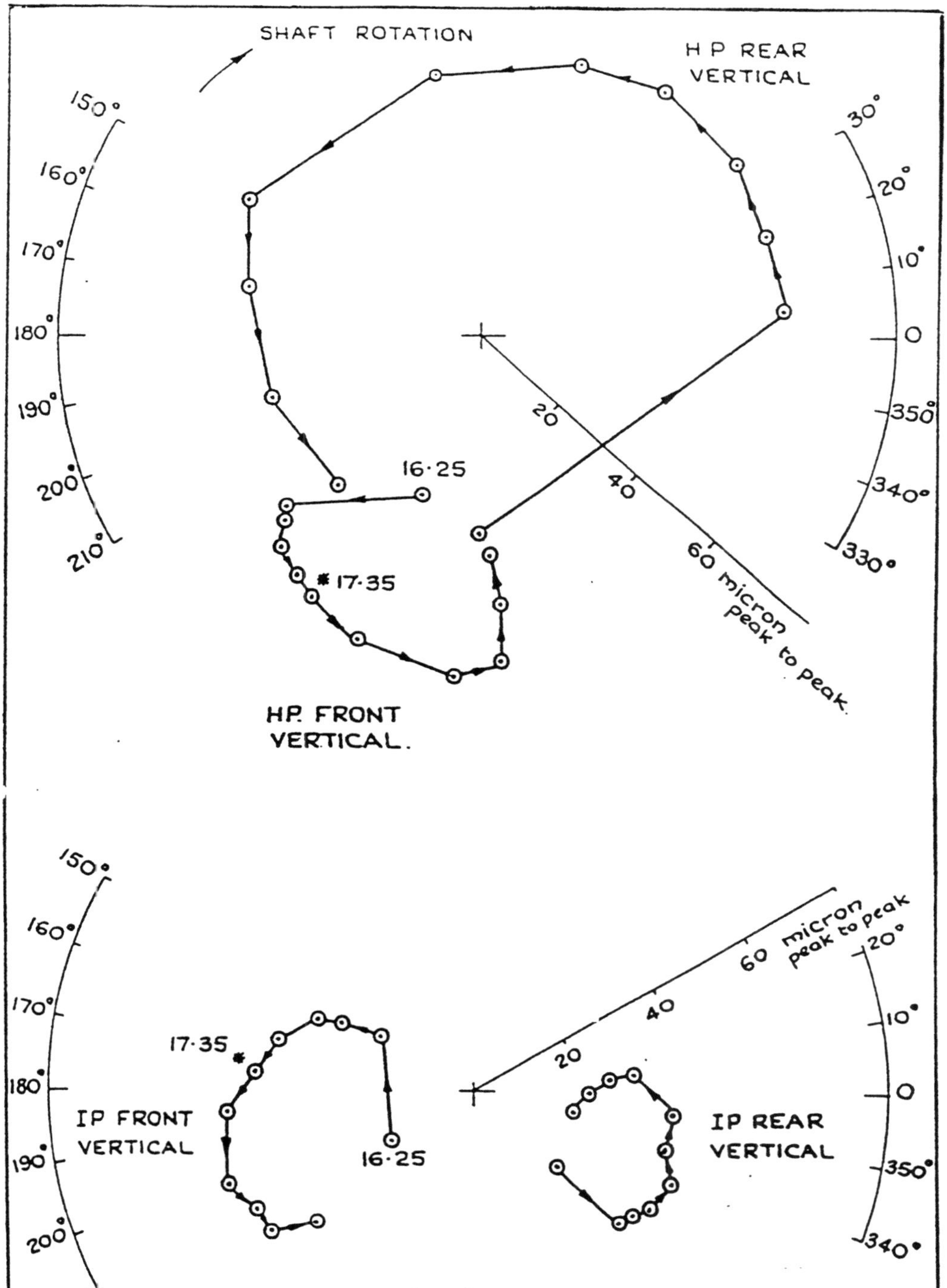

SHAFT ROTATION
H P REAR
VERTICAL
150°
160°
170°
180°
190°
200°
210°
30°
20°
10°
0
350°
340°
330°
20
40
60 micron
peak to peak
16·25
17·35
HP. FRONT
VERTICAL.
IP FRONT
VERTICAL
IP REAR
VERTICAL

Fig. 6

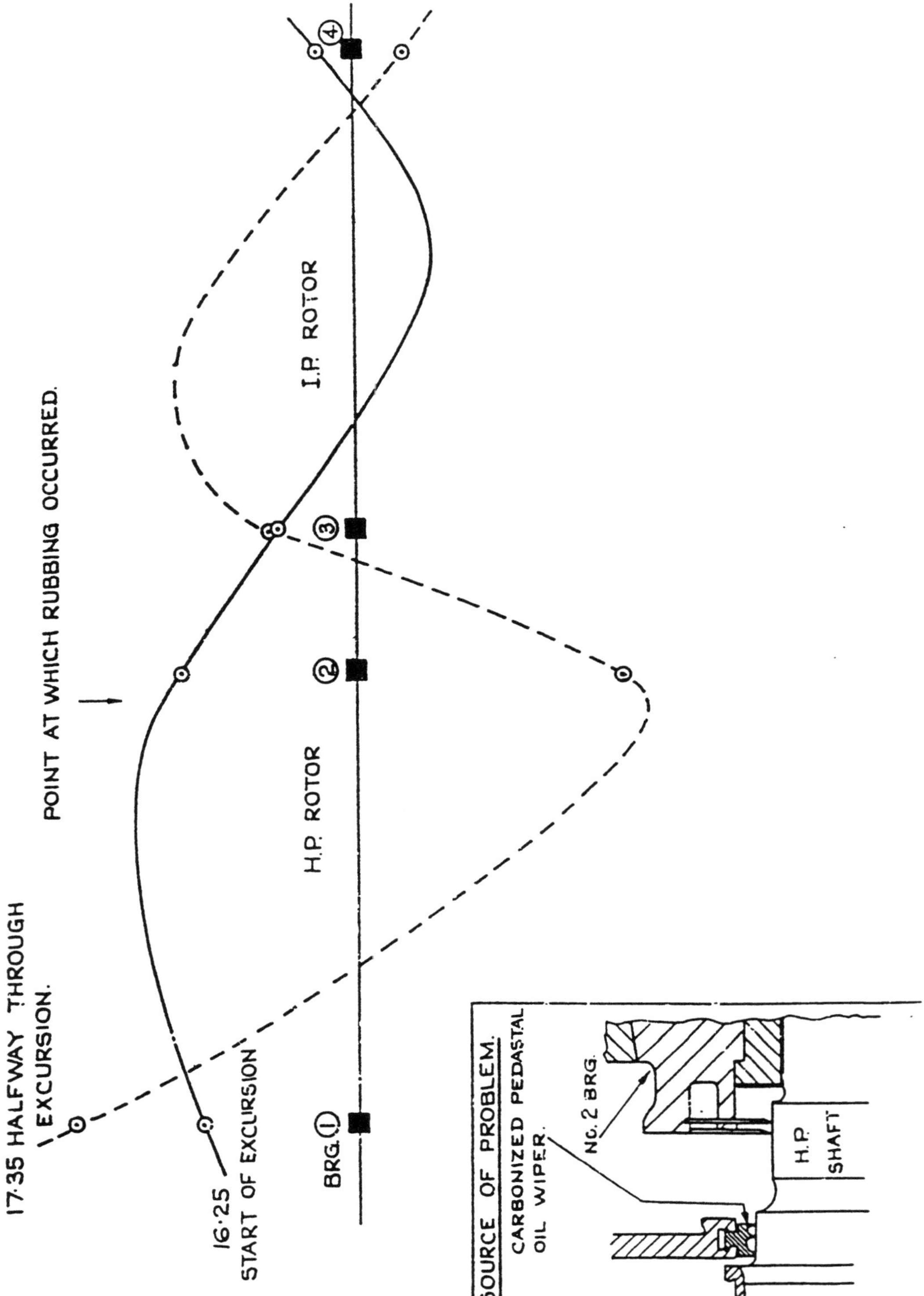
17·35 HALFWAY THROUGH
EXCURSION.
16·25
START OF EXCURSION
POINT AT WHICH RUBBING OCCURRED.
BRG. 1
2
3
4
H.P. ROTOR
I.P. ROTOR
SOURCE OF PROBLEM.
CARBONIZED PEDASTAL
OIL WIPER.
No. 2 BRG.
H.P.
SHAFT

Fig. 7

4.2 Loss of IP Shrouding

A turbogenerator unit was operating at a steady load of around 430 MW. Suddenly there was a dramatic increase in the overall level of HP and IP eccentricity, which was observed on the station eccentricity trace. The machine load was immediately reduced to 200 MW but there was no accompanying reduction in the level of IP eccentricity, which remained steady. Over the next hour governor valves were closed in sequence in an attempt to reduce the levels. This exercise caused a variation in the HP eccentricity but the IP eccentricity was not significantly affected. Several sets of data were acquired in quick succession. This was done in order to observe any trends in the data caused by for instance a transient rub, where the levels would return to normal after a time interval. No such trends were observed and it was concluded that material had been lost from the rotor so the machine was taken out of service immediately., The signals from the four vertical shaft probes were then recorded during the rundown.

The once per revolution vibration vectors of the shaft as recorded before and after the first event are shown in [8a]. The vectors representing the resultant changes in shaft vibration are plotted in [8b]. It can be clearly seen that the vector changes that occurred at the ends of the IP rotor were in phase and the largest change was at bearing three. The vector changes at the ends of the HP rotor were out of phase indicating that the source of excitation was closer to the coupling end, and likely to be on the IP rotor. When the vectors, before and after the first event were plotted as "projected" modes as in [9], it appeared that the defect was in the IP front flow.

The once per revolution component of shaft vibration was recorded during two rundowns. When plotted in vector form as in [10] this confirmed the on load diagnosis. Amplitudes at both bearings had increased, with the increase at three being greater than at four, and they had also moved together through an angle of 90°. The unit was taken out of service and it was discovered that shrouding had been lost from the first two blades rows in the IP front flow. The bearing pedestals on these units are very stiff. So several months earlier when pedestal vibrations were being monitored on rundown after shrouding had been lost from the first stages of the IP front flow of another unit, the changes observed in the pedestal vibration were nothing like as clear and conclusive as the changes in shaft vibration observed on this unit.

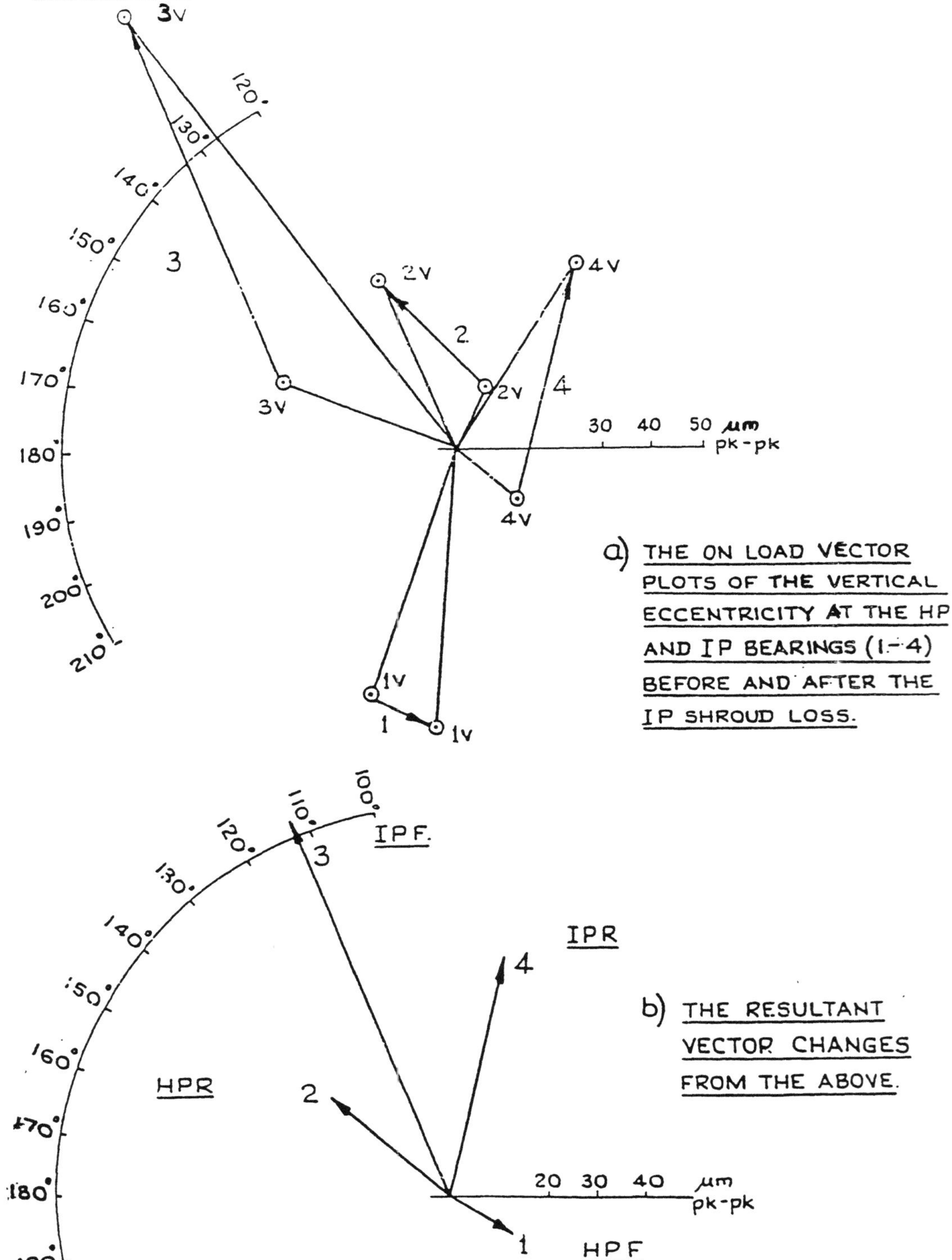

3V
120°
130°
140°
150°
160°
170°
180°
190°
200°
210°
3
2V
2
4V
4
2V
3V
30 40 50 μm
pk-pk
4V
a) THE ON LOAD VECTOR PLOTS OF THE VERTICAL ECCENTRICITY AT THE HP AND IP BEARINGS (1–4) BEFORE AND AFTER THE IP SHROUD LOSS.
1V
1
1V
100°
110°
120°
130°
140°
150°
160°
170°
180°
190°
IPF.
3
IPR
4
b) THE RESULTANT VECTOR CHANGES FROM THE ABOVE.
HPR
2
20 30 40 μm
pk-pk
1
HPF

Fig. 8

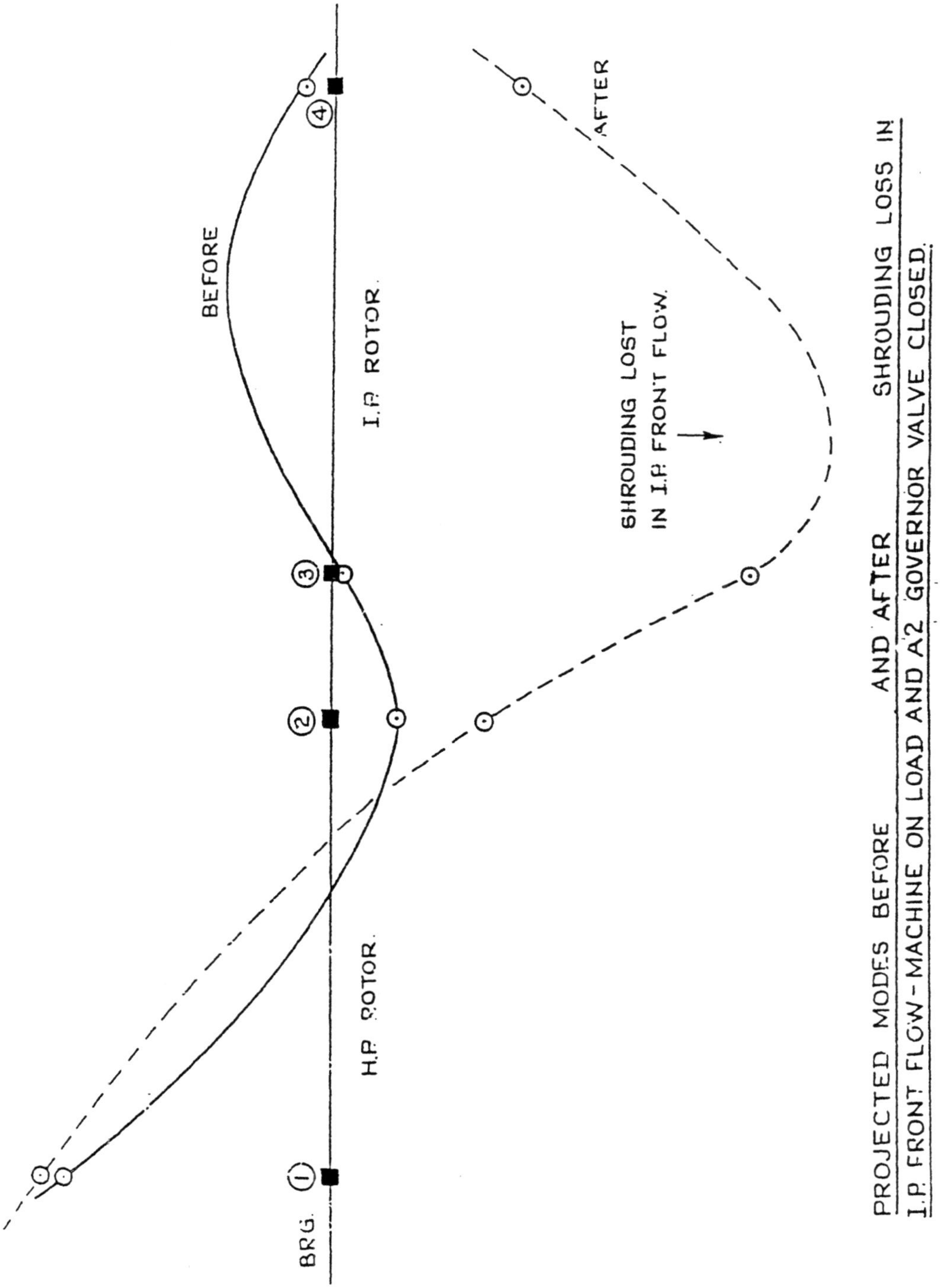
BEFORE
AFTER
I.P. ROTOR
H.P. ROTOR
SHROUDING LOST
IN I.P. FRONT FLOW.
BRG.
1
2
3
4
PROJECTED MODES BEFORE AND AFTER SHROUDING LOSS IN
I.P. FRONT FLOW - MACHINE ON LOAD AND A2 GOVERNOR VALVE CLOSED.

Fig. 9

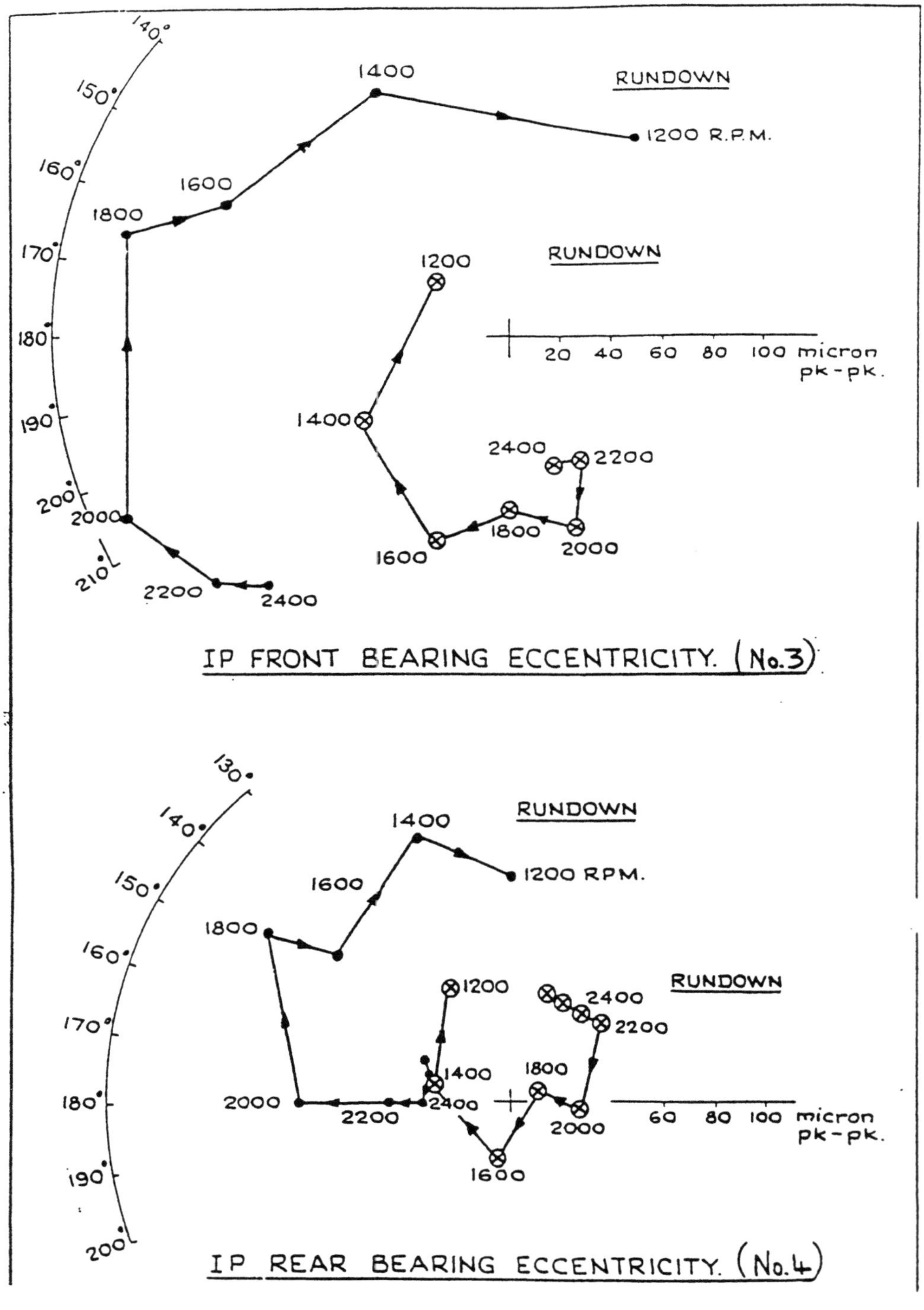

RUNDOWN
1200 R.P.M.
1400
1600
1800
2000
2200
2400
140°
150°
160°
170°
180°
190°
200°
210°
RUNDOWN
1200
1400
1600
1800
2000
2200
2400
20 40 60 80 100 micron pk-pk.
IP FRONT BEARING ECCENTRICITY. (No.3)
RUNDOWN
1200 RPM.
1400
1600
1800
2000
2200
2400
130°
140°
150°
160°
170°
180°
190°
200°
RUNDOWN
1200
1400
1600
1800
2000
2200
2400
60 80 100 micron pk-pk.
IP REAR BEARING ECCENTRICITY (No.4)

Fig. 10

4.3 Mechanical Slip-Stick in Generator Rotor Bars

A turbogenerator unit was returned to service after a major overhaul with a new generator rotor. On return to service pedestal vibration levels at the front and rear generator bearing were entirely satisfactory, at values between 0.5 and 1.5 mm/s rms.

A series of tests were carried out at a generator load of 480MW, in order to examine the electrical integrity of the rotor; these involved changing the rotor current and generator reactive load. The reactive load was increased from 100 to 200MVAR with little effect on the vibration level. When the load was further increased to 300MVAR the vibration on the generator pedestal, both front and rear, increased sharply and remained at the new level even when the reactive load was reduced to 100MVAR. The vertical pedestal vibration levels finally settled out at approximately 5mm/s rms at the front and 3.5 mm/s at the rear where they remained until the next outage. When the machine was next loaded up, the vibration levels were found to be at the post-overhaul values of 0.5 to 1.5mm/s. In addition to the specific information acquired from the turbovisory and other instrumentation, the turbogenerator was being monitored, as is now the normal procedure, by an on-line vibration monitoring system. This study briefly describes the information gathered by the on-line monitoring system before, during and after the electrical tests.

The front and rear main generator bearings (the subject of this study) provide eight of the ac signals monitored (4 pedestal and 4 shaft) Recorded together with these are static signals relevant to the generator operation including: generator load (MW), generator reactive load (MVARS), generator rotor current, hydrogen gas temperature and pressure.

The generator front and rear vertical pedestal vibrations have been chosen for describing the dynamic changes associated with this generator rotor incident. The other six available generator ac signals (horizontal and vertical shaft and horizontal pedestal) have been analysed and concur with this data. The amplitude of the once per revolution component of these generator front and rear vertical vibrations (GENFV_VIB and GENRV_VIB) is plotted along with load (LOAD (MW), reactive load (MVAR) and rotor current (ROTOR_AMP) in figures 11 to 14 against time, for separate periods covering between 3 and 7 days. The figures are presented in colour to assist with interpretation and time synchronisation of the traces. The relationship between the scale associated with each label and the common 0 to 100% Y-axis is given at the left hand side of each figure, i.e. vibration is 0 to 10 mm/s rms, load is 0 to 600MW, reactive load is -200 to +400MVAR and rotor current is 0 to 5000A.

When the unit first returned to service from major overhaul the generator pedestal vibration levels settled out at values between 0.5 and 1.5 mm/s rms. The reactive power however did not exceed +130MVARS. [11] shows this initial situation. What is clearly evident from [11] is the sudden and permanent change in vibration at the generator pedestals associated with the increase in reactive power to +300MVAR. Other points to note from [11]:

1. Although the reactive power was increased to approximately +170MVAR this had no significant effect on generator pedestal vibration.

2. The dependency of the vibrations on reactive power and rotor current (not shown in the figure) seems more pronounced after the tests than before.

[12] again shows the chosen variables over a different period. At 20.30 the machine came out of service and was returned to service at 11.46 on the following day. The most important feature of [12] to note is the return of the generator pedestal vibrations to a lower level after a short outage. These levels are similar to those measured post overhaul. [12] shows that the operating regimes of the reactive power and rotor current parameters, before and after the short outage, are similar.

[13] presents the data acquired and shown in [11]. This shows the test data in much more detail. Associated with the changes in generator pedestal vibration amplitudes are changes in the phase of the vibration. This is best represented by plotting the vibration data in polar form as in [14], the arrows show the direction of movement of the measured points. A large in-phase component of vibration is evident indicating a bend in the rotor.

[15] shows the amplitude and phase of the once per revolution component of vibration for the generator front and rear vertical vibrations (GENFV_VIB and GENRV_VIB) i.e. bearings 11 and 12, during rundown. The dashed curve is the rundown soon after the electrical tests when the levels were high. This rundown is compared with a later rundown, when the vibration levels had reverted to their lower operational level and can be considered "normal". The earlier rundown is showing a bend in the generator rotor. The phases of the two signals remain apart for the rundowns until a speed of around 800 rpm is reached when they suddenly come together ([15] bearing 12). This may be due to the release of the bend at that point on the curve.

Conclusions

From [11], [12], [14] and [15] it appears that changing the reactive power during the electrical tests induced a temporary bend in the generator rotor which was sustained and remained unchanged until the next outage when the rotor straightened either at a speed of 800 rpm during rundown or on barring. Such symptoms must be associated with sticking components e.g. wedge bars on the generator rotor.

From [11] the previous highest value of reactive power was +170MVAR and this caused no significant changes in the pedestal vibrations. In [13] the small drop in the GENFV_VIB level began when the reactive power had reached +180MVAR whilst rising up to +200MVAR where there was a pause. The upturn in the GENFV_VIB did not begin until the reactive power had reached +300MVAR. When the reactive power was held at a steady +300MVAR there was a rapid upward trend in both front and rear generator pedestal vibrations which then appears, from [13], to be settling at levels of 3.3 mm/s and 2.1 mm/s rms respectively. On subsequent reduction of the reactive power towards +100MVAR the generator vibrations turned upwards again, when passing through +170MVAR [13]. It must be noted that, as the signals are scanned sequentially, there is a time lapse of approximately 12 seconds between measurement of the vibration signal and the reactive power, the vibration being measured first. This time lapse is not considered to be significant here.

To understand why the GENFV_VIB amplitude dips before increasing on [13] a closer examination of the polar plot [14] was made. [16] shows those points presented in [14] which were measured between the times 10.00 hr and 13.00 hr when the most significant change took place. The picture can be broken into four general movements [16].

1. An in-phase movement of both front and rear vectors by a small amount in a direction of 280° as the generator reactive power was increased from +100 to +170MVAR (from 10.00 to 11.00), represented on the plot as point 'A'.

2. As the reactive power was increased from +200 to +300 MVAR, the vectors moved out of phase by an equally small amount to the point 'B'. This is manifested as a reduction in the GENFV_VIB amplitude [13] between 11.00 and 11.40 hours.

3. When the reactive power is constant at +300 MVAR the vectors move a significant amount from 'B' to 'C'.

4. As the reactive power is reduced the vectors continue to move still further in the same direction from 'C' to 'D'.

An explanation of these events is that the initial increase of the reactive power to +170MVAR causes a small thermal bend (to 'A') in a direction of 280°, producing the vibration vectors at point 'A' in [16]. A further rise in the reactive power to +300MVAR results in an increased thermal bend, and also a slight slippage of a rotor bar at one end, moving the vectors to point 'B'. This slippage increases rapidly with the power steady at +300 MVAR, as the current further heats and expands the rotor bars, inducing a more serious bend in the 90° direction. Having completed the slippage both the bend and vibration stabilise until the reactive power is reduced to +100MVAR again. The resulting contraction of the stuck rotor bar now causes further bending of the shaft, taking the vectors to point 'D' in the figure.

Following these dramatic changes in the vibration amplitudes on the generator bearings, there was an increased sensitivity to changes in the reactive power. [17] summarises this relationship for both the abnormal and normal states.

The evidence suggests there is a mechanical slip-stick action associated with the components of this generator rotor which can be induced when operating at high reactive power levels.

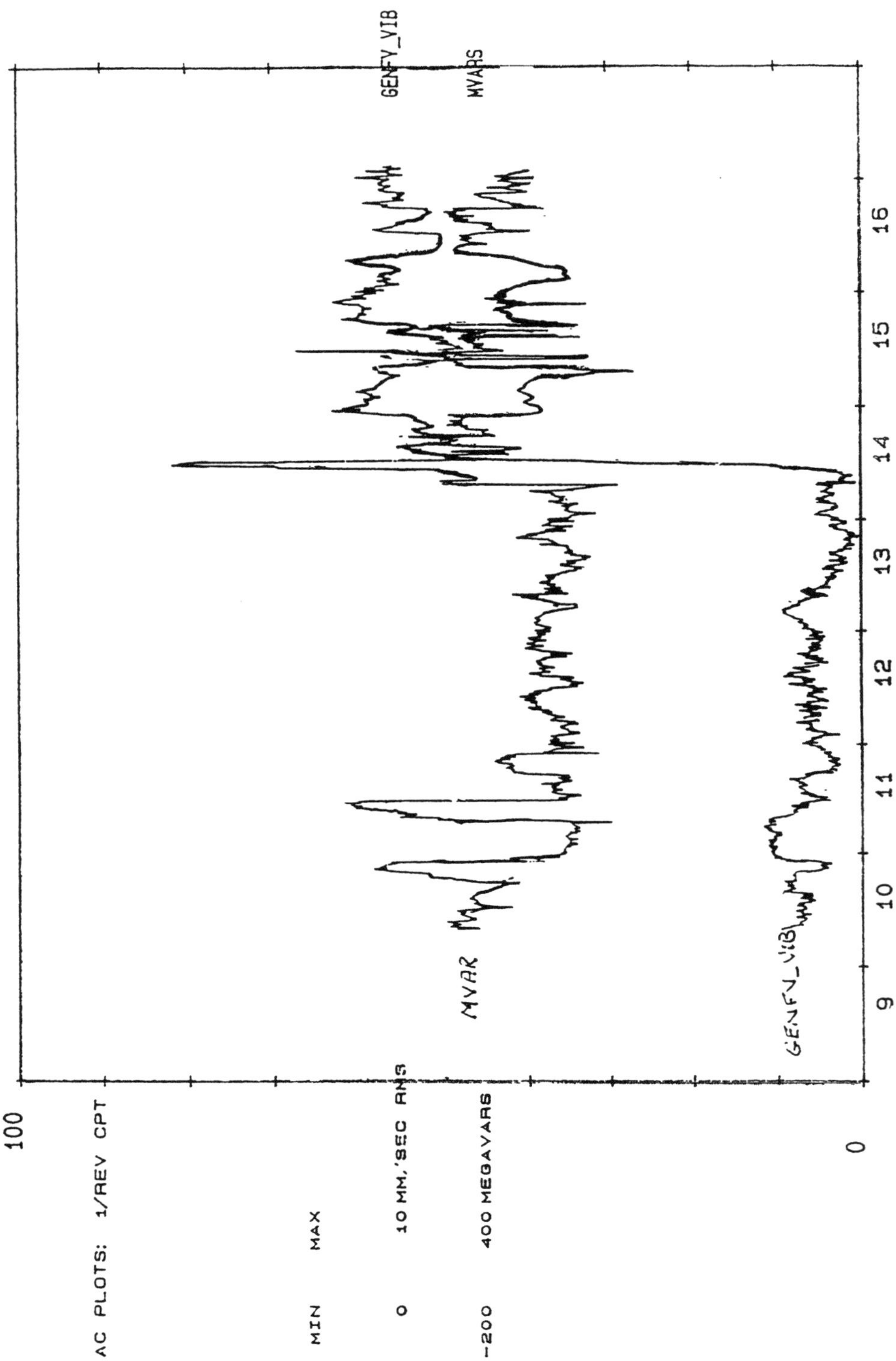
AC PLOTS: 1/REV CPT
MIN MAX
0 10 MM/SEC RMS
-200 400 MEGAVARS
100
0
GENFV_VIB
MVARS
MVAR
GENFV_VIB
9
10
11
12
13
14
15
16

Fig. 11

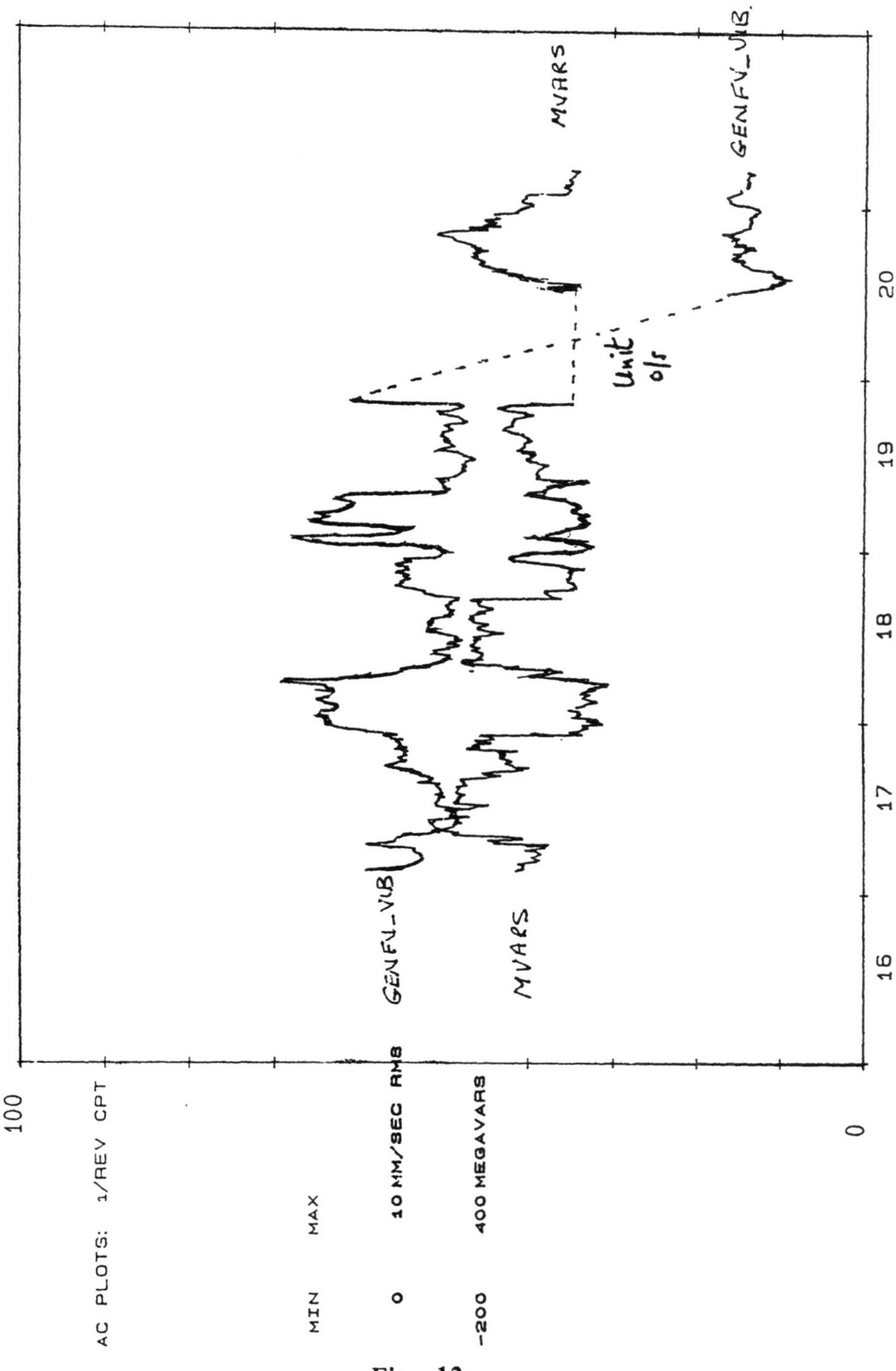
AC PLOTS: 1/REV CPT
MIN MAX
0 10 MM/SEC RMS
-200 400 MEGAVARS
100
0
16
17
18
19
20
GENFV_VIB
MVARS
MVARS
GENFV_VIB
Unit o/s

Fig. 12

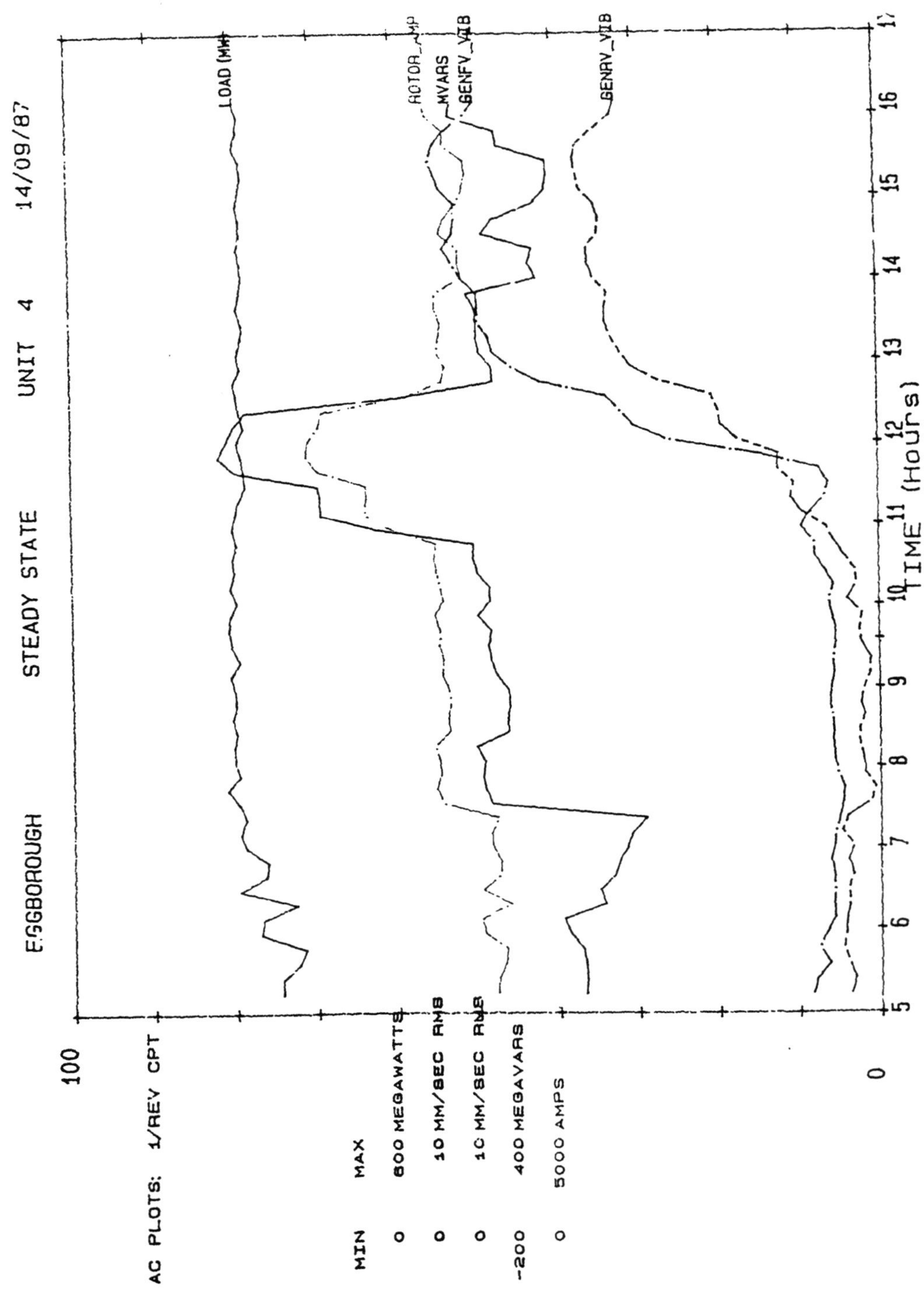
ESGBOROUGH
STEADY STATE
UNIT 4
14/09/87
LOAD (MW)
ROTOR_AMP
MVARS
GENFV_VIB
GENAV_VIB
TIME (Hours)
5
6
7
8
9
10
11
12
13
14
15
16
17
100
0
AC PLOTS: 1/REV CPT
MIN MAX
0 600 MEGAWATTS
0 10 MM/SEC RMS
0 10 MM/SEC RMS
-200 400 MEGAVARS
0 5000 AMPS

Fig. 13

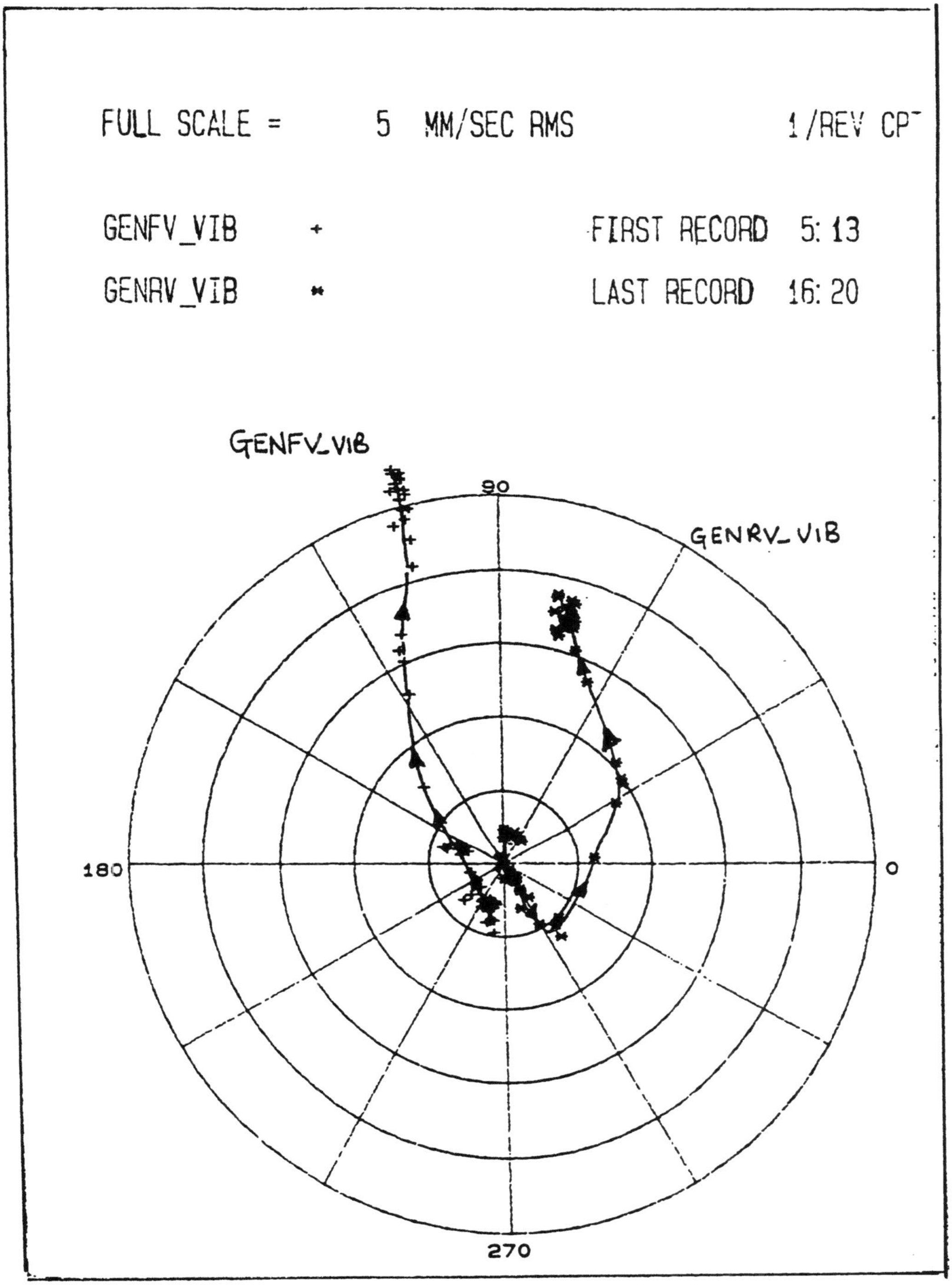

PROGRESSION OF GENERATOR FRONT AND REAR VIBRATION VECTORS

Fig. 14

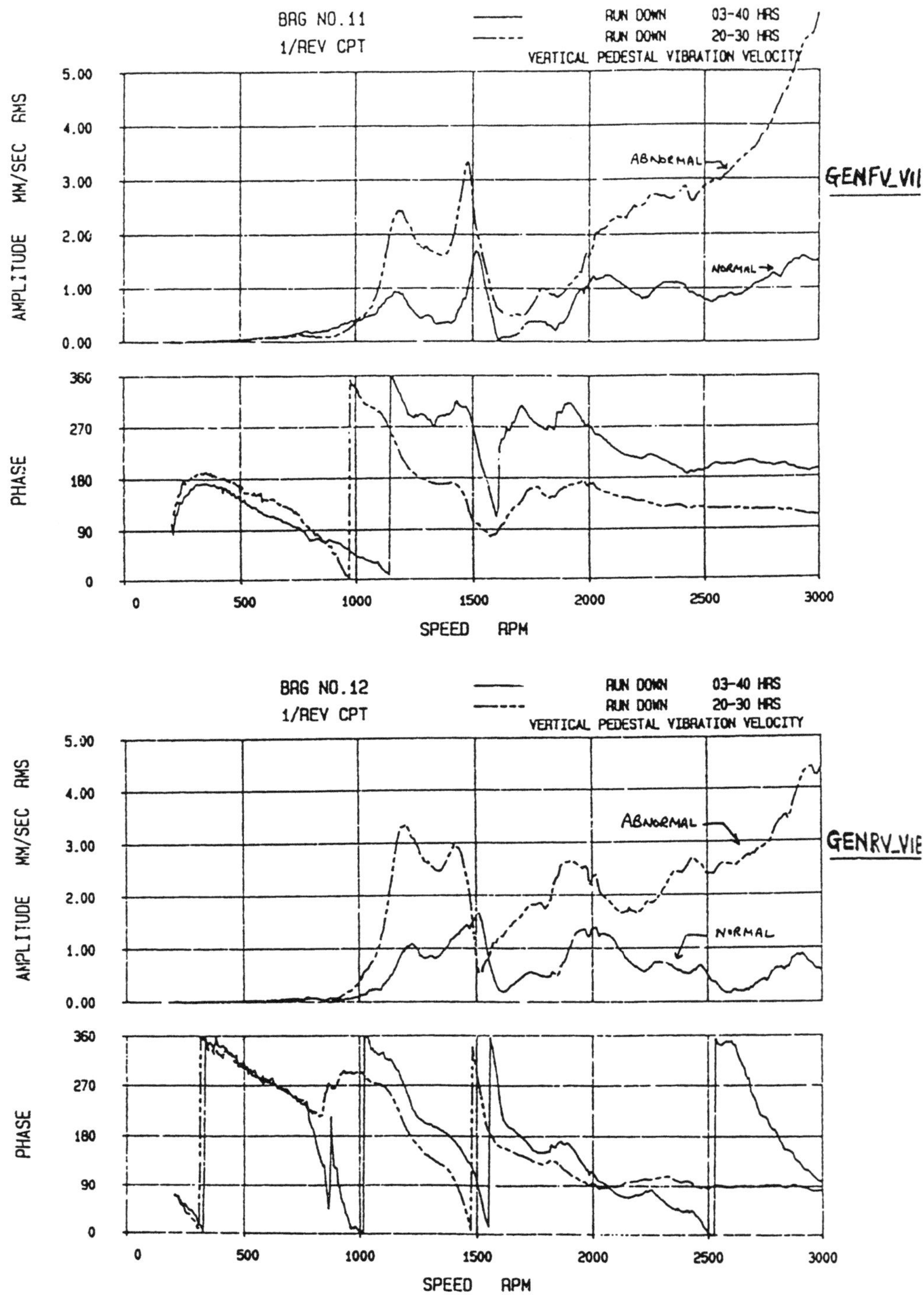
BRG NO.11
1/REV CPT
RUN DOWN 03-40 HRS
RUN DOWN 20-30 HRS
VERTICAL PEDESTAL VIBRATION VELOCITY
AMPLITUDE MM/SEC RMS
5.00
4.00
3.00
2.00
1.00
0.00
ABNORMAL
NORMAL
GENFV_VI
PHASE
360
270
180
90
0
0
500
1000
1500
2000
2500
3000
SPEED RPM
BRG NO.12
1/REV CPT
RUN DOWN 03-40 HRS
RUN DOWN 20-30 HRS
VERTICAL PEDESTAL VIBRATION VELOCITY
AMPLITUDE MM/SEC RMS
5.00
4.00
3.00
2.00
1.00
0.00
ABNORMAL
NORMAL
GENRV_VI
PHASE
360
270
180
90
0
0
500
1000
1500
2000
2500
3000
SPEED RPM

Fig. 15

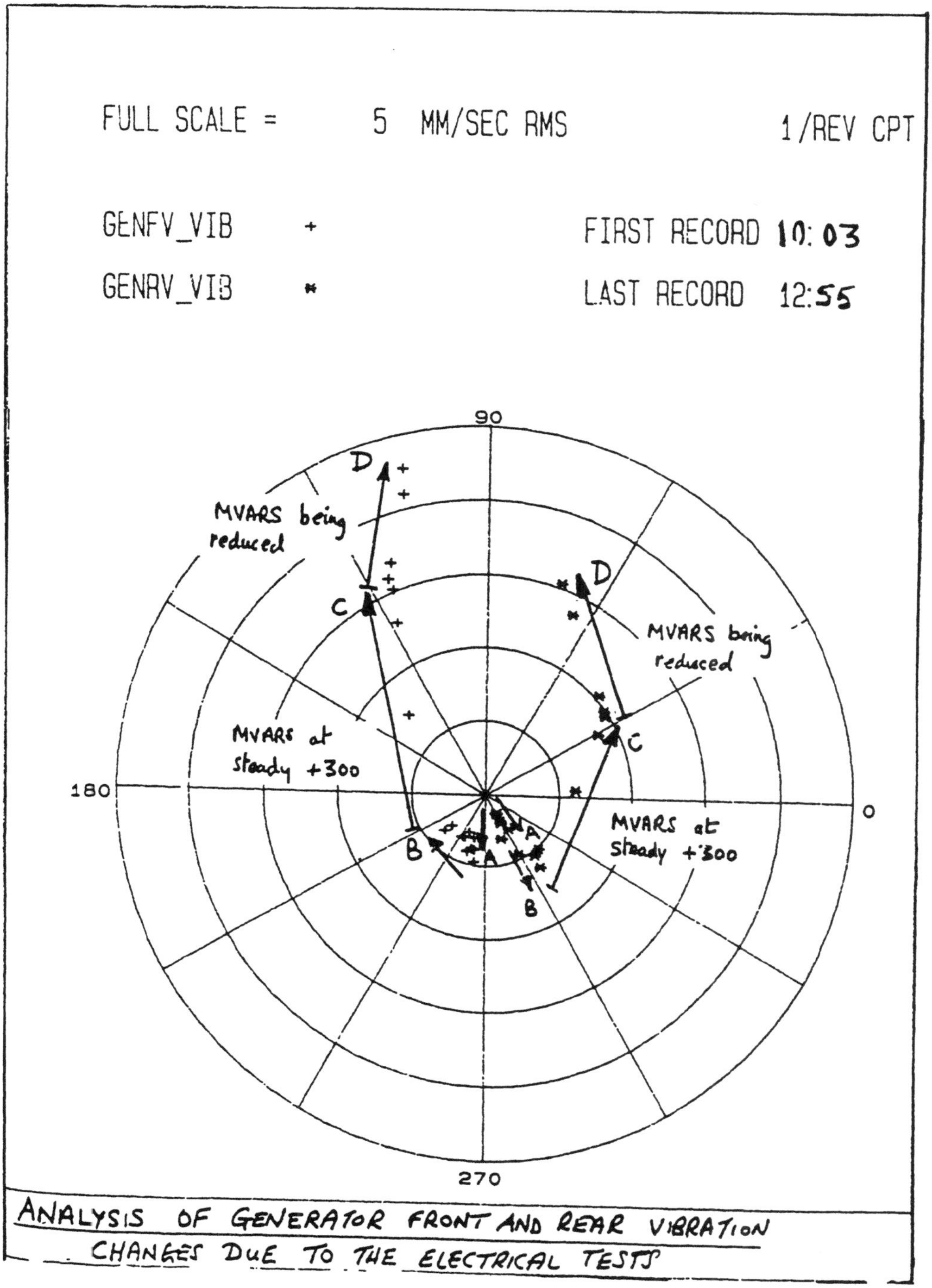
FULL SCALE = 5 MM/SEC RMS
1/REV CPT
GENFV_VIB +
GENRV_VIB *
FIRST RECORD 10:03
LAST RECORD 12:55
90
180
0
270
D
C
B
A
MVARS being reduced
MVARS at steady +300
ANALYSIS OF GENERATOR FRONT AND REAR VIBRATION CHANGES DUE TO THE ELECTRICAL TESTS

Fig. 16

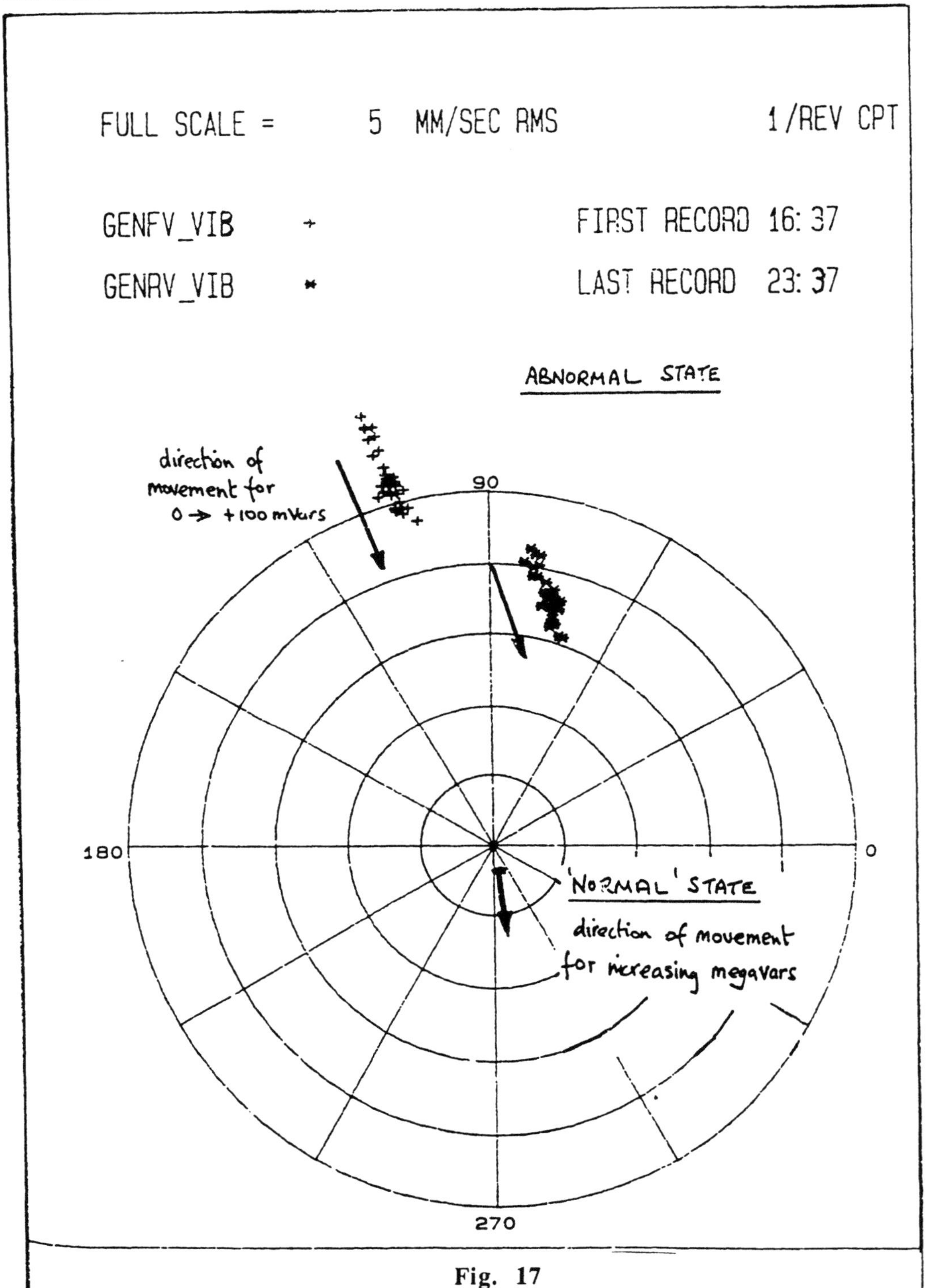
FULL SCALE = 5 MM/SEC RMS
1/REV CPT
GENFV_VIB +
GENRV_VIB *
FIRST RECORD 16: 37
LAST RECORD 23: 37
ABNORMAL STATE
direction of movement for 0 → +100 mVars
90
180
0
270
'NORMAL' STATE
direction of movement for increasing megavars

Fig. 17

5.0 SUMMARY

The data analysis system as described has shown how easy it is to set up an automatic data analysis facility for load monitoring using standard commercial hardware. Much more difficulty was encountered at other stations, where previously used manually operated equipment was converted to computer control. It has also been shown that sophisticated data analysis techniques normally associated with a central laboratory facility can be applied close to the turbine, and that more intensive monitoring of a turbine on load can be of assistance when diagnosing machine problems. It is necessary to examine established practices and equipment in the light of the new microcomputer technology.

If future monitoring is to be successful then diagnosis must lean towards the recognition of vibration patterns between many vibration transducers and the relationship of the patterns to other operating parameters, with less emphasis placed on absolute values.

When monitoring on-load with no apparent fault, variations in vibration levels can be individually correlated with plant parameters using statistical techniques to establish trends. This allows the reliance which had to be placed on the accurate determination of small vector changes of the spectral components, usually from a limited number of pedestal transducers to give earlier and more accurate diagnosis of problems such as cracked rotors.

In a known fault condition, the patterns that emerge from monitoring many vibration transducers, when correlated with theoretical models, will allow the models to be improved so that they may be used in the future with more confidence in the identification of a variety of machine problems.

INVESTIGATIONS ON THE VIBRATIONAL BEHAVIOUR OF GAS TURBOGENERATOR UNITS AT THE FIRST START-UP: AIMS AND METHODOLOGY. CASE HISTORIES ON 120 MW UNITS

C. Frigeri
ENEL S.p.A. DSR/CRA, Cologno Monzese, Italy

M. Pastorino
ENEL S.p.A. DCO, Rome, Italy

G.A. Zanetta
ENEL S.p.A. DSR/CRA, Cologno Monzese, Italy

ABSTRACT

The paper illustrates methods and results of an investigation on the dynamic behaviour of gas turbine units, about which there was no previous operation experience in the Company.

Deep insight into the dynamic response was gained. Vibration problems were early detected and their solution could be promptly devised by the manufacturers.

The use of experimental and numerical tools is described and the advantages of performing such activity on new turbogenerator units are discussed.

1. INTRODUCTION

The installation of a large number of 120 MW gas turbogenerator units for continuous operation was started at ENEL, as a part of an important repowering program and for use in combined cycles of steam thermal plants.

As a part of the commissioning program, the definition of the dynamic baseline of a machine is performed to provide a reference situation for the operative staff.

Moreover, gas turbines of such size were never operated before in the Company.

Therefore du e to the comparative novelty of such units, it was decided to perform a thourough investigation on the dynamic characteristics of those turbogenerators, in order to point out potential improvements and to devise the appropriate corrective actions in due time. Experimental influence coefficients for the balancing planes accessible in field were measured.

The investigation was performed by means of an integrated use of experimental and numerical tools, like a multichannel digital acquisition system, modal analysis identification techniques, rotordynamic mathematical models and a multiplane balancing code.

The paper details methods and results obtained so far by this work.

2. MACHINE DESCRIPTION

The 120MW units under exam consist of the gas turbine rotor and the generator rotor, rigidly connected by an intermediate shaft and supported on four bearings (Fig. 1).

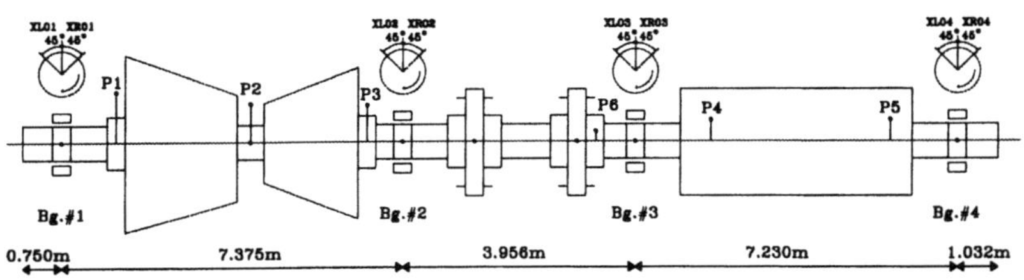

Fig. 01 - Vibration measurement points and balancing plane locations.

The relevant weights are approximately 56 Tons, 26 Tons and 2.5 Tons.

The blade rows of the gas turbine are installed on four separate disks, tightened by axial bolts. In turn, the turbine shaft is bolted to the nineteen stage compressor, where the disk of each stage is shrunk on the shaft. The whole shaft is equipped with tilting pad bearings and, together with the casing, it is supported on a flexible foundation.

The generator, hydrogen cooled, is provided with partial arc bearings and the bearing supports are separate from the casing. Instability of the rotor, experienced in the past on a smaller unit, was among the main concerns, from the point of view of the dynamic behaviour.

3. INSTRUMENTATION AND ANALYSIS TOOLS

For vibration measurement and continuous monitoring, the machines are equipped with two proximity transducers at each bearing, 45 degrees apart from the vertical direction (Fig. 1).

Since the very beginning, it was decided to add a velocity meter on top the cover of each proximity probe, in order to be able to pick-up the absolute shaft vibration (Fig. 2).

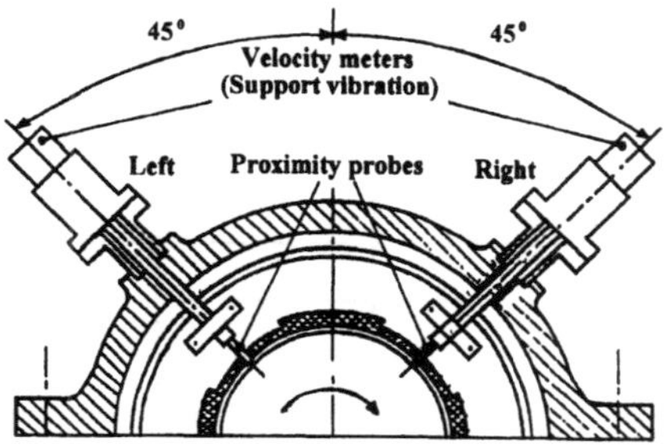

Fig.02 - Trasducer arrangement for vibration measurement at bearings.

Acquisition and data processing were performed by a multichannel digital system installed on a mobile laboratory, described in details in [1] and [2].

For a special identification task, a commercial modal analysis system [3] was used, for acquisition and processing of frequency response data resulting from impact tests.

Critical speeds, mode shapes and forced responses of the shaft line were computed by means of a rotordynamic computer code [4].

Experimental influence coefficients and balancing masses were obtained by a multiplane balancing code [5], [6].

Full integration of these experimental and analytical tools provided not only qualitative interpretation of the vibration data but also quantitative results.

4. EXPERIMENTAL INVESTIGATIONS AND NUMERICAL ANALYSES

The first goal of the job was to draw a baseline of a single unit, in order to identify the frequency range of the critical speed, the relevant mode shapes, margins against instability, sensitivity to thermal and, in general, operational conditions.

In addition to the vibration signals, a number of operation quantities, to be correlated wit vibration measurements,was recorded.

As for the vibration behaviour, the prominent features emerging from the exam of the recordings, and that could be attached to any such unit,were:

- High margins against instability, from tests with different oil-film temperature.
- Insufficiency of the proximity transducers to monitor the vibration behaviour of the turbine on its first critical speed, due to the flexibility of the foundation
- High axial vibrations at both generator supports, possibly influencing also radial relative shaft-support vibration readings.
- Turbine sensitivity to thermal conditions, in particular on the first critical speed (Fig.3).
- A little damped critical speed on the generator in the vertical plane, not far above the nominal speed (Fig.4a)
- A critical speed, more damped, in the horizontal plane all the bearings and just over the nominal speed (Fig.4b).

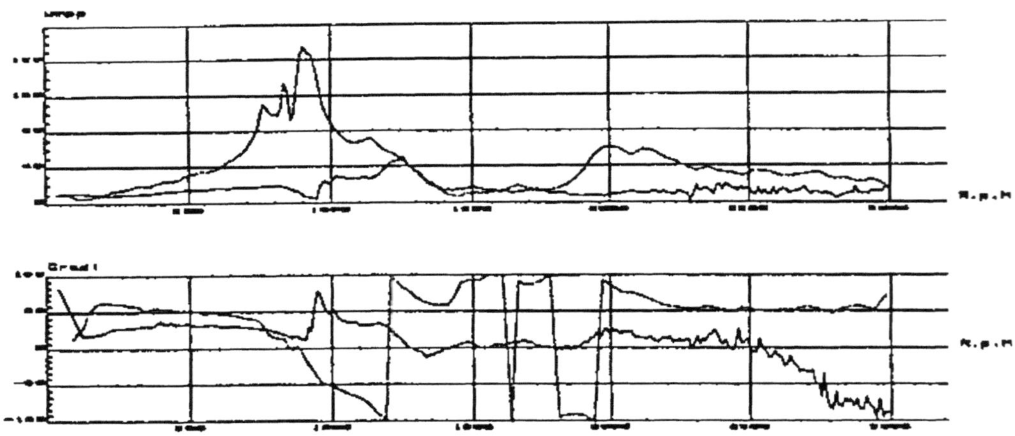

Fig.03 - Absolute shaft vibration on a turbine bearing during a cold run-up and the subsequent hot coast-down (dashed line).

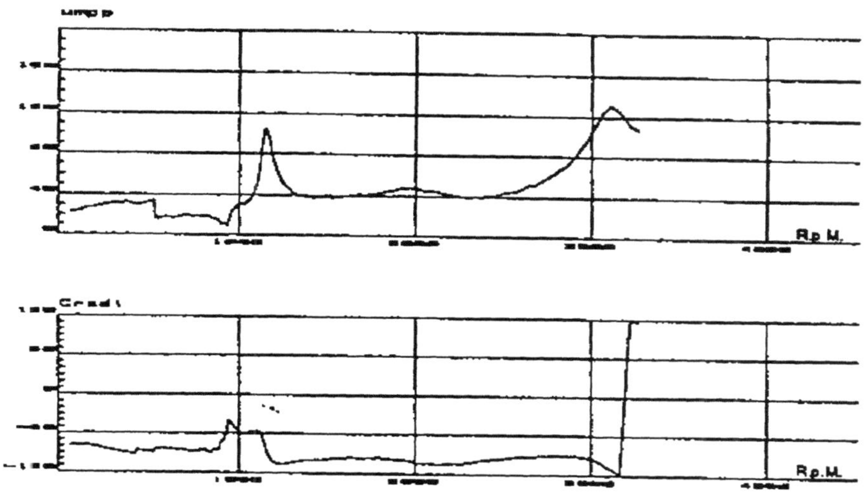

Fig.04a-Generator vibration in the vertical plane during an over speed test.

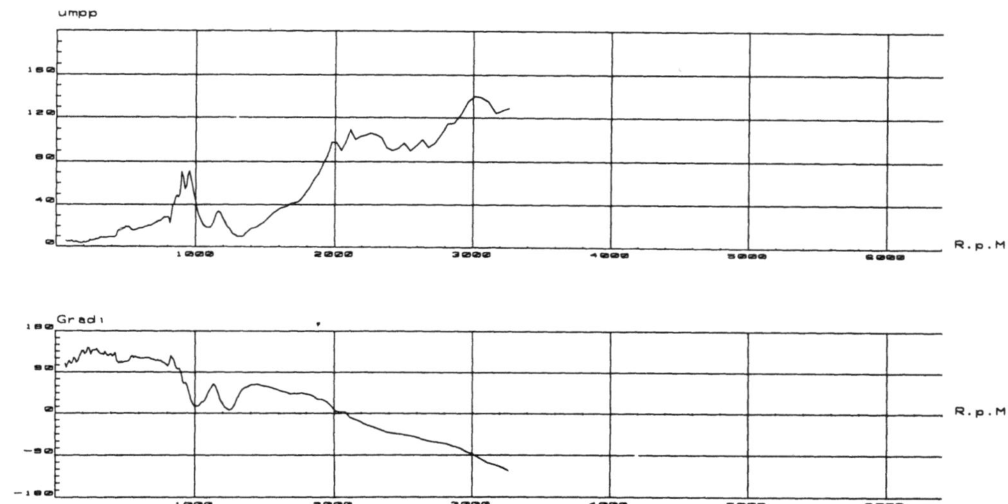

Fig.04b- Vibration at one bearing in the horizontal plane during an overspeed test.

In subsequent measurements on other units, these two critical speeds were found to be spread over a range of 100÷150 R.p.M.

The investigation was then specialized, in order to ascertain the possibility to remove the potential sources of high vibrations.

An impact test was performed on the generator supports and data were analyzed by means of a modal analysis system.

In the axial direction, a resonant frequency at 60.5 Hz was responsible for high vibrations (Fig. 5a). On other units this value was found to be even lower. The corresponding mode shape was defined (Fig. 5b). A residual contribution of the foundation beam flexibility was also highlited.

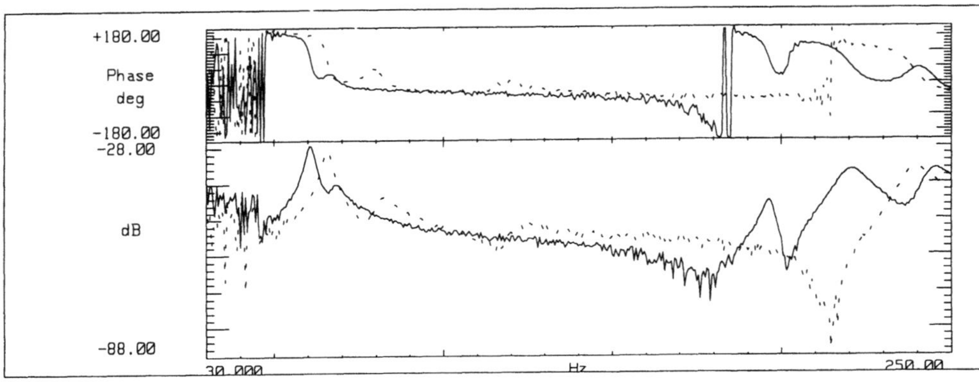

Fig.05a- Frequency response in the axial direction rom the impact test on the original (continuous line) and on the modified support.

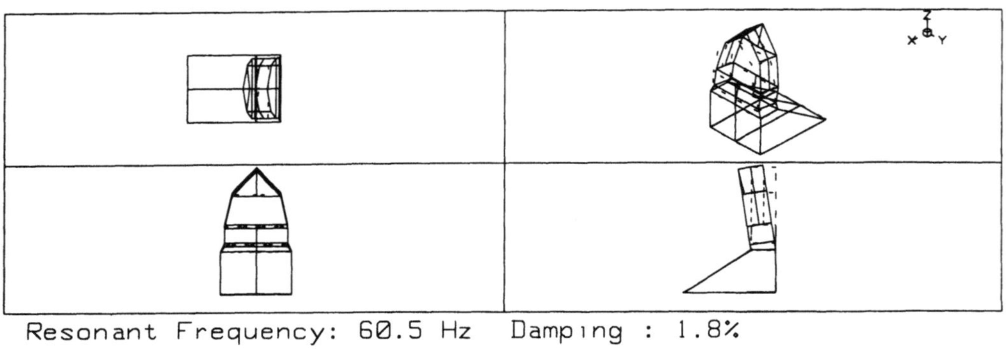

Fig.05b- Vibration mode shape for the resonant frequency in the axial direction.

In the vertical and horizontal directions, the resonant frequencies were far above the nominal speed. Those frequency responses were curve-fitted to identify reliable compliance values of the generator supports, for subsequent use in the computations of the shaft line dynamics (Fig. 6).

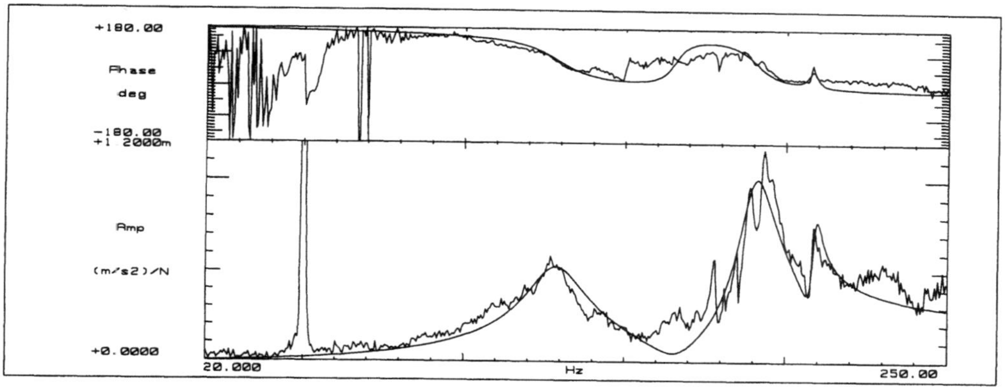

Fig.06a- Support frequency response in the vertical direction fron impact test. Experimental curve and curve fitting results.

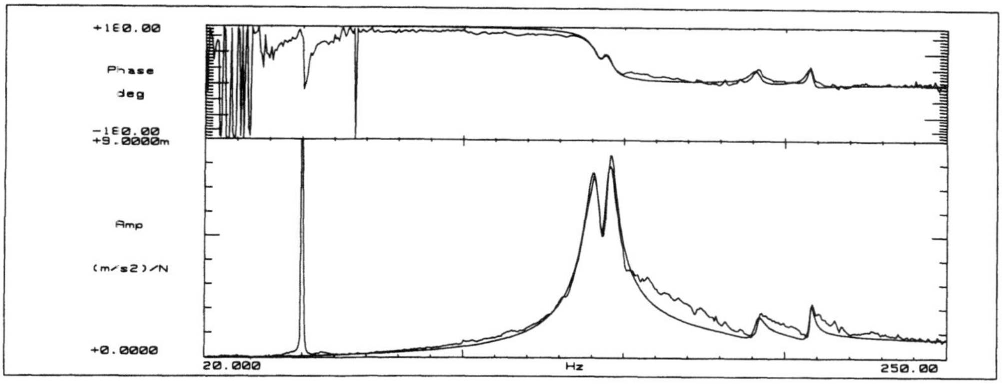

Fig.06b- Support frequency response in the horizontal direction from impact test. Experiment al curve and curve fitting results.

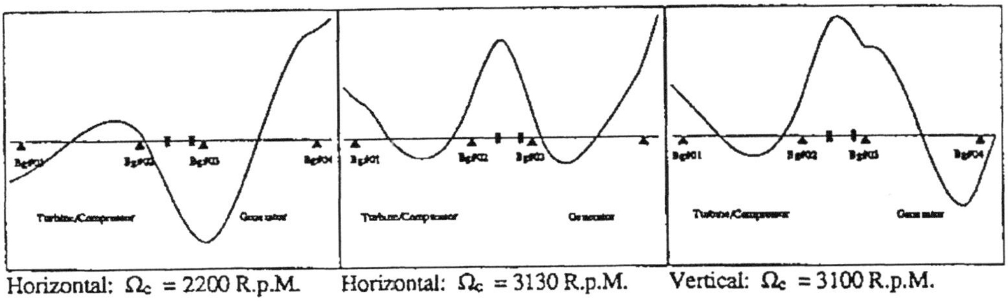

Fig.07- Dominant mode shapes and corresponding critical from rotordynamic model computation.

Moreover this data were compared with the deformation shape drawn from measurements in operation at different locations on the supports, to single out residual flexibilities.

A rotordynamic model of the shaft line was tuned in to the experimental data. The parameter values for the generator foundation were taken from the above mentioned curve fitting. The study gave a clear insight in the dynamic response of the shaft and the dominant mode shapes were identified (Fig. 7).

Apparently, the generator was working below its second critical speed in the vertical plane. A sensitivity analysis was undertaken to verify the possibility to move away the second critical speed of the generator.

That was found to be sensitive to the oil film stiffness. Also, combining experimental data with compliance values from modal analysis tests, identification of the oil film stiffness was tried. The results indicated values remarkably lower than the computed ones.

In the horizontal plane, the critical speed was due to the dynamics of the coupling shaft, in connection with a third mode of turbine and generator. Skew or eccentricity errors in the coupling can give rise to high excitation, owing to the high correlation between this mode shape and the rotating bow induced in the shaft (Fig. 8).

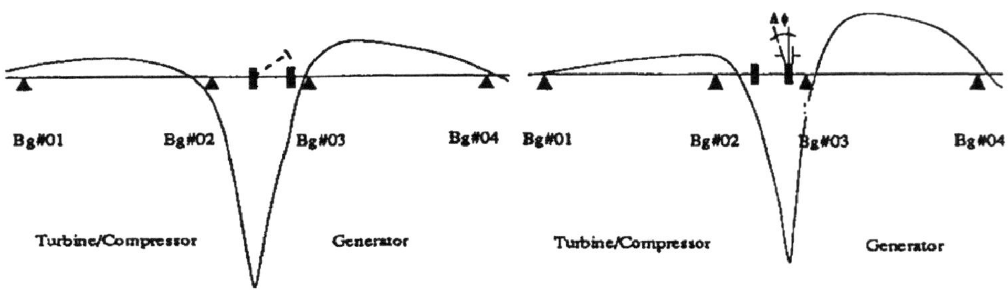

Fig.08- Rotating bow induced by eccentricity or skew errors at the coupling faces.

Basing on such considerations, it was predicted that the coupling location would have been essential for the balancing of the whole shaft line in most cases.

In the light of the experience and information gained on that unit, other similar, more specialized investigations were undertaken on those units that showed peculiar vibration problems.

Different faults were correctly identified. Two of them were inherent to a particular machine. On the contrary, other faults could potentially affect any other unit, even though they appeared clearly on one or two units only.

They can be summarized as follows:

- Oversensitivity to excitation current on a generator rotor. The rotor was substituted.
- Oversensitivity of a turbine rotor to initial thermal conditions. A compromise is being sought by properly balancing.
- Rubs at the hydrogen seal rings on the generator. The occurrence of the rub was also influenced by the scaling oil temperature.
- Rubs at a floating seal on the generator casing near the coupling.

These rubs were erratic and particularly severe on two units, but marginally present also on other units.

An unusual feature was the vibration increase with only minor changes in the phase angles (Fig. 9).

That was explained by the fact that at 3000 RpM the rotor was working below the dominant critical speeds.

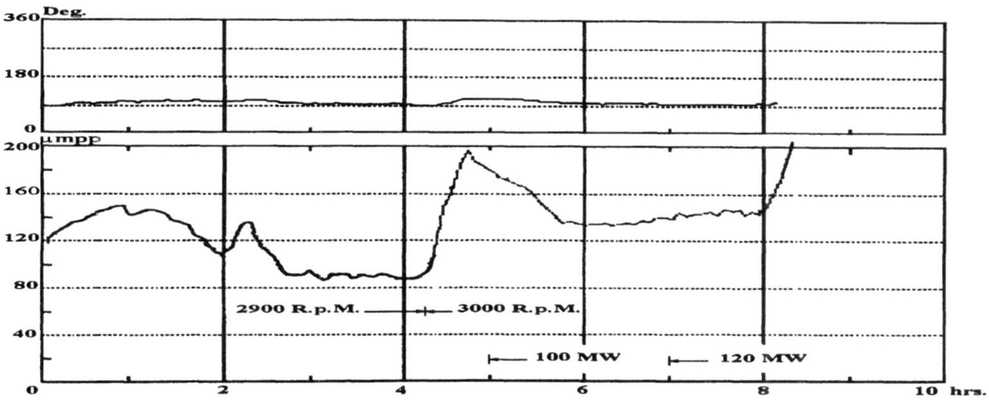

Fig.09- 1X component vibration recordings with on the seals

5. REMEDIAL ACTIONS

The previous diagnoses correctly addressed the remedial actions, worked out by the manufacturers. For the most part, those changes have already been implemented on the machines and demonstrated to be effective.

In order to improve axial vibration response, the supports were stiffened, thus increasing the resonant frequency and lowering the static deflection (Fig. 5a). For future plants, the concrete support beam will be modified to reduce its torsion and the lay out of the lub oil piping will be redesigned, thus allowing the installation of a even stiffer support. This was already done in the last installation.

A critical speed is not itself a problem, but it increases the sensitivity to changes in the unbalance state.

Since the second critical speed of the generator in the vertical plane was little damped, it was thought that a good improvement could be obtained by moving it a little toward higher frequencies. A first result was obtained by substituting insulating glassfiber supporting elements, which were found to be somewhat flexible, for steel elements, while reducing the insulation to a thin film. The determinant factor should be the increase of the oil film stiffness. Work is still in progress on the subject. The analysis also showed that stiffening of the rotor itself and reducing the weight in the area of the retaining rings gives appreciable increase of such critical speed. Future rotors will be manufactured accordingly.

Positioning and clearances of the hydrogen seal rings were revised. Moreover heaters have been added into the oil circuit and specifications were given about oil temperature.

The floating seal on the turning gear casing was in practice eliminated, after changing the suction circuit of the oil drops from the coupling.

6. INFLUENCE COEFFICIENT MEASUREMENT AND BALANCING

A significant investigation task was the measurement of the influence coefficients (IC) for the balancing planes accessible in field, in order to able to apply multiplane balancing codes based on the IC method.

In fact, from the first trim balancing tests, it was apparent that the balancing of the turbine only could remarkably affect the generator and viceversa, in most cases. This situation was also confirmed and explained by the results of the rotordynamic model (see Par.4).

According to the design of the rotors, five balancing planes were accessible on the plant, three on the turbine compressor rotor, two on the generator rotor. In consideration of the influence of the coupling section discussed above, also mass addition on this section was tested, taking advantage of four screwed holes in the coupling face.

Prescriptions were given to properly control the machine operation from one run to another, in order to obtain the same thermal conditions as far as possible.

For the coupling and inner generator planes, amplitude and phase of the trial masses were decided by means of the rotordynamic model. That greatly helped to go toward a balancing of the shaft line during each trial run, thus improving the quality of the results. In fact, during installation, the first trial mass on the generator was put just 180° apart from the prescribed position. The resulting IC were three or four times the true ones, supposedly because high vibrations gave rise to rubs against the seals.

Comparing experimental results and numerical predictions for the IC of those three planes it was shown that amplitude errors were less than 20% and phase errors ranged in between $\pm 20^{\circ}$.

The IC were obtained for both nominal speed and hot condition run-down transients. In general, no matter how much care was taken in the operating condition control, coherent IC values were obtained only for speeds above the 1500÷2000 RpM. Below this speed range, the results were not always reliable, owing to non repeatable behaviour of the rotors and week effect of the added masses.

Four units have been successfully balanced so far by means of a multiplane balancing code (Fig. 10).

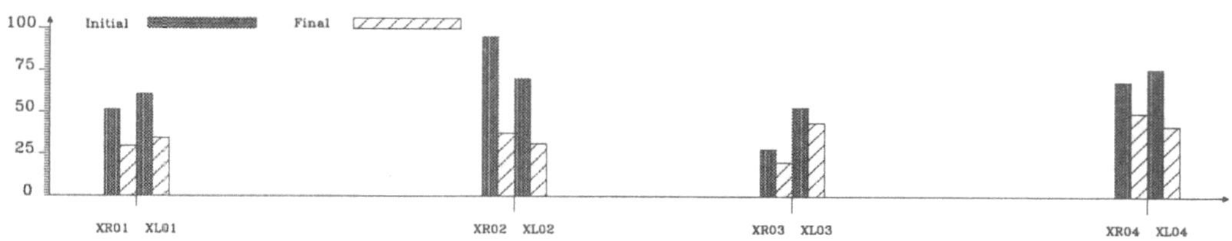

Fig.10- R.M.S. values of corresponding vibrations before and after trim balancing on four units

Two of them were balanced at first shot; the other two requested two runs. Data were processed by a multiplane balancing code: in three cases, three balancing planes were used at one time, after appropriate choice; in a fourth case the coupling plane only was used. Of course, those balancing results were also processed to update and to improve IC values.

In all four cases, the coupling plane was involved and turned out to be essential, as it was guessed. Therefore balancing holes will be provided on the coupling face in future rotors.

Again, the experience gained so far demonstrates the need for overcoming more traditional balancing procedures in favour of more powerful and, by now, well established numerical tools.

7. CONCLUSIONS

A comprehensive investigation on the vibrational behaviour of new gas turbine units was performed at the first start-up. An effective integration of experimental and analytical approaches was put into practice. The outcomes of this activity have been manyfold.

Generator critical speeds close to the operating speed were detected and possibile non-repeatable vibration response on the gas turbine was identified, while candidate components for implementation of remedial actions were devised. Locations of potential sources of rub were singled out and removed. Improvements in the vibration monitoring instrumentation and the opportunity for a new balancing plane accessible in field were also suggested. Experimental influence coefficients were obtained for the balancing planes accessible in field and effectively used in a multiplane balancing code.

On the basis of the information gathered by means of this experimental and numerical investigation, the manufacturers have been able to properly devise the remedial actions.

The results were already used and implemented on more than six units now and, for all of them, at the commissioning stage.

This experience proves the timeliness of such activity and, therefore, the suitability of introducing the identification of the vibratory response baseline among the usual activities carried out at the first start-up of a new turbogenerator unit, especially when a machine of a particular type was never operated before.

REFERENCES

1. A.Benanti, C.Frigeri, G.Silvestri: Vibration Monitoring and Mechanical Faults Diagnostics on Large Rotating Machinery in ENEL Power Plants, IFToMM/CISM Symposium on "Diagnostic of Rotating Machines in Power Plants" - Udine(ITALY) - October 27÷29.1993
2. R.Bernante, C.Frigeri, A.Vallini, G.A.Zanetta: Diagnostic Oriented Flexural Vibration Monitoring Systems for LargeTurbogenerator Units. AIPnD Int.Conf. on"Monitoring and Predictive Maintenance of Plants and Structures" Florence (ITALY) 17÷20 May 1992
3. LMS CADA-X. General Concepts Manual-CADA-X-0200-UM-02
4. DYTS04 USER'S GUIDE-CEGB-RD/L/P-15/80-JOB No.VE316
5. G.A.Zanetta: An Interactive Version of the Balancing Program "EKILB" and Remarks on the Tests Performed on Models, ENEL Internal Report No. 330.012-6 rapp. 649/86-Nov 1986
6. C.Frigeri, M.Tonsi, G.A.Zanetta: Theoretical and Practical Aspects in the Field Balanc ing of Large Turbogenerator Rotors. Int. Conf. "ROTORDYNAMICS '92"-Venice (ITALY) 28÷30 April 1992- Springer- Verlag 1992

EXPERIMENTAL TESTS AND MODEL BASED CALCULATIONS FOR THE DIAGNOSIS OF A CRACK IN 320 MW GENERATOR

G.L. Lapini and M. Zippo
CISE SpA, Segrate, Italy

N. Bachschmid and A. Collina
Technical University of Milan, Milan, Italy

A. Vallini
ENEL S.p.A. DPT-VDT-STE, Pisa, Italy

ABSTRACT

The paper reports the results of an extensive experimental test program for the diagnosis of a transverse crack in a generator rotor of a 320 MW steam power plant.
An adequate vibration monitoring equipment was installed on the machine in order to supply the items for the diagnosis of the malfunction.
The experimental dynamic behaviour of the rotor in-service and during the off-line tests in the factory was simulated by means of mathematical model of the rotor including the cracked effect: a satisfactory agreement was found with the experimental results.
The performed analysis strongly indicated the presence of a crack, which was later found by ultrasonic internal inspection.

1. INTRODUCTION

During the last period of normal service the 320 MW steam turbine-generator, whose schematic line axis is shown in Fig.1, revealed a progressive increase in total absolute shaft vibration levels and a high sensitivity to load variation.
The potentially dangerous situation indicated by this behaviour was initially underestimated by the plant personnel because of the poor quality of the standard installed vibration monitoring system. Only after the alarm levels were exceeded vibration analyses were carried out by vibration specialists.
Additional sensors were installed on the generator bearings (eddy current probes and accelerometers) in accordance to the typical VIPROX configuration, see Fig.2, and a PC based acquisition and analysis system was used for vibration monitoring and analysis.

2. GENERATOR DYNAMIC BEHAVIOUR IN SERVICE

2.1. Experimental tests (rated speed and run-down)

In order to analyse the in-service dynamic behaviour of the generator rotor the following tests were carried out at rated speed: load variations, exciting current variations, generator cooling H_2 temperature variations.
Figs.3-4 and Fig.5, which are relative to the different tests, show a very high sensitivity of the vibrations and particularly of its first harmonic components, to the variations of the three above-mentioned parameters.
In particular, the sensitivity to the load and to the exciting current variations could indicate either electrical faults (short circuit) or structural defects, while the sensitivity to H_2 temperature was consistent with the hypothesis of structural defects in the rotor (crack).
In any case, the stop of the machine was suggested, in order to complete also the dynamic analysis of the rotor during the run-down.
No significant effects on vibration levels were pointed out when the generator was disexcited while during the run-down extremely high vibration levels were reached at the first critical speed (Fig.6) producing rubs on the bearings and seals.
On the contrary, because of the low eccentricity values at the turning gear speed, no permanent rotor bending was evidenced, at least at bearings location.
A contribution to the very high vibration levels measured at the first critical speed came probably also from an irregular hydrodynamic condition inside the bearings induced by rubs. Fig.7 shows, for instance, the strong decrease in oil film pressure of one of the generator bearings in correspondence of the critical speed.
The amplitude of the second harmonic component (2X) at half the first critical speed was relatively small in comparison with the first harmonic component (1X), but significantly higher than the value measured during a run-down recorded two years before, Fig.8.
The increase of the amplitude in correspondence of the second harmonic first critical speed, with no phase variation at the critical speed and at lower speeds, indicated a loss of stiffness in the weaker principal stiffness axis, which in the examined rotor was opposite to the polar axis, allowing to identify the angular position of the crack, if any.

A slight decrease of the value of the first critical speed was also pointed out.
The in-service dynamic behaviour of the generator was compared with the behaviour of a cracked rotor, simulated by an analytical model (described in the following paragraph).
Since the results of this simulation were consistent with the experimental data, it was suggested to take the rotor out of service.
A first inspection carried out in the plant revealed no crack, so that it was decided to transport the rotor to the factory for deeper analysis.

2.2. Analytical simulation of generator dynamic behaviour

Analyzing the static deflection measurement results, described in par.3.2, the crack location and its depth were identified by trial and error procedure: the most probable location was found by an automatic program utilizing a simple crack model (two external equal and opposite moments), and a least square procedure.
The depth of the crack was estimated using the rather sophisticated crack model, described in [1] and [2], and looking for the best agreement between experimental and theoretical results (analysing different values of the depth of the crack).
With the above identified location and depth of the crack, which is shown in Fig.9, the model of Fig.10 was used for calculating, by means of a computer code, the dynamic behaviour of the rotor, in a constant temperature distribution condition (thermal steady state condition) and in the thermal stress condition which is obtained after half an hour cooling transient with an external rotor temperature gradient of -0.5 °C/min.This latter condition could roughly represent the in-service situation of the hydrogen temperature lowering.
As known, the external rotor negative temperature gradient causes an opening of the crack, and a subsequent increase in the bow and therefore in the first harmonic vibration component.
Table 1 shows the comparison between theoretical and experimental results, obtained by calculating the variation due to the thermal transient.
Although the agreement is more qualitative than quantitative (probably due to inexact assumptions relative to the thermal situation and to a not completely reliable mathematical model of the rotor-bearing-supporting structure) the results strongly confirm the suspect of the crack.
In Fig.11 the calculated first and second harmonic vibration components during a run-down transient are shown, in which a cooling transient with an external rotor temperature gradient of -0.8 °C/min was supposed in order to take into account the hydrogen cooling effect (since the generator is disexcited).
The theoretical results do not fit completely the experimental ones due to the presence of an unknown unbalance on the rotors (for the first harmonic) and of a not well defined residual axial unsymmetry in the generator (for the second harmonic).
Furthermore, the thermal transient in the cracked area, where the temperature and stress distribution can only be estimated very roughly, is certainly changing during the run-down (which lasts approx. fifty minutes) and lastly the mathematical model has some uncertainties with regards to the oil film parameters and the supporting structure modal parameters, which may have a strong influence on the behaviour at rated speed (3000 rpm).
A tuning of the model with appropriate experimental data was not possible.

Nevertheless the qualitative good agreement between theoretical and experimental data confirms again the presence of the crack.
The very high vibration level in passing the first critical speed can be explained by the loss of damping properties of oil film in presence of high amplitude orbits (non-linear effects which were not considered in the above described model).
The apparently low residual bow at lower speeds is confirmed by the theoretical results: at bearing location the bow due to the crack (which is rather far away from the bearings) effects only very little first harmonic vibrations.

3. OFF-LINE TESTS IN THE FACTORY

When the rotor was brought to the factory the theoretical and experimental research program was continued in order to confirm the presence of a transverse crack, which could not be identified either by a visual inspection nor by dye penetrant tests because, as will be shown later, its location was below the windings end. To reveal the crack it was decided to use different techniques in order to evaluate their effectiveness.

3.1. Modal analysis bump tests

Modal analysis bump tests were carried out in order to verify if significant variations of modal parameters occurred by changing the rotor angular position.
This technique had been tested in the past for a cracked turbine rotor and was applied to evaluate its potentialities with a generator rotor (which is axially unsymmetrical).
The rotor, put on double straight rollers, was excited by an instrumented hammer at twenty equally spaced angular positions (Fig.12) and the vibrational response was measured simultaneously by six accelérometers located along the shaft. The first 6 bending modes were identified. Tab.2 reports the frequencies of the 1^ and 2^ bending modes as measured at the longitudinal rotor position 3Z.
As it can be seen, the frequency of the first bending mode has always a single value which changes with rotor angular positions as shown in Fig.13.
On the contrary, split peaks appear for the second and higher frequencies at some rotor angular positions.
Experimental laboratory tests performed in the past on a symmetric rotor with a transverse crack had revealed a similar behaviour.
It is apparent from Fig.13 that the axis of maximum stiffness did not coincide with the rotor pole axis (Fig.12) and that there was a significant stiffness variation between two perpendicular axes and also on the same axis as the bending moment was reversed.
That was probably due to the action of self-weight bending which could open and close the crack.
Yet, the graphical display on Fig.13 points out a poor repeatability (after a rotation of 360° a different frequency value was measured, also if the rotor was rotated back in the initial position) which was probably related to hysteresis and non-linearity effects (in part due to rotor support system); that made the understanding of results more complicated.

3.2. Static deflection measurements and comparison with analytical results

The generator rotor was also placed on a lathe and its deflection in the two opposite horizontal directions was measured during a slow rotation.
This permitted to evaluate the deflection separately from measuring section ovalization: in particular from the measured deflections the first harmonic components (due to a permanent bow and to the breathing mechanism of the crack) and the second harmonic components (due to its distributed residual axial unsymmetry and to the localized crack) have been calculated. The results were compared with similar ones obtained several years before on the same rotor. The residual axial unsymmetry was identified from former measurements and applied to the actual rotor: the increment of the actual second harmonic component in the deflection was attributed to the crack and permitted its identification, as already said.
The location identified by the second harmonic components was coincident with the one identified by the first harmonic component.
Fig.14 shows the above evaluated experimental second harmonic components compared with those calculated with the crack model.
The good agreement strongly confirms the presence and location of the crack.
The generator rotor was then placed in the balancing cell, where heating and cooling facilities were available and its statical displacements during heating and cooling transients were measured.
Fig.15 shows the experimental and theoretical first harmonic component variations due to the cooling transient. Again a satisfactory agreement was found.
These results confirm the validity of the statical deflection measurement method for detecting a structural defect, also with respect to modal parameter measurement methods.
The latter method is based on natural frequencies measurements which are proportional to the square root of the stiffness; the statical deflection instead is inversely proportional to the stiffness and therefore more sensitive to the stiffness variation.

3.3. Vibrational response of the rotor in the balancing cell

In addition to the experimental data collected in the field it was judged to be very interesting to study the dynamic behaviour of the generator rotor in the balancing cell of the factory, where the rotor could be also thermally stressed.
By temperature changes of the rotor it was tried to cause the opening and closing of the supposed transverse crack. The subsequent dynamic behaviour was analyzed both at rated speed and during run-up and run-down. Rotor vibrations were measured at different longitudinal positions by couples of eddy current probes (Fig.16).
For safety reasons the rotor was never turned at speeds higher than the first critical speed (about 750 rpm); since the rotor was bent, to reach this speed it was necessary to balance the rotor with a 17 Kg weight.
Fig.17 shows the first harmonic vibrations measured at point 4 (retaining ring) during run-downs at three different rotor thermal conditions:
- during a heating transient (the outer rotor surface temperature was 62 °C),
- during a cooling transient (the outer rotor surface temperature was 50 °C),

- nearly stationary thermal condition at end of the cooling transient (the outer rotor surface temperature was 40 °C).

The graphs show a considerable eccentricity at low speed especially at the retaining ring and higher vibration levels during the cooling transients.

This behaviour was consistent with the presence of a transverse crack.

As already said in par.2.2, an external negative rotor temperature gradient induces an opening of the crack causing an increase of the rotor bow and therefore of the amplitude of the first harmonic vibration (the opposite happens during the heating process).

Fig.18 shows the second harmonic components for the same tests.

As it can be seen, there are high amplitude levels at half the first critical speed (confirming a rather high axial rotor unsymmetry) but no significant thermal effects connected to the presence of a crack.

3.4. Comparison with theoretical results

Assuming the previous thermal rotor condition at 40 °C as a constant temperature situation and the warm condition at 62 °C still as a heating transient with an outer rotor surface temperature gradient of + 0.12 °C/min, which was the mean rotor temperature gradient during the heating process, and taking into account the additional unbalance of 17 Kg on the rotor body, the diagrams of Figs.18-19 were obtained.

The agreement with the experimental results seems to be satisfactory, both with respect to the second and first harmonic components, confirming the validity and the reliability of the crack model and of the computer code for calculating the dynamic behaviour of a rotor with cracks, unbalances and residual unsymmetries.

3.5. Considerations about the different experimental techniques used during the off-line test

The different techniques used during the off-line tests in the factory showed a different effectiveness.

In particular, more satisfactory results were obtained from static deflection measurements and from vibrational response measurements in the balancing cell with rotor running at different thermal conditions.

Some uncertainties arised, on the contrary, analyzing the results of the modal analysis bump tests; that was due both to the test setup (in particular the rotor support system) and to the natural axial unsymmetry of the rotor generator which partially masks the crack effects.

3.6. Ultrasonic rotor inspection

As a consequence of the experimental and analytical results, the rotor was examined ultrasonically.

The ultrasonic inspection of a generator rotor is generally a difficult and time consuming operation because of its structural complexity.

The analytical results allowed to restrict the inspection area near the retaining ring (coupling flange side).

The poor accessibility of the rotor under the electrical windings required the use of a special

ultrasonic beam rebound technique.
Ultrasonic inspection was performed with the ARGO automatic system [3], developed by CISE, which was partially modified for this particular application.
The inspection confirmed the presence of a transverse crack located near the relief between the shaft and the active rotor portion whose dimensions are shown in Fig.20.
Successive dye penetrant tests, performed after the removal of the retaining ring and of a few rotor windings, showed clearly the presence of the crack (see Fig.21).

4. CONCLUSIONS

Accurate measurements and the comparison of the experimental results with theoretical results, obtained by means of a mathematical model, allowed to identify the presence of a crack and its approximative location and extension.
This crack, which extention was already close to a 50% of the cross sectional area, was impossible to detect with standard dye penetrant or ultrasonic tests, due to its location under the windings ends.
From the case history described in this paper, the presence of an adequate vibration monitoring system in the power plants clearly appears to be strictly necessary at least for avoiding catastrophic failures.

REFERENCES

[1] G.Diana, N.Bachschmid, F.Angeli "An on-line crack detection method for turbogenerator rotors" Int.Conf. on Rotordynamics, Tokyo, sept. 1986, pp.385-390.
[2] N.Bachschmid, D.Del Vescovo "Thermal bow in an axisymmetrical shaft containing a transverse notch and related vibrational behaviour" III Int.Conf. on Rotordynamics, Lyon, sept. 1990, pp.239-244.
[3] S.Ghia, A.Sala "Ultrasonic examination of large forging"
Int.Conf.Materials for Power Engineering Components,
Cost 501, Julich-Germany, 13-14 oct.92.

Tab.1

Variation (experimental)

	I harm. ampl.	I harm. ph.	II harm. ampl.	II harm. ph.
XA08	36	159	9	162
XB08	23	21	1	35
XA09	133	98	42	20
XB09	99	166	40	165
XA10	87	15	7	84
XB10	33	47	9	21

Variation (numerical)

	I harm. ampl.	I harm. ph.	II harm. ampl.	II harm. ph.
XA08	7	-6	0	0
XB08	8	-120	0	0
XA09	75	98	9	20
XB09	52	145	10	-161
XA10	65	154	1	-135
XB10	52	8	1	68

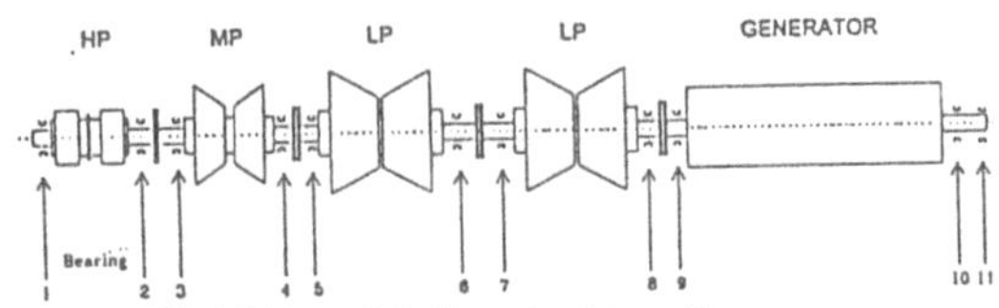

Fig.1 Scheme of the line axis of the turbine-generator with bearings denomination.

Tab.2

BUMP-TEST ON CRACKED GENERATOR

Angular position (polar angulus)	Experimental FRF's results [Hz] 1st mode	2nd mode
0 (0°)	20.87	46.5
1 (24°)	20.75	46.14
2 (40.5°)	20.75	46.02
3 (57°)	20.75	45.9
4 (73.5°)	20.51	45.65
5 (90°)	20.39	45.53
6 (106.5°)	20.51	45.65
7 (123°)	20.75	46.02
8 (139.5°)	20.75	46.02/46.94
9 (156°)	21.11	46.51
10 (180°)	21.36	47
11 (204°)	21.48	46.51/47.61
12 (220.5°)	21.24	46.63/48.46
13 (237°)	21.12	46.51
14 (253.5°)	21	46.39/46.68
15 (270°)	20.9	46.14/49.07
16 (286.5°)	21	46.26
17 (303°)	21.12	46.51
18 (319.5°)	21.12	46.75
19 (336°)	21.24	46.87
0 (360°)	21.24	46.87

Excitation point is 3z
Measurement point is 3z

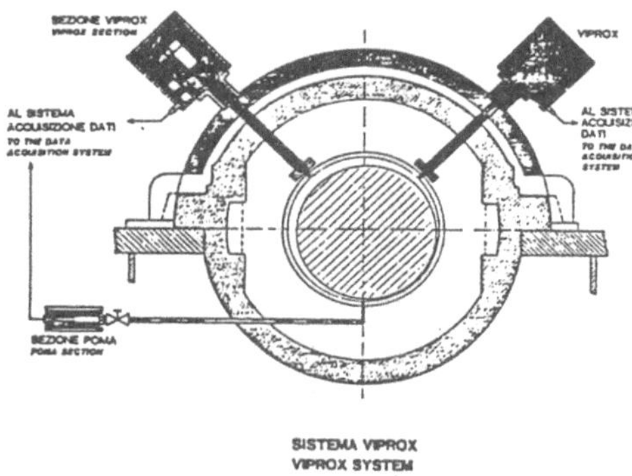

Fig.2 Transverse bearing section with a typical VIPROX sensors arrangement.

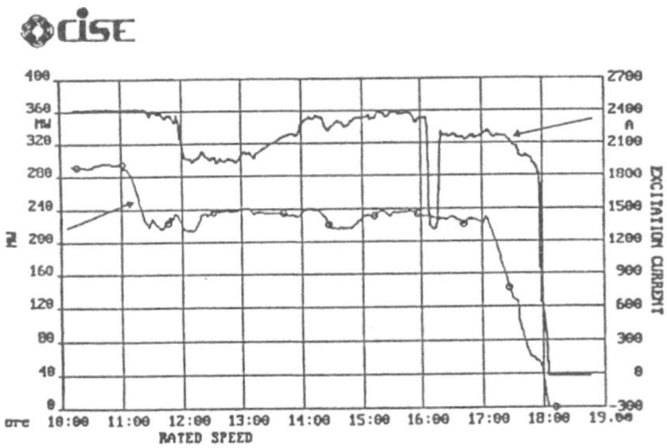

Fig.3 Load and excitation current transient.

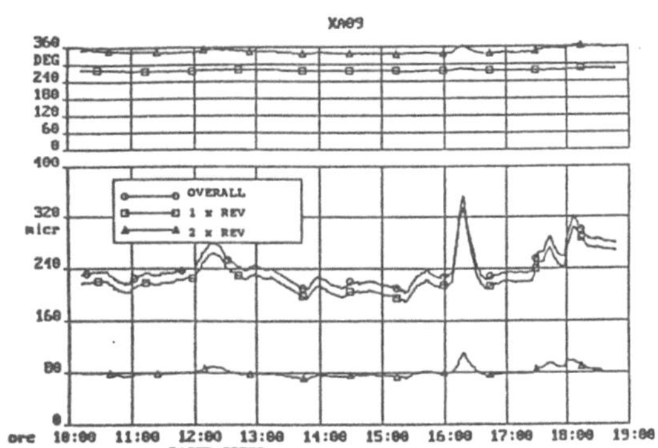

Fig.4 Relative shaft vibration measured on the 1st generator brg.(no 9) along load and excitation current variations.

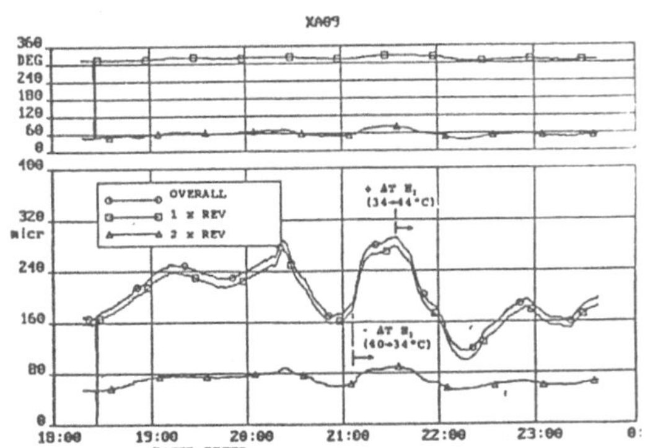

Fig.5 Relative shaft vibration measured on the 1st generator brg.(no 9) along load, excitation current and H_2 temperature variations.

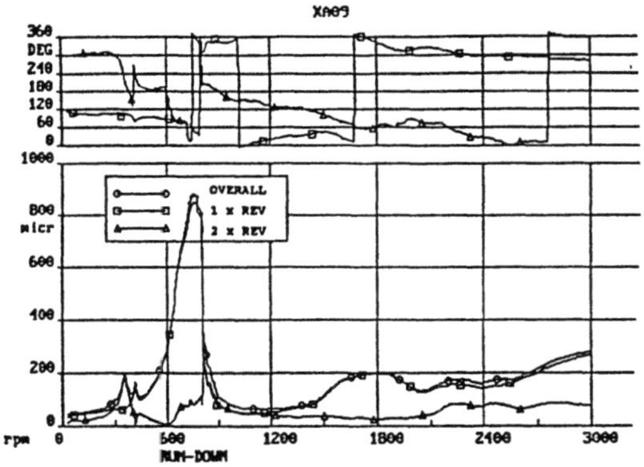

Fig.6 Relative shaft vibration measured on the 1st generator bearing (no 9) during the run-down before generator out of service.

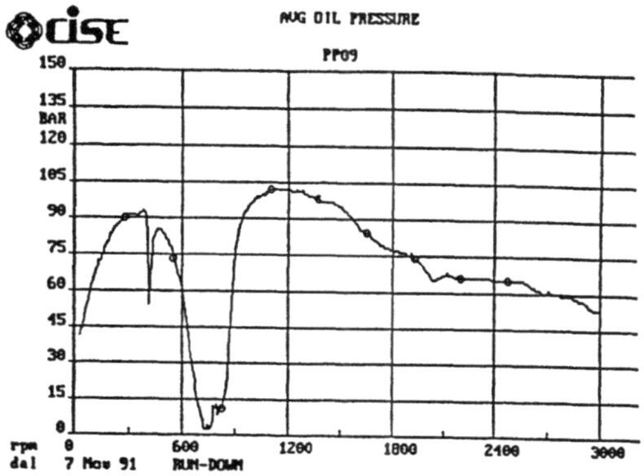

Fig.7 Average oil film pressure measured in the brg.no 9 during the run-down before generator out of service.

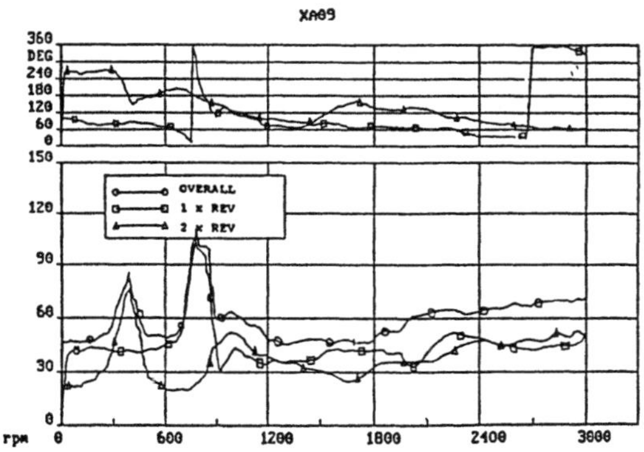

Fig.8 Relative shaft vibration measured on the 1st generator brg. (no 9) during a previously run-down.

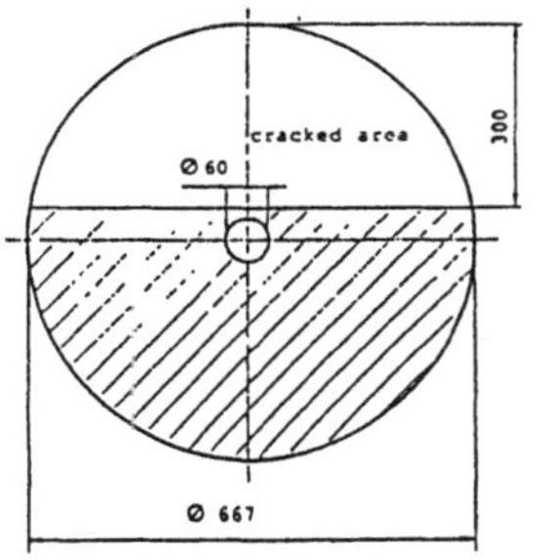

Fig.9 The cracked section model.

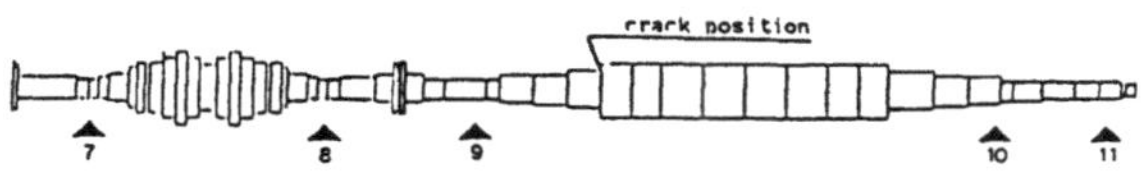

Fig.10 The rotor model (only generator plus LP turbine).

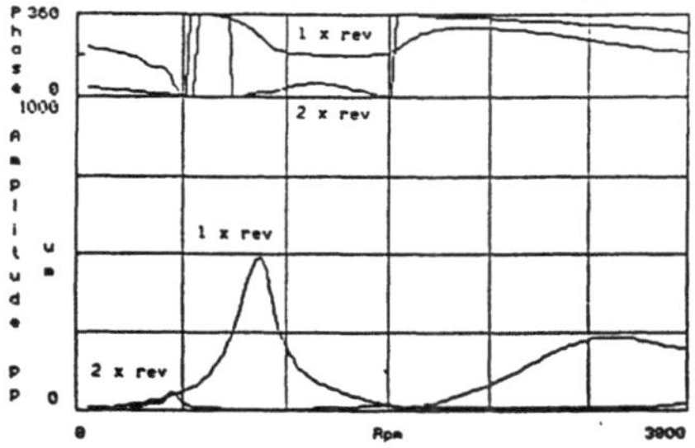

Fig.11 Run-down transient in brg.no 9 to be compared with Fig.6.

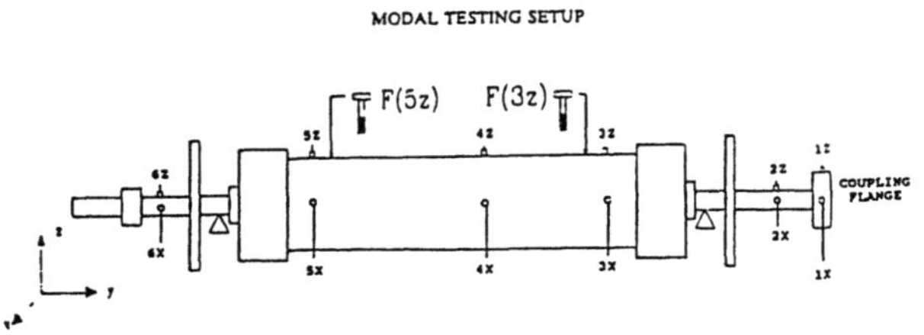

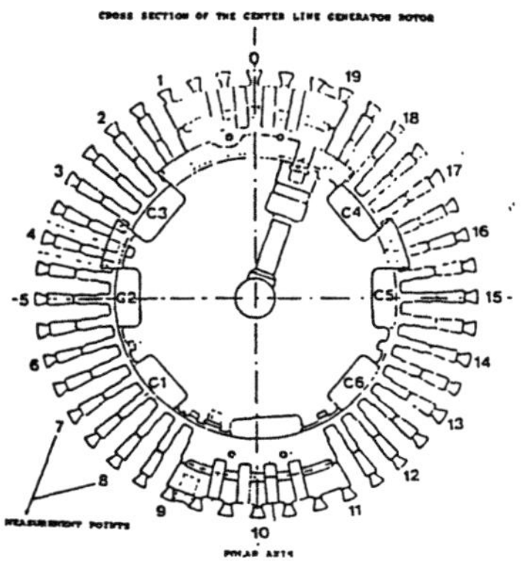

Fig.12 Modal analysis bump test setup.

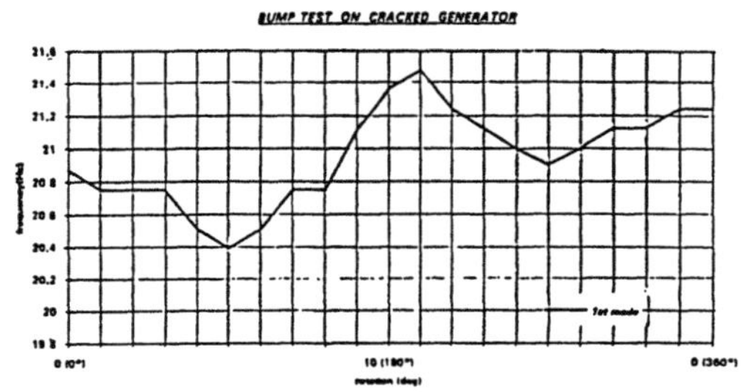

Fig.13 First natural frequency variation relative to angular position of rotor generator.

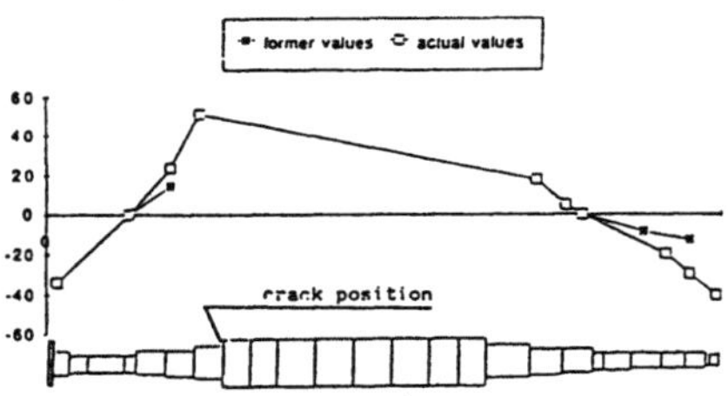

Fig.14 Second harmonic statical deflection variations due to a crack.

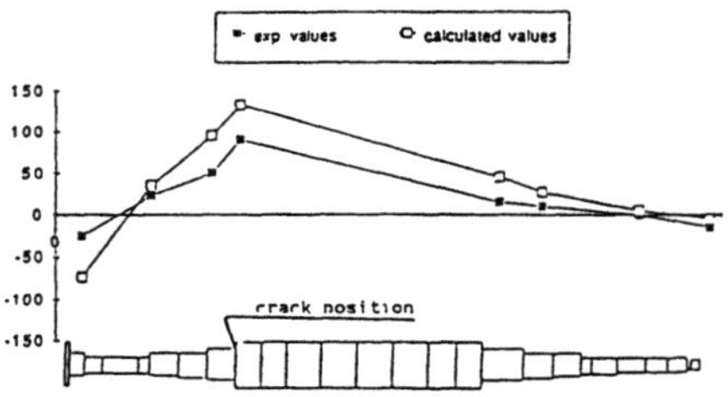

Fig.15 First harmonic statical deflection variations due to a cooling transient.

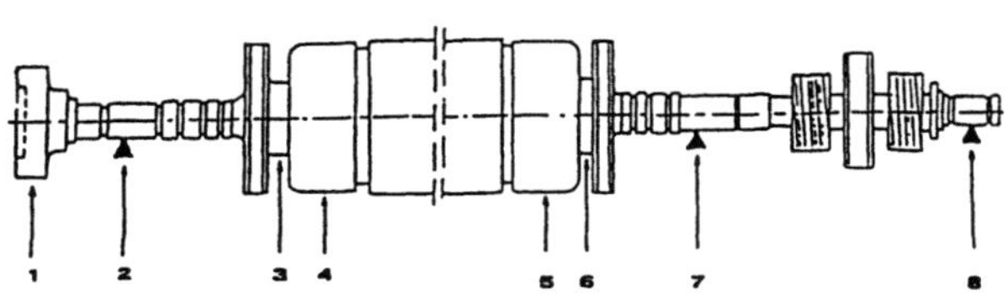

Fig.16 Eddy-current probes location in the balancing cell.

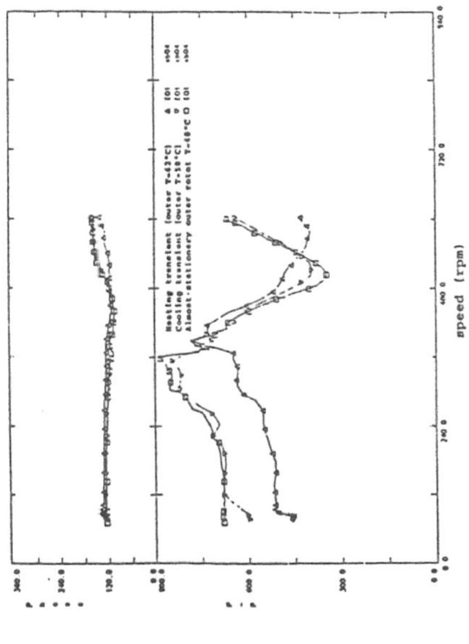

Fig.17 First harmonic components measured on the location 4 during run-downs at different rotor thermal conditions.

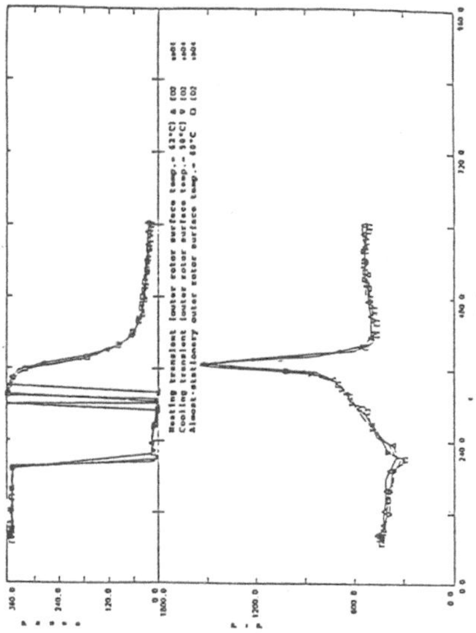

Fig.18 Second harmonic components measured on the location 4 during run-downs at different rotor thermal conditions.

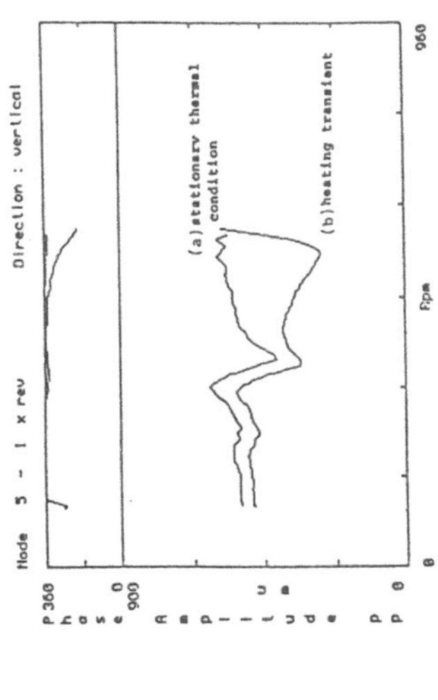

Fig.19 First harmonic component at measuring point 4 in different thermal conditions.

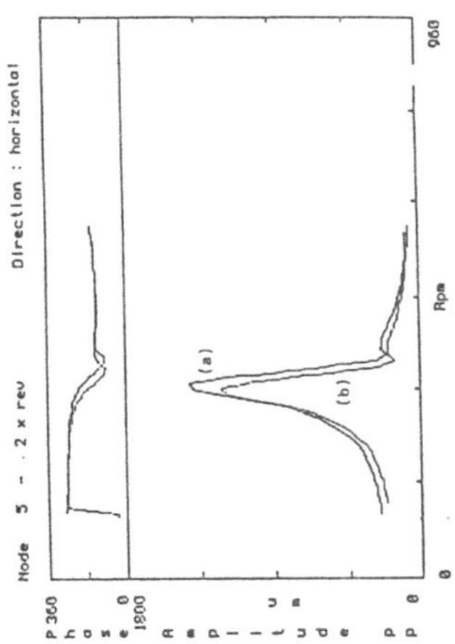

Fig.20 Second harmonic component at measuring point 4 in different thermal conditions.

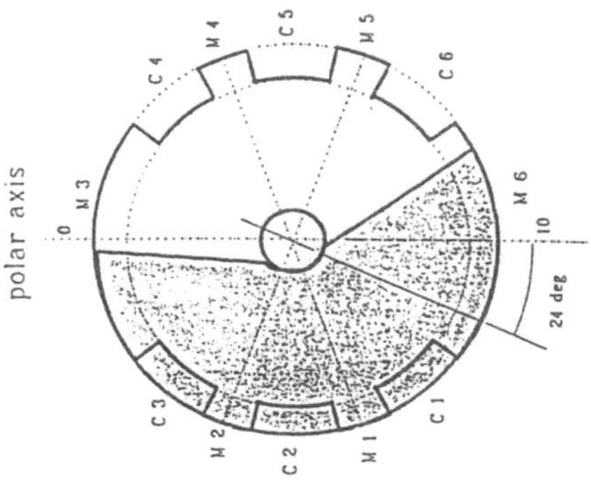

Fig.21 Schematic transverse section of generator rotor showing the dimensions of the crack (dark area).

EXPERT DIAGNOSTIC SYSTEM FOR GAS TURBINES

G. Sisti and G. Ferrari Aggradi
Nuovo Pignone, Florence, Italy

INTRODUCTION.

In recent years users of turbomachinery have shown growing interest for the use of computer based systems for health monitoring and diagnosis. These systems use both the thermodynamic and the mechanical parameters measured, and their combination, to detect incipient damage before major failures. In the beginning of the 1980's Nuovo Pignone has designed and built a Turbine Diagnostic System (**TDS**) for gas turbines. Both hardware and software are largely unrelated to the type of driven machinery; the system may therefore be applied to centrifugal compressors, pumps, electrical generators. Main features of TDS are user friendly in operation and reliability; it is flexible and can be constantly updated with the innovation called for by a continuously evolving state-of-art.

Special attention has been paid to Main Machine Interface, to be operated and used in data interpretation also from not computer skilled staff.

Special attention has been paid to Main Machine Interface, to be operated and used in data interpretation also from not computer skilled staff.

The first application of TDS has been in a natural gas compression station. Then it has been installed in seven oil pumping stations in Saudi Arabia and in Iraq, on a turbogroup for energy recovery in a F.C.C.plant, in four gas compression stations in Iran, and in three gas compression stations in Venezuela. A second-generation diagnostic system has been meanwhile developed, using the expert system's technology.

TDS PURPOSE.

The purpose of the system is to provide a diagnostic instrument that can be used to plan maintenance procedures in relation to actual needs, either in alternative to or besides routine maintenance. The system provides information used to prevent possible failures that could cause severe secondary damage. This results in increased availability of the machines (longer intervals between scheduled maintenance procedures), lower costs of maintenance, effected predominantly on condition, and thus effective improvement of plant efficiency. The system also supplies files of data that can be used for analysis after any damage that may occur; in addition, they provide a history of the life of the machine. In fact, the hourly average values of vibrations and operating parameters of the group for a year can be stored. All the stored data can be transferred by the operator to a floppy disk at any moment.

The system has two primary functions, checking the mechanical characteristics of the machine by comparing them with those stored in the data bank and monitoring the performance of the machine and verifying any decrease in its efficiency. The system uses both the thermodynamic and the mechanical parameters measured, as well as combinations of these, serving the dual purpose of:

- detecting incipient failures immediately;
- checking the efficiency of the installation (energy savings).

METHODOLOGY ADOPTED.

The diagnostic system has been developed from the GPA methodology, whose basic premises have been postulated by L.A. Urban.

Analysis of the thermodynamic operation of the turbogroup is based on measurement, at different positions on the machine, of many of physical quantities (temperature, pressure, flow rate), whose variation from the preestablished reference conditions can be correlated to the state of deterioration of the various components. For this to be possible in unequivocal manner, it is necessary to have a mathematical model of the operation of the machine ("Gas Path Analysis" model), consisting of a set of equations that link the values assumed by some thermodynamic quantities in coincidence with certain positions of the machine with the value of the parameters assumed as index of the health status of each component (efficiency, pressure drop, flow coefficient, etc.)

MATHEMATICAL MODEL.

A machine, which is an assembly of elements interacting with each other, is endowed by its designers with a set of special characteristics that are reflected in the performance of the machine itself and are, in final analysis, linked to geometric factors (shapes, dimensions, clearances between mobile parts and in movement, etc.), kinematic factors (speed, acceleration) and many other physical or physical/chemical factors such as the nature of the materials used for the various components, their internal and surface conditions, etc. During the life of the machine, many parameters that determine its performance may undergo change, for various reasons (wear, plastic creeps of hot parts etc.). Obviously, it is impossible to directly check the evolution of the factors determining the performance of the various parts, as it would require continuous arrest and disassembly of the machine. Consequently, it becomes necessary to investigate the evolution in performance of the various components (performance that we will say as "non-measurable" values) through the analysis and processing of suitable "measurable values" (pressure, temperature, speed, etc.) which are related to the former values. To do this, we must know the relationship between the "non-measurable values" and the "measurable" ones, a relationship represented by the mathematical model for the machine. If we indicate by x_1, x_2,.....x_n the above mentioned values measurable and non-measurable, the operation of the machine is regulated by the aggregation of a system of m equations in n unknowns, which in differential form becomes:

$$\begin{cases} \dfrac{\partial \varphi_1}{\partial x_1} dx_1 + \dfrac{\partial \varphi_1}{\partial x_2} dx_2 + \ldots\ldots \dfrac{\partial \varphi_1}{\partial x_n} dx_n = 0 \\ \dfrac{\partial \varphi_2}{\partial x_1} dx_1 + \dfrac{\partial \varphi_2}{\partial x_2} dx_2 + \ldots\ldots \dfrac{\partial \varphi_2}{\partial x_n} dx_n = 0 \\ \ldots\ldots\ldots\ldots\ldots\ldots\ldots\ldots\ldots\ldots \\ \dfrac{\partial \varphi_m}{\partial x_1} dx_1 + \dfrac{\partial \varphi_m}{\partial x_2} dx_2 + \ldots\ldots \dfrac{\partial \varphi_m}{\partial x_n} dx_n = 0 \end{cases}$$

The deviations in the various quantities x_1, x_2,.....x_n around an operating point P_o of the machine can be evaluated by linearizing the system.The operation around point P_o is thus defined by a linear system of the type:

$$(1)\quad \begin{cases} a_{11}\,\Delta x_1 + a_{12}\,\Delta x_2 + \ldots\ldots a_{1n}\,\Delta x_n = 0 \\ a_{21}\,\Delta x_1 + a_{22}\,\Delta x_2 + \ldots\ldots a_{2n}\,\Delta x_n = 0 \\ \ldots\ldots\ldots\ldots\ldots\ldots\ldots\ldots\ldots \\ a_{m1}\,\Delta x_1 + a_{m2}\,\Delta x_2 + \ldots\ldots a_{mn}\,\Delta x_n = 0 \end{cases}$$

whose coefficients (i.e. the partial derivative ($\delta\varphi_i/\delta x_i$) evaluated in P_o) form the matrix $\mathbf{A} = [a_{ij}]$. Matrix A is determined on the basis of the design characteristics of the machine. Obviously, it is necessary to know the operation of the machine in all the zones of the utilization map. Matrix A being known it is possible, by establishing a number (n-m) of "measurable" quantities, to render the systems determinate and to calculate the value of the "non-measurable" m quantities.
Taking as reference a one-shaft gas turbine, with variable compressor stator vanes, directly coupled to a generator, according to Fig.1 the installation is subdivided into four "basic modules": axial compressor, combustion chamber, turbine and generator.

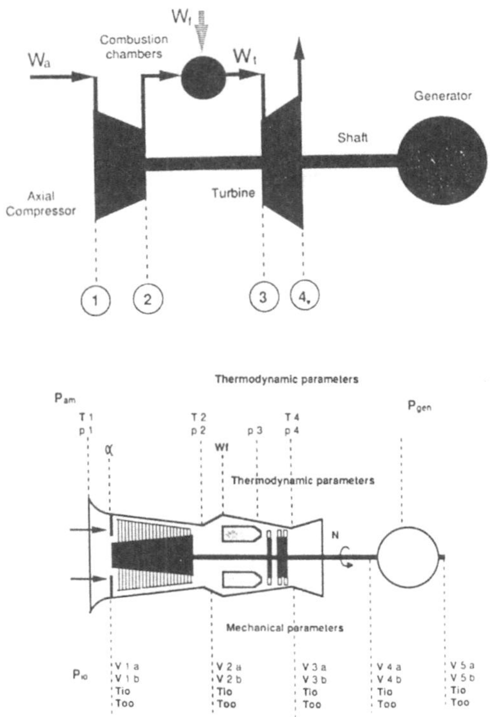

Fig. 1

The quantities measured to render determinate the system of equations (1) are:

T_1, P_1, T_2, P_2	axial compressor suction/discharge temperature/pressure
α	IGV position
N	shaft speed
W_f	fuel flow rate
P_3	turbine inlet pressure
T_4, P_4	turbine discharge pressure and temperature
P_{gen}	generator electric power

All the quantities measured are given in ISO conditions. The deviations from the stored reference conditions are calculated and, with the matrix of the coefficients $A = [a_{ij}]$ in the operating point, the system of equations (1) is resolved, giving the variations in the "non-measurable" quantities, that are axial compressor flow-rate, efficiency, flow coefficient and power, combustion chamber efficiency and loss coefficient, turbine inlet temperature, efficiency, flow coefficient and power.
If there is no decreasing in the machine performance the variations of these "non measurable" quantities are null or negligible. To store a set of reference conditions able to cover the whole potential operating range of the turbogroup, the possible zones of operation have been subdivided into "domains" of preestablished size; the reference conditions are stored in the field, either with a new machine or after a mayor overhaul; they can be stored by the operator or automatically by the system, when the machine is operating in a "non-signed" domain. The size of the various "domains" has been limited so that when the real operating points falls at the extreme edge of a domain, the error due respect the barycentre point is negligible .

DESCRIPTION OF TURBINE DIAGNOSTICS SYSTEM.
The TDS is composed of hardware modules and well proven software packages, which are configurable and expandable according to the type of turbo-machinery concerned and to the required analyses. The TDS hardware architecture (Fig. 2) consists of a Front End Processor (for data acquisition and processing) and a Workstation (for calculation and man-machine interface function) connected on a LAN Ethernet.
P6009 Front End Processor (FEP) consists of the following:
- 32 bit Motorola computer, series 68000, developed in VME-bus technology, mass storage devices, real time OS9 operating system.
- Microprocessed RTU for data acquisition from the turbogroup.
- VAS (Vibration Acquisition System) provided with the firmware for frequency analysis by means of the FFT function.
- RS 232C serial lines to communicate with machinery control subsystems.

P6009 Workstation consists of the following:
- 32 bit Motorola Computer, series 68000, developed in VME-bus technology, mass storage devices, UNIX operating system, X-windows graphic environment.
- high resolution 19" Monitor
- Hard copy for colour printing of video display images
- Alphanumeric Printer for alarm summaries and event messages report.

Basic Software. TDS supports the most widely used software standards:
- UNIX System V operating system for general use.
- OS9 for the management of real-time acquisition programs.

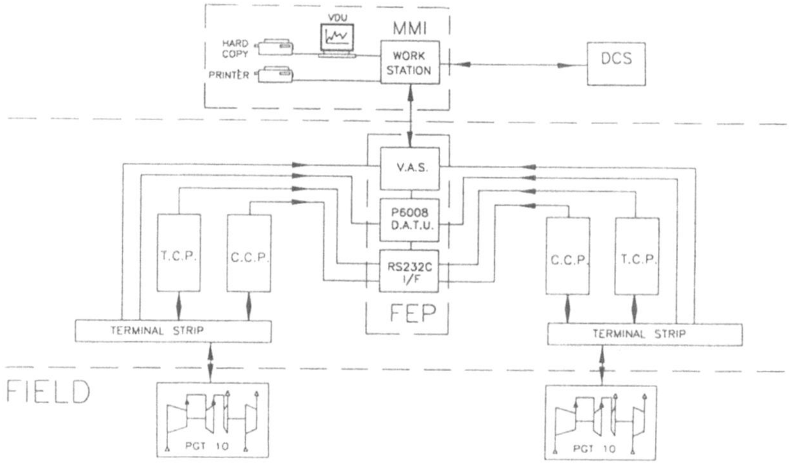
VDU
MMI
HARD COPY
PRINTER
WORK STATION
DCS
V.A.S.
P6008 D.A.T.U.
RS232C I/F
FEP
T.C.P.
C.C.P.
C.C.P.
T.C.P.
TERMINAL STRIP
TERMINAL STRIP
FIELD
PGT 10
PGT 10

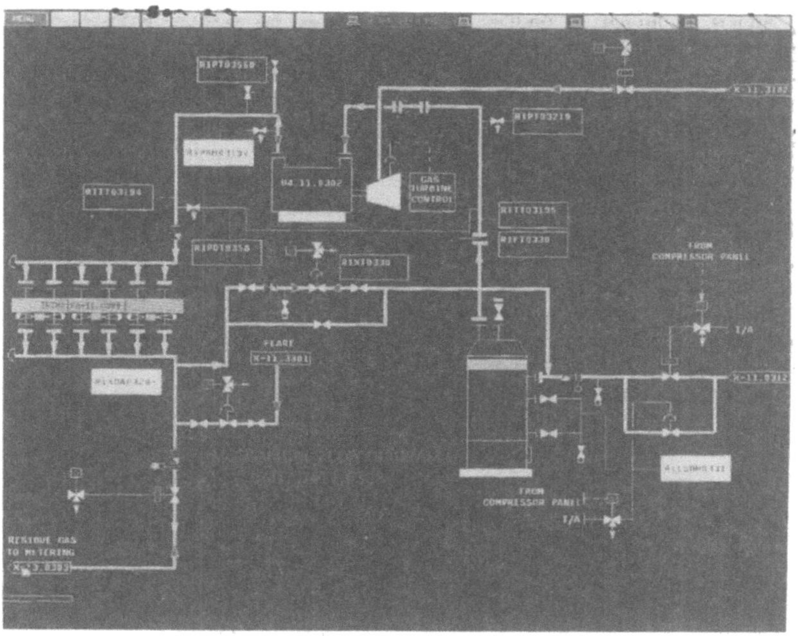
GAS TURBINE CONTROL
FROM COMPRESSOR PANEL
I/A
FLARE
FROM COMPRESSOR PANEL
I/A
RESIDUE GAS TO METERING

Fig. 3

- X window, multitask window graphic environment for data display
- Ethernet LAN for the communication between subsystems.

Application software. Developed in "C" language, includes:

- System configurator: it defines the turbogroup synoptic diagrams and system database.
- TDS diagnostic program: it detects the malfunctions in the TDS system and suggests remedial maintenance interventions.

Process Software is structured on 4 software packages: Data Acquisition Trend Monitoring, Performance Analysis, and Vibration Analysis

DATA ACQUISITION.

It is the basic function of the TDS system that provides configuration, data acquisition and processing, data base, man-machine interface and system diagnostic functions.

Analog variables. They are periodically gathered from transducers fitted to the turbogroup or from its control system, and converted into engineering units. They are sampled on a periodic basis using master/remote message exchange sequences initiated by the computer. The main variables are pressures, flows, temperatures and speed. Error checking and recovery from error in data transmission are provided.

Analog data processing. Data checking and digital filtering are applied to ensure the validity of received data. The analog values are converted to engineering units. Current data is compared with the previous set of data for significant changes, and is compared against predefined alarm limits and rate of change limits, adjustable on a per point basis.

Digital data. The digital data relevant to points in ON or OFF state are scanned at high speed. The detected changes of state initiate the event or the alarming procedure according to the type of point.

Turbogroup displays. The monitoring of the turbogroup is carried by the TDS through two levels of graphic displays: the Overview level and the Detailed level. The turbogroup Overview Display shows the current value of the most significant variables and the status of the main devices. The detailed information about each turbogroup section is presented in the second level graphic displays. Proper friendly procedures are provided to make the retrieval of the information quick and easy (fig. 3).

Alarm handling. All the alarms, both analog and digital, are presented on the Alarm Summary. Each alarm message contains date and time of occurrence, point name, tag, status. For the analog alarms the current value and exceeded limit are reported too. Proper operator procedures are provided for alarm handling, acknowledgement and detection.

Summary of analog variables. All the analog data are reported in summaries. The information presented for each point includes the point name, tag, current value, engineering units and alarm status. Summaries are arranged either on a turbogroup section or on type of variable basis.

Point displays. A point display is provided for each point. It includes all the attributes of the point, such as name, tag, current value, raw date, instrument range, alarm limits, alarm inhibition flag, etc...

System configuration. TDS ensures full capability to add, modify and delete points, to perform displays and logs generation and modification. The configurator procedures easily lend themselves to editing. The modifications are tested before system running.

TREND MONITORING

Trend Graph. TDS provides full graphic curve display to present up to three current values of selected variables simultaneously. Any variable in the data base is selectable by the operator for trending.The time axis resolution is 700 alues. The amplitude axis has an adjustable zero and span per variable. Resolution of the amplitude axis is 500.

Trend Graph Displays include Real Time, Short Term and Historical Trends.

Real Time Trend. The display of the variables is updated at each scan cycle. It is a valuable tool during the unit start-up

Short term trend. Snapshot data for the selected parameters are stored and displayed. The time period range from the last hour to last few seconds. Zoom facility for amplitude is provided (fig.4).

Historical trend. This graph is quite similar to the short term trend but it covers a long period, up to one year. The values of the selected parameters are averaged to allow compression of data.

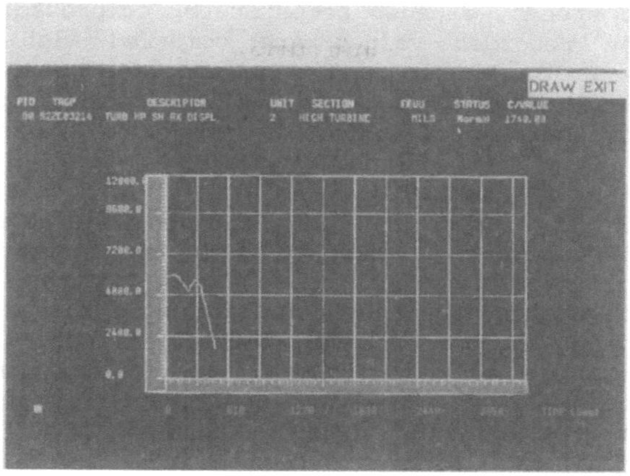

Fig. 4

Sequence Monitoring. The Sequence Monitoring graphs help to investigate the behaviour of the engine during the transients. The values of selected engine parameters are compared with start-up/shut-down baseline data "signed" when the engine is new or after overhaul. Pre-trip data are available to support the postmortem analysis.

Combustion Monitoring. A feature of TDS is to show in a polar form the temperature distributions around the exhaust chamber. Uneven temperatures are detected as deviation from the averaged values and by computation of max spread, resulting in identification of combustion chamber shortcomings.

Printouts. TDS fully supports printing of short term and historical trends, sequence snapshots, periodic and on request logs of any turbogroup section.

PERFORMANCE ANALYSIS

The thermodynamic model applied to a gas turbine and its load provides the operator with truly exceptional assistance. It provides a "Photograph" of the machine for following the evolution of its performance, an evolution that is calculated by the TDS, virtually in real time. The operator can thus promptly detect the problems (fouling or wear on blades, wear on seals, abnormality in combustion chambers), and identify the component in which the problem is arising.

Some parameters to be processed cannot be gotten directly from the field, but are calculated from other parameters by means of special algorithms (integrated values, averaged values, peak values, flow-rates and others as required). All measured and calculated parameters are normalized to ISO standards (15 °C, 101.2 kPa).

Load map. It is an outline display of the performances. For compressors the map shows the current situation in terms of pressure ratio and volumetric flow against a background of manufacturer's performances curves. In any condition the current value is continuously updated. The background curves can be adjusted by proper operator procedures .

Axial compressor fouling monitoring. One of the "non-measurable" quantities that TDS calculates is the axial compressor efficiency.

Should the deviation between the computed efficiency in the current domain and the efficiency stored be more than a predefined quantity a proper alarm is notified, to advise axial compressor fouling.

Turbogroup performance. Power absorbed by the load, the turbine power and the heat rate are recorded when machine is new, against baseline parameters as speeds, exhaust temperature, compressor flow. During the normal running the recorded values are compared with the "signed" values. Should a deviation be detected, TDS outputs a relevant alarm.

Turbine map. The map shows the current operating point in terms of power and shaft speed against a background of manufacturer's performance curves

VIBRATION ANALYSIS

It is based on the frequency analysis of the vibration signal. Displacements, velocity and acceleration are the dynamic parameters collected from the transducers mounted in the rotating machinery. The conversion from time-domain signals output from A/D converters to Frequency domain is performed by FFT algorithms. Control of sampling is internal, with a frequency dependant on the frequency range selected for analysis. The mechanical analysis charts make it possible to analyze each component of the dynamic signal, thus obtaining information on its amplitude, frequency, phase, RPM, etc. For each vibration sensor, a Frequency Analysis is carried at preset time intervals and on event such as start-ups, shutdowns and alarms. When one of the vibration signals is in alarm, the analysis is started and the dynamic data of the variable concerned are "frozen". Both the signal and the spectrum are stored and graphically analyzed. Software filtering and validation techniques are applied to analog signals to reduce noise and transducers malfunctions.

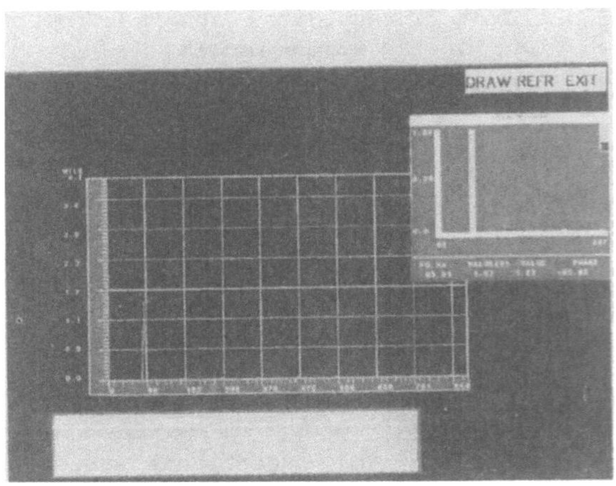

Fig. 5

Vibration waveforms. The amplitude of up three signals is shown vs. time.

Spectrum. The spectrum shows the frequency components of the vibration signal in terms of the amplitude on y axis versus frequency. The frequency analysis can be carried out both at preset time intervals and on event such as start-ups and shutdowns (fig.5).

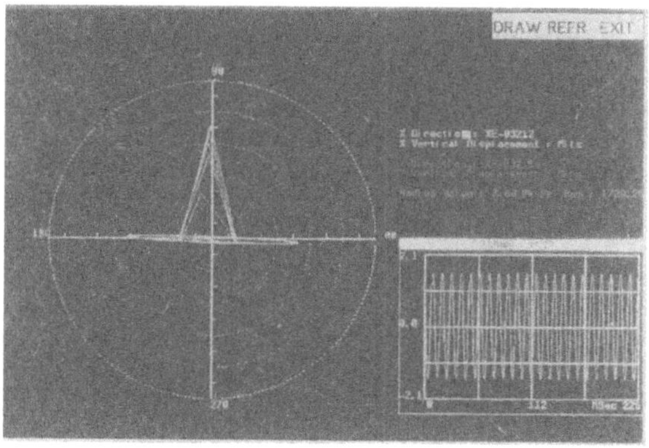

Fig. 6

Orbit diagram. It shows the shape of the shaft centerline motion detected by means of two shaft proximity probes, placed at 90° (fig.6).

Waterfall plot. The frequency spectrum of any selected vibration is shown vs. speed and time. It is useful during start-ups and shutdowns to investigate abnormal vibration components and critical speeds (fig.7).

Bode diagram. It represents the amplitude and phase of a selected harmonic vs. the shaft speed (fig.8).

Nyquist diagram. It represents, in a polar form, the amplitude and the phase of a selected harmonic vs. the shaft rotation speed (fig.9).

FURTHER DEVELOPMENTS.

The diagnostic system described until now simplifies proper on-condition maintenance and the prevention of oncoming failures, it can give the user a picture not only of the complete machine but also of all various its subunits. The system keeps the various modules into which the turbounit is subdivided under control, signals where there are variations in the operating parameters in a module in relation to the machine when new, gives alarm messages but leaves the operator to interpret messages and to decide the action to take. Only an expert on the machine can interpret the data, correlate thermodynamic information with mechanical data and decide the best course of action. Experts are not always readily available, and this is especially true in the case of complex plants. Therefore, having industrially

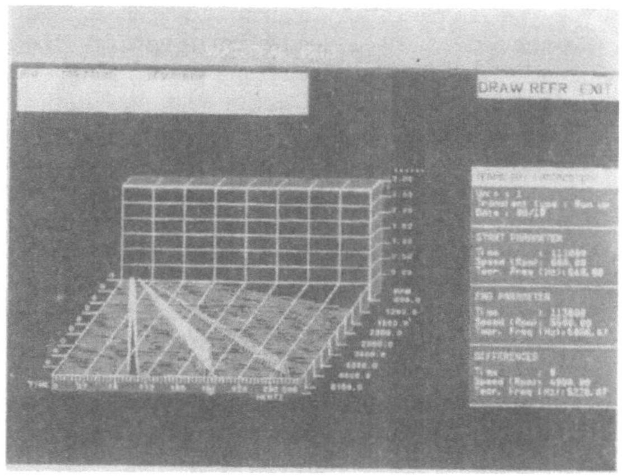

Fig. 7

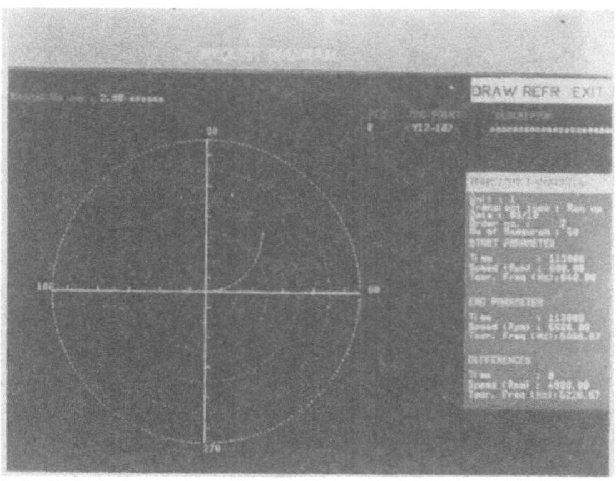

Fig. 8

set up the TDS, naturally Nuovo Pignone started to develop a second generation diagnostic system, using the technology of expert systems. Expert systems came into being towards the mid-seventies to designate a new programme typology set up within the scope of Artificial Intelligence, distinguished by their capacity to solve complex interpretation problems with performances up to the standard of an expert in the field. To date, expert systems capable of solving all types of complex problems have been developed. There are all sorts of fields of application ranging from medicine to biochemistry, from genetics to mineral research, and from the financial field to the diagnosis of complex equipment.

Purpose of the expert system

The system was conceived as a tool for the plant operator when tackling faultily functioning of the turbine during the various turbine conditions, such as outage, ready-for-startup, firing, in operation and tripping. In particular, the target was to put the operator in a position to carry out initial checks in order to troubleshoot. Anyway, the operator was to be put into action to draw up a detailed diagnostic report, supplemented with as much information gleaned during the diagnostic session as possible. The expert system interfaces with the turbine through its TDS. This supplies alarms and values that can be gotten automatically from the turbine package control, supplementing them with thermodynamic and mechanical alarms that it itself is capable of generating. Final aim of the system is a predictive maintenance strategy, based on real time knowledge of the machinery health conditions; using the expert system technology the knowledge, the experience and the intuition of machinery specialists have been added to routine analysis of diagnostic system. Computer has stored a great amount of technical knowledge and can select intelligent judgments on likely causes of trouble or poor working. The large amount of data base knowledge that expert system requires has been built by interviewing experts and leaders in the field of plant design. Information has been written in the form of rules: rules of judgment and rules of good guessing, typical of expert-level decision. These rules have been placed in an empty shell expert system, which has been integrated with TDS. TDS transmits to expert system all data available, both measured or computed: the expert system scans the knowledge base until a satisfactory evaluation is made; then gives a report to the operator. To give operator good decision support, the system produces extensive explanation of how the different diagnosis have been concluded; so operators and maintenance people will have an actual help. The expert system can also operate in stand alone mode in order to be able to simulate data supplied by the TDS in order to test new diagnostic situations to train new operators.

On-line operation

Periodically the system retrieves from the field the state of alarms and signals acquired automatically. Data is represented graphically on the screen, featuring an instrumentation panel for measured variables and three alarm panels, divided up into types: auxiliary panel, thermodynamic panel and mechanical panel.

Every field acquisition is checked by the system to identify the existence of alarms or abnormal values. Then, the operator's intervention is signaled by the diagnosis button turning red so as to be able to start up the troubleshooting session. The operator can start diagnosis by working on the relative button. By putting the diagnostic session into operation, the system starts troubleshooting.

Troubleshooting consists of formulating hypotheses supported by data known at the time. Hypotheses are then checked by asking the operator attendance in the field: simple measurements of data not acquired automatically or checking of plant components. Checks can just be visual to acquire the state of a component (for istance the running condition of an electric motor) or more complex tests capable of determining whether it works properly or not. Upon every manual acquisition request, the operator can interrogate the system as to the reason for its request and ask greater details on how to carry out the necessary check . An explanation of the hypothesis in progress is given as a cause-effect linkage of faulty operating conditions. The diagnosis will continue until for each failure (alarms present or abnormal variables) at least a justifiable cause has been identified or there are several hypotheses that can immediately be checked. At this stage, the system presents a summary of the conclusions it has drawn. For each of them, it presents a window (result) featuring a description of the conclusion drawn, recommendations and an explanation about how the signaled malfunctioning justifies the detected field situation .

Results can be broken into:

Diagnosis: identifies ascertained malfunctioning of a component.

Failure :identifies faulty functioning of the plant (for istance a pump out of operation) ascertained without finding a verifiable self-explanatory cause.

Hypothesis: identifies possible malfunctioning of a component part capable of explaining, wholly or partially, faulty functioning of the plant that it is not possible to justify with retrieved data. Hypotheses are put forward as an alternative to diagnosis when the checking of a part is impossible because the operator is incapable of carrying it out and answering the question.

Stand-alone mode

In stand-alone mode, the diagnostic system enables the operator to stand in or the TDS giving him a chance to set with a mouse on the graphic interface the state of alarms and values of variables that can be acquired automatically at his discretion. Having completed the setting stage, it is possible to start the diagnosis session as in the on-line mode. If needed, the variables detected automatically that have not yet been defined shall be requested by the operator only when used by the system.

FUTURE DEVELOPMENTS OF THE SYSTEM

At present, the expert system is built for a PGT 10 gas turbine, driving a centrifugal compressor for pumping natural gas, installed at the SNAM station in Cortemaggiore. At that station, it has been installed a TDS, combined with an expert system for industrial experimentation and for checking any necessary refining.

ROTORDYNAMICS OF INDUSTRIAL GAS TURBINES BY FE MODELS

S. Salvano and M. Bertotti
Fiat Avio, Turin, Italy

ABSTRACT

The development trend of advanced industrial gas turbines is towards larger sizes and higher power outputs improving performance and thermal efficiencies to compete in the international market.

This implies, inter alia, that rotating parts are much more stressed having the same and often more stretched requirements in terms of life and reliability.

Among the technical disciplines related to the design of Gas Turbines, Rotordynamics is playing a more critical role along the engine development from the prediction phase through experimental results to the service operation.

As a consequence, the mathematical modelling is becoming an indispensable tool and efforts to improve its suitability to reproduce the physics are continuously pursued.

This paper describes the achievements obtained applying an improved FE model, using SAMCEF code, to the new advanced Gas Turbine 701F, the first 230 MW entered service, comparing critical speed prediction and experimental test results gathered in the Shop Tests.

1. INTRODUCTION

The development trend of advanced industrial gas turbines is towards larger sizes and higher power outputs improving performance and thermal efficiencies to compete in the international market.
This implies, inter alia, that rotating parts are much more stressed having the same and often more stretched requirements in terms of life and reliability.
Among the technical disciplines related to the design of Gas Turbines, Rotordynamics is playing a more critical role along the engine development from the prediction phase through experimental results to the service operation.
As a consequence, the mathematical modelling is becoming an indispensable tool and efforts to improve its suitability to reproduce the physics are continuously pursued.
This paper describes the achievements obtained applying an improved FE model, using SAMCEF code, to the new advanced Gas Turbine 701F, the first 230 MW entered service, comparing critical speed prediction and experimental test results gathered in the Shop Tests.

2. PROBLEMS RELATED TO ROTORDYNAMICS ON INDUSTRIAL GAS TURBINES

Several kind of problems related to rotordynamics are faced during the different phases on the life cycle of an industrial gas turbine.
In the initial design phase, evaluation of critical speeds is a very critical item : since these rotors are generally supercritical, very precise evaluation up to the third or fourth critical speed is necessary to guarantee an adequate margin with respect to the operating speed. Besides rotor stability has to be carefully checked, as well as rotor / stator coupling.
In the end, choice of many design parameters is affected by rotordynamics:
– bearing span and related compressor and turbine scantling ;
– bearing stiffness and damping characteristics ;
– support and casing stiffness ;
– inertial data of rotating parts.
In the detailed design phase, definition of design tolerances for different components like rotors and bearings can be affected by rotordynamics related problems such as maximum allowable initial unbalance or bearing behavior.
This definition is established through a concurrent engineering approach with which is made a trade–off between manufacturing, design and structural needs.
Shop tests, either on a rig facility without load or directly on the power plant, are the first check–point of validation of the design. In this phase, balancing of the machine at full speed and power may be required to take the vibrations below the allowable limits.

Allowable vibration limits themselves are defined through a proper understanding of the rotordynamics of the machine.
Finally, vibration monitoring during operation is continuously required ; they can be the symptom of any malfunction of the machine.
All these problems can be managed by using the mathematical FE model of the gas turbine, opportunely tuned.

3. *PHYSICAL AND MATHEMATICAL MODEL*

The turbomachinery components are generally classified in three categories:
– rotating parts such as disks and shafts with the relevant connection elements like flanges, curvic couplings;
– static parts like casings and pedestals;
– connection elements between rotor and stator : these include bearings with particular reference to aerodynamic and hydrodynamic forces induced by rotor motion.
Rotordynamics basic equations are derived by virtual work principle using Lagrange's equations:

$$\frac{d}{dt}\left(\frac{\delta T}{\delta q_i}\right) - \frac{\delta T}{\delta q_i} + \frac{\delta U}{\delta q_i} = Fq_i$$

FE technique permits to treat them in the matricial form.

Rotating parts are characterized by kinetic energy *T* discretization forming mass and gyroscopic matrices, by strain energy *U* setting up the relevant stiffness matrix and by dissipative forces *F* that go into damping one.

Respect to rotating parts, fixed parts do not have gyroscopic contribution.

Connection elements, whose behavior is generally considered linear, can be described by appropriate stiffness and damping matrices able to represent the fluid–structure interaction forces.
Finally structural damping *B* is assumed proportional to *K* stiffness and *M* mass matrices:

$$B = \alpha K + \delta M$$

Besides, both α and δ are related to critical damping ratio c/c_{cr} and speed *w* by the following relation:

$$\alpha = 2\,(c/c_{cr})\,/\,w$$

$$\delta = 2\,(c/c_{cr}) \times w$$

4. FE MODEL

4.1 APPROACH

Several methods are being used to perform critical speed calculation among which the most significant are the Transfer matrix method and the FE approach, since closed form solutions are available only for simple configurations and not of practical interest.

Transfer matrix approach is based on Myklestad–Prohl iterating or modified Holzer method, but it is subjected to some limitations like beam type formalism, problem size limitation, numerical instabilities in case of complex models, linear analysis and poor flexibility.

On the contrary, FE method is more versatile and allows the modelling of complex rotors and the fixed parts of the structure.

With SAMCEF code, both beam and axysimmetric formalisms can be used.

The beam model is the classical approach and the most useful in the preliminary design phase.

The axysimmetric model is appropriate in the detailed analysis phase since it allows to obtain finer models and so detailed local information on the stress levels, having the possibility to perform in parallel a structural analysis on the same kind of model.

Axysimmetric model is coupled with a Fourier expansion of the displacement field in circumferential direction. Assuming that bending deformation is of major interest, the analysis is associated to first harmonic with sin and cosine components.

In this case, ovalization of the cross section of the rotor can be taken into account for anisotropic rotors or supports as well as the gyroscopic effect of introducing a coupling between the two harmonics components.

4.2 MODELLING

While characterization of stiffness and mass of the structure, both stator and rotor, can be easily evaluated , stiffness properties of the connection elements and damping of all the system are a much more difficult task.

For what concerns bearing film characteristics, they are derived by a data base on which experimental curves are given as function of fluid dynamics parameters like Sommerfield and Reynolds number for different bearing types.

Starting from these data, used in the initial definition phase, the matrix coefficients are tuned basing on experimental critical speeds values.

Damping of the model is tuned through experimental value of Q–Factor.

Practically, to do that , viscous damping coefficients of journal bearings are maintained unchanged and a structural damping, with a different value for each critical speed,is put on the model.

Being rotor and static parts treated as separated superelements, two α and δ are available to act on in order to get the experimental Q–Factor of the first two critical speeds.

Calibrating the model, the response of the engine to an external force like an unbalance is much closer to reality.

5. AN APPLICATION CASE: THE 701F GAS TURBINE

The example here reported is referred to the ***701F***, the advanced 50 Hz industrial gas turbine of the new generation with a power output of 230 MW, designed within the Trilateral Cooperative Agreement between FIAT Avio, MHI and Westinghouse. Cross–section is shown in fig. 1.

Adopted FE model is a 2D plain axysimmetric, made of isoparametric elements having three or four nodes (fig. 2). The model can be extended in circumferential direction for 3D visualization (fig. 3).

Input data used to build the model are :

– electronic EUCLID file of the cross–section of the machine ;

– inertia data of rotating parts non–axysimmetric (essentially spindle bolts and blades of which it is also necessary the number per stage);

– mass data of static parts non–axysimmetric (combustor and vanes).

Main features of the FE model are :

– the subdivision in two superelements, one for the rotor and one for the stator, in order to manage the problem size ;

– the introduction for the non–symmetric parts of artificial material properties , acting on E and density, in order to restore proper stiffness and inertia ;

– the use of lumped masses for complex static parts like combustor and bellmouth;

– stiffness and damping matrix coefficients of bearing oil film are given as a function of speed, while bearing supports, compressor radial struts and turbine exhaust tangential ones are modelled by FE.

6 – TYPE OF ANALYSIS AND MAIN OUTPUTS

6.1 CRITICAL SPEED ANALYSIS

Different steps have been followed on this phase of the analysis :

– define the Campbell Diagram and determine critical speeds of interest;

– define the stability diagram and evaluate eventual instability problem;

– perform a sensitivity analysis on connection elements stiffness coefficients to evaluate critical speed range of variability.

In order to completely verify the good behavior of the model and the relevant superelements, the analysis has been conducted in three different steps:

a – rotor only with bearing film and connection to ground through bearing supports but without pedestals;

b – stator only connected to ground through pedestals ;

c – whole model .

In the first case , being the bearing film characteristics symmetric, Campbell diagram is the typical case on which at w=0, each critical speed corresponds to a couple of coincident frequencies that diverge with speed respectively in a forward and in a backward one, being the last one of low practical interest since not excited by rotor unbalance (fig. 4).

The case of the stator only is very simple, being the frequencies independent from the speed (fig. 5).

The last case is the real one on which rotor/stator coupling can be noticed. In fact, having introduced the pedestal stiffness, that is different in horizontal plane respect to vertical one, the anisotropy in the model generates a Campbell with two separate frequencies , horizontal and vertical, already distinguished at no speed (fig. 6).

Table 1 reports the above results.

Mode shapes of the first three critical speeds are shown in figg 8 to 10.

For what concerns stability of the rotor after crossing a critical speed, diagram of fig. 7 shows the value of the real part of each complex solution versus speed. Negative values in the whole range mean that no instability problem is foresee.

6.2 FREQUENCY RESPONSE

Subsequent step on the analysis is the response of the rotor subjected to an external load.

Most interesting case is that of a load caused by an unbalance, corresponding to a first harmonic excitation.

Both the cases of static and dynamic unbalance distributed along the rotor or acting separately an the three balancing plane have been analyzed, providing the response of the rotor in frequency range of interest (0 to 55 Hz).

For each case, a sensitivity analysis has been conducted by acting on the structural damping in order to get experimental Q–Factors for the first two critical speeds crossed before reaching operating speed .

Analysis results are given in terms of amplitude and phase of vibration at the bearings and the balancing planes.

These outputs, organized as polar diagrams and referred to each balancing plane load case, provides the so called "Effect Diagrams", used in the balancing activity during the commissioning phase.

On fig. 11 , it is shown the vibratory response of both compressor and turbine bearing in the vertical plane to a 10 Kgmm unbalance put on the compressor balancing plane, while on fig. 12 the same response is given in the polar diagram.

Besides damping of the system for each critical speed and relevant Q–Factor are evaluated.

6.3 TRANSIENT ANALYSIS

A transient analysis has been carried out considering the same unbalance as for the frequency response.

A typical case of 18 min to reach 3000 RPM was examined to evaluate the effective dynamic response of the system and compare it with the frequency response case , although a gas turbine start–up transient cannot be considered a so sudden phenomenon.

Responses are given in fig. 13 and are in the order of the steady–state case confirming what above mentioned, while in fig. 14 frequency spectra of the above case is shown.

The evolution of the orbits (fig. 15) permits to notice the phase change through a critical speed as well as the whirl sense change across Backward critical speeds.

7. COMPARISON WITH EXPERIMENTAL DATA AND DISCUSSION

The FE model was tuned on the basis of Shop Test results carried out during 1992 and 1993, at MHI facility of Yokohama (Japan), where the first 701F unit successfully ran at 1350 C TIT and then connected to the external grid for supplying electric power to the local utility company Tokyo Electric.

For the purpose of this paper, vibration pick–up positioned on the vertical plane has been taken into consideration for analytical–experimental comparison.

Most interesting experimental results are vibration amplitude of compressor and turbine bearing during start–up to rated speed and relevant polar diagram amplitude/ phase (fig. 16).

According to that, first two critical speeds in the vertical plane are detected as well as the relevant maximum vibration level and Q–factor.

Comparison with analytical results are depicted on table 2 and shows good agreement both in terms of critical speed values and vibration response.

8. CONCLUSIONS

In this document, a description of the main capabilities of an improved FE model of an industrial Gas Turbine has been given.

This tool can be extensively used during all the phases of the project, to support the solution of different problems related to rotordynamics.

For what concerns the diagnostics during operation in Power Plants, the use of the model permits, inter alia, to establish allowable vibration limits; to derive effect diagrams for balancing purposes; to study, on the basis of the vibration pick–up response, the unbalance distribution along the rotor, individuating any localized vibration source; to carry out the transient response of the machine to a sudden unbalance derived by any mechanical malfunction and to perform local detailed structural analysis in order to investigate any kind of rotordynamics related phenomenon.

Table 1: ANALYTICAL Critical speeds (Hz)

case 1 : ROTOR only	*case 2: STATOR only*	*case 3: whole model*
	7. 1HS	*6.1 1HS*
18.2 1HR		*12.2 1HR*
18.8 1VR		*17.2 1VR*
	21. 1VS	*21.9 1VS*
	22.5 2HS	*23.2 2HS*
41.7 2HR		*27.5 2HR*
	36. 2VS	*34.7 2VS*
43.4 2VR		*36.8 2VR*
	43.2 3HS	*42.3 3HS*
	46.7 3VS	*44. 3VS*
55.5 3HR		*46.9 3HR*
56.5 3VR		*53. 3VR*

H : HORIZONTAL V : VERTICAL
R : ROTOR S : STATOR

Table 2: Experimental vs Analytical comparison

	Vertical Critical Speeds (Hz)		*Q – FACTOR*		
	ANALYTICAL	*EXPERIMENTAL*		*ANALYTICAL*	*EXPERIMENTAL*
*1^*	*17.2*	*15.3*	$(a = 0.001)$	*6.4*	*5.5 – 8*
*2^*	*36.8*	*38.*	$(a = 0.001)$	*3.5*	*6*

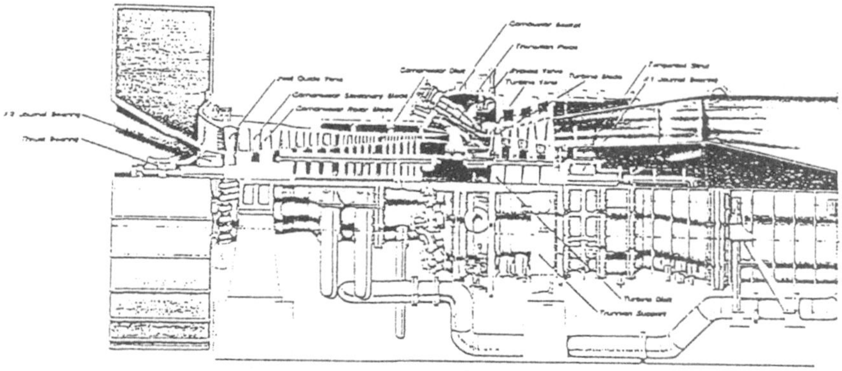

FIG. 1 : 701F GAS TURBINE CROSS-SECTION

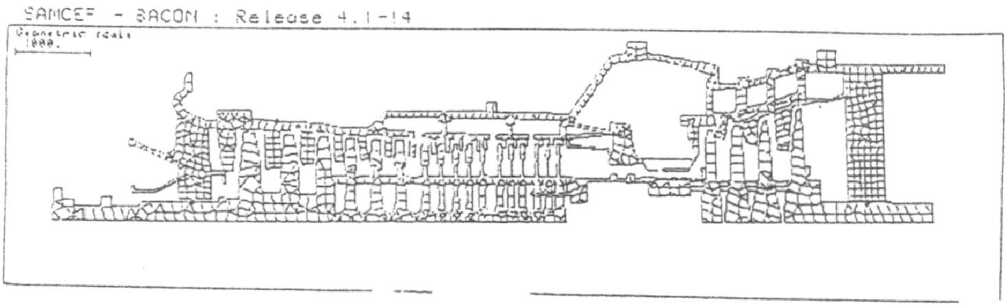

FIG. 2 : 701F FE MODEL

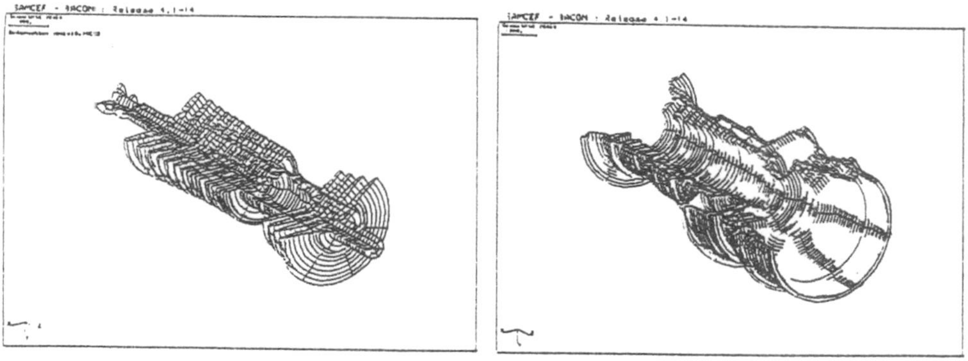

FIG. 3 : 3D VIEW OF ROTOR AND STATOR

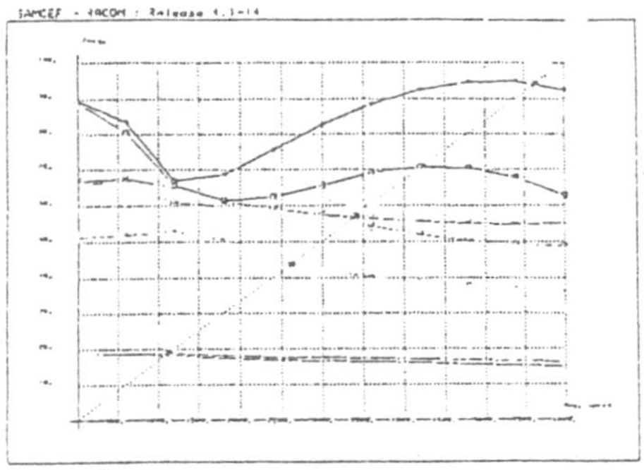

FIG. 4 : CAMPBELL DIAGRAM: CASE 1 ROTOR ONLY

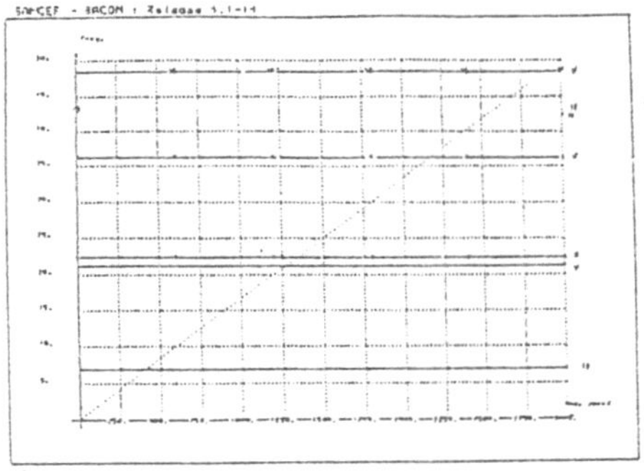

FIG. 5 : CAMPBELL DIAGRAM: CASE 2 STATOR ONLY

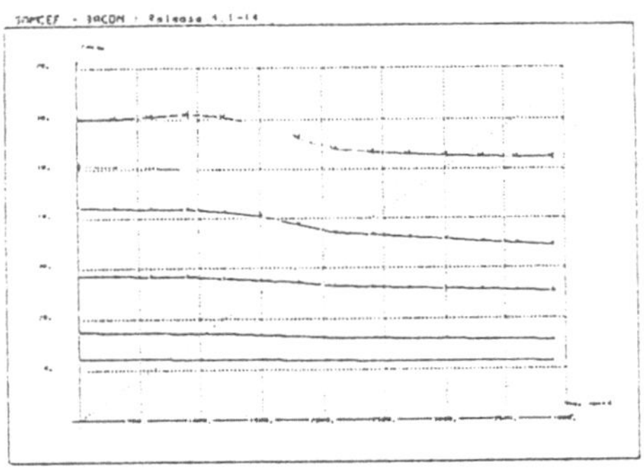

FIG. 6 : CAMPBELL DIAGRAM: CASE 3 WHOLE MODEL

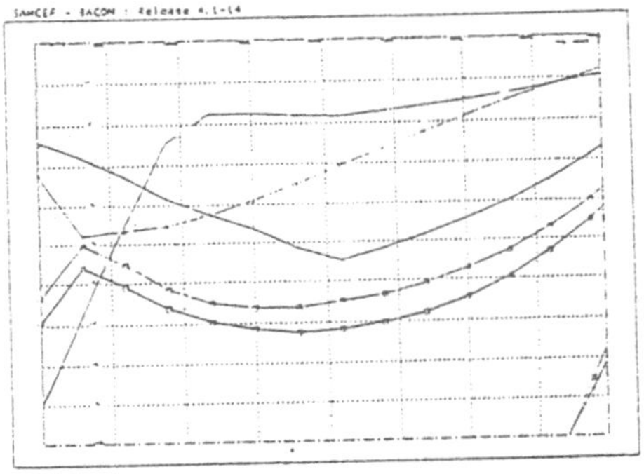

FIG. 7 : STABILITY DIAGRAM

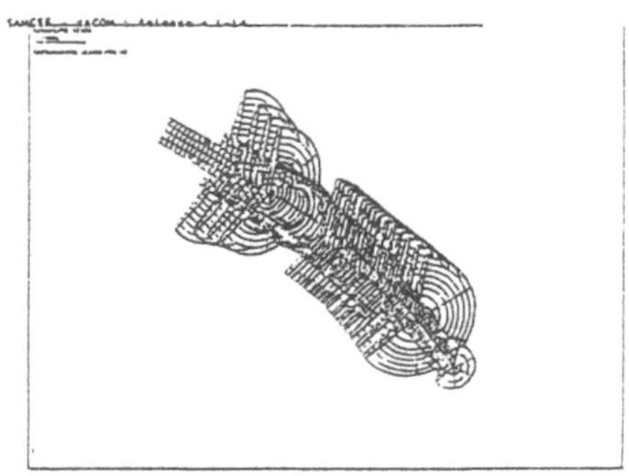

FIG. 8 : 1ST CRITICAL SPEED DEFORMED SHAPE

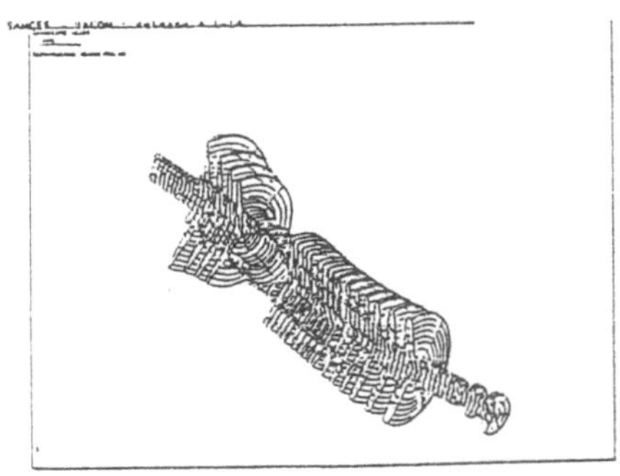

FIG. 9 : 2ND CRITICAL SPEED DEFORMED SHAPE

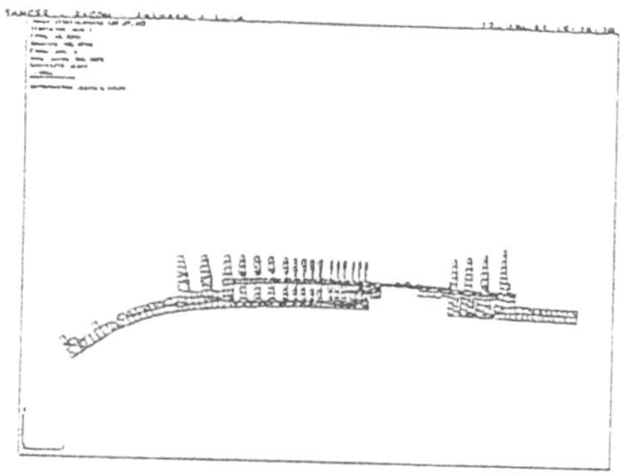

FIG. 10 : 3RD CRITICAL SPEED DEFORMED SHAPE

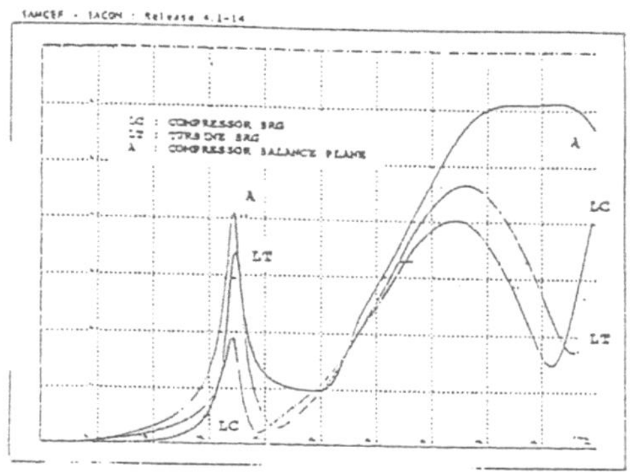

FIG. 11: FREQUENCY RESPONSE AT BEARINGS TO UNBALANCE ON COMPRESSOR PLANE

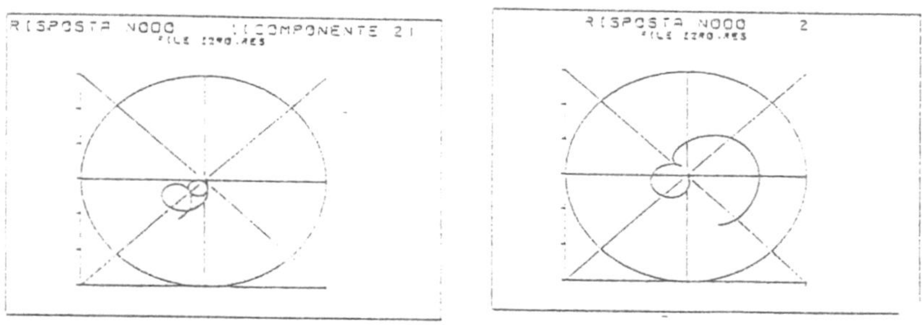

FIG. 12: EFFECT DIAGRAM

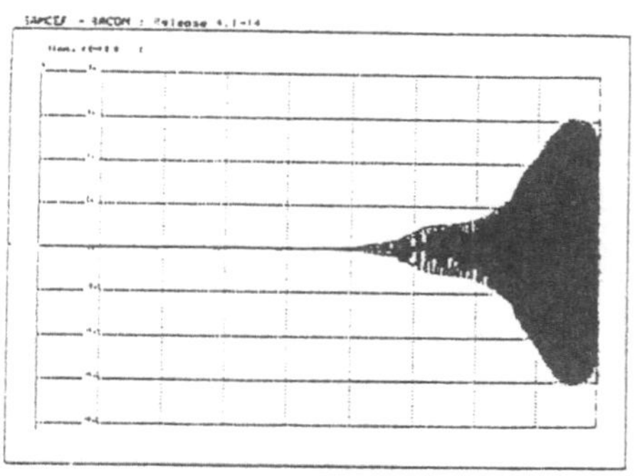

FIG. 13a: TRANSIENT RESPONSE AT TURBINE BEARING TO UNBALANCE ON COMPRESSOR PLANE

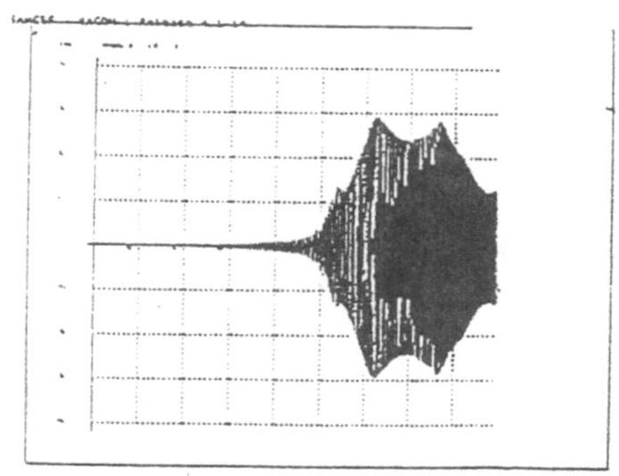

FIG. 13b: TRANSIENT RESPONSE AT MID SPAN TO UNBALANCE ON COMPRESSOR PLANE

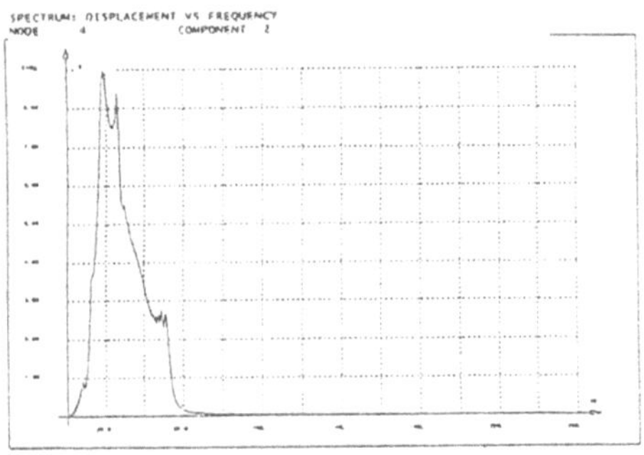

FIG. 14: FREQUENCY SPECTRA AT MID SPAN TO UNBALANCE ON COMPRESSOR PLANE

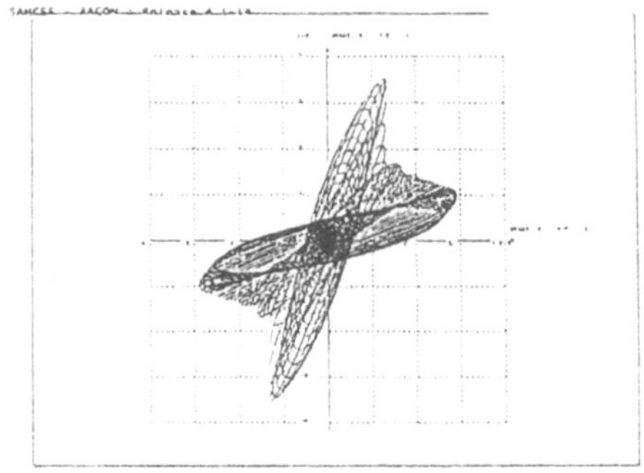

FIG. 15: TURBINE BEARING ORBIT

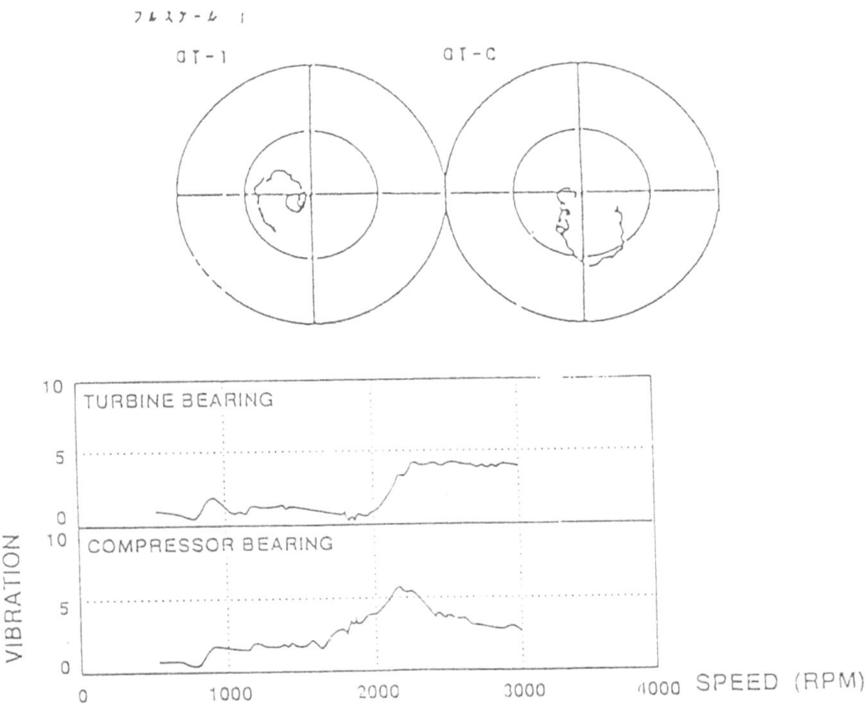

FIG. 16 : 701F SHOP TESTS AT YOKOHAMA: EXPERIMENTAL VIBRATIONS AT BEARINGS

RIGHT AND LEFT-HAND NATURAL MODES OF INDUSTRIAL ROTATING MACHINES. MODEL STUDY

R. Bigret, S. Vialard and P. Karajani
I.S.M.C.M., Saint Ouen, France

R. Chevalier
E.D.F. - D.E.R., Clamart, France

ABSTRACT

The rotors of rotating machines are guided by links. The matrixes representing links are, in general, not symmetrical. Right and left-hand natural modes differ. Left-hand modes are particularly used to define modal unbalances.

The high pressure rotor of a turbine generator of 1000 MW at 1500 rpm of a nuclear plant is here being studied. Three models are described ; their first flexural natural pulsation, while on fixed supports, is equal to 1116 rpm, close to the first critical speed. The structure of the first two models is that of a De Laval-Jeffcott rotor, the third rotor concords with a high pressure rotor. For each of the models, two kinds of bearings have been teamed : cylindrical and 3 lobes.

Natural pulsations as well as right and left-hand modes are determined by literal formulas for the first model, by special software for the other two. Comparisons are set up between amplitudes and the phases of the right and left-hand natural modes ; the differences are related to the assymetries of impedance matrixes.

The quality of the first two models, in reference to the concordant model, is shown through the comparison between the natural modes.

1. INTRODUCTION

The rotors of rotating machines are guided in their stators by links which may consist of fluids, ball-bearings or magnetic fields. In certain rotating machines, particularly turbines, the tightness systems, situated between the stator and the rotor, reduce or even eliminate, leakage between the compartments.

The links and the tightness systems are characterized by impedance matrixes which associate radial forces to radial displacements. The impedance matrixes depend, among other factors, on the design, the position of the rotor in relation to the non-rotating structure and the rotating speed. They may not necessarily be symmetrical. Gyroscopic moments are expressed by a non symmetrical matrix.

In general, the associated natural pulsations and the natural modes are determined by using the impedance matrixes, symmetrical or not ; on the other hand, responses to forces related to the rotor, particularly to forces due to unbalances, by a simplifying procedure, are often shown by modal decomposition, which do not take into account the assymmetry in the impedance matrixes, i.e. which accept the equality of the so-called right and left-hand modes (!). Annex 1 gives the definition of the right and left-hand modes.

Essentially, this note brings into relation the differences between natural pulsations, natural right and left-hand modes and the assymmetries of the impedance matrixes of two bearings for the model of a high pressure rotor of a turbine generator of 1000 MW at 1500 rpm and two simplified models of this rotor.

2. HIGH PRESSURE ROTOR OF A TURBINE GENERATOR OF 1000 MW/1500rpm

In France, 75% of the electrical energy is supplied by 53 power units, spread over 17 nuclear sites equipped with turbine generators of 1500 rpm : 34 of ≈ 900 MW, 19 of ≈ 1300MW ; in 1995 a turbine generator of 1500 MW will be put into service.

• Figure 1 describes a rotor similar to a high pressure rotor of a 1000 MW, 1500 rpm turbine generator. Bearings 1 and 2, identical, are characterized by an impedance matrix of displacement ; at the contact/bearings, the modules of the terms of the impedance matrix of the non-rotating structure are very large ; owing to this fact, its vibrations amplitudes are nil.

• The natural pulsations and the natural right and left-hand modes are determined by software that uses a finite-element method ; it takes into account the rotation inertia of the elements, the influence of a shear force affecting the deflection of the rotor and the gyroscopic moments. The mesh system makes it possible to define 36 elements attributed with mass and bending stiffness.

• Table 1 shows the first natural pulsations $\omega_{\cdot}$ (complex pulsation $p_{\cdot} = \delta_{\cdot} + j\omega_{\cdot}$ with $j^2 = -1$, $\delta_{\cdot} = 0$: conservative system), and the corresponding natural modes for the impedance matrixes where the terms are reduced to equal direct stiffness which, owing to their high value, (5.10^{15} N/m), represent the fixed swivel supports ; for 10^{17} N/m natural pulsations and modes are quite identical. Table 1 shows that natural pulsation at 0 rpm gives rise to two families of neighboured pulsations (18,9 Hz → 19,2 Hz, 19,3 Hz and 18,6 Hz, 18,5Hz) which manifest themselves by forward and reverse precessions ; they are influenced by gyroscopic moments.

The natural pulsation equal to 1116 rpm (18,6 Hz) is used in all that follows, in order to define the two models called De Laval-Jeffcott, where the pulsations and natural modes will be compared to those of the high pressure rotor. The critical speeds defined by the maximum in amplitudes of forced regime by an unbalance located on point C (fig. 1 and 2), are close to 1116 rpm.

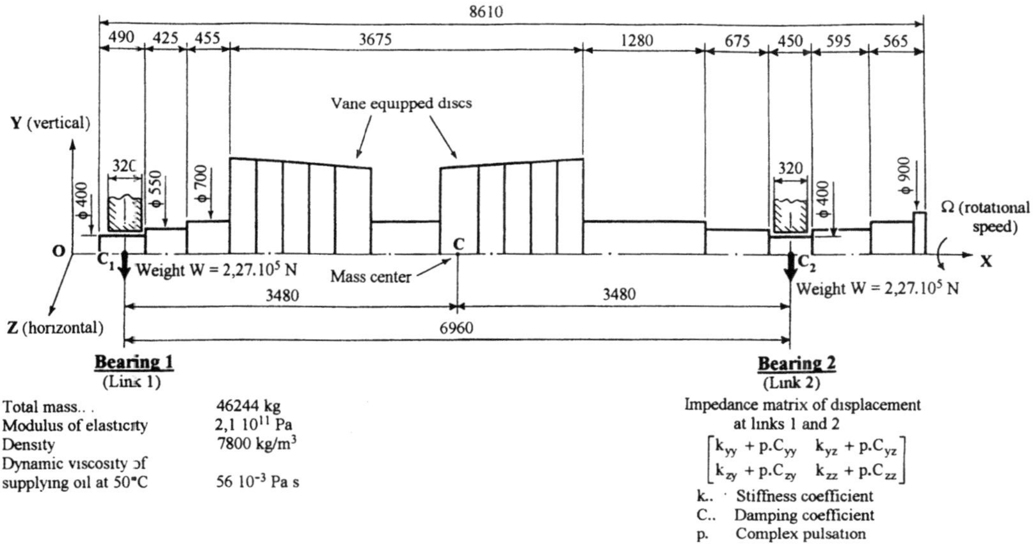

Fig. 1 Rotor similar to a high pressure rotor of a steam turbine rated at 1000 MW at 1500 rpm. Geometry and characteristics.

Table 1 High pressure rotor Natural pulsations - Natural mode

Links at C_1 and C_2 Rigid swivel	Rotational speed 0 rpm	Rotational speed 1116 rpm (18,6Hz) (≈ Critical speed)	Rotational speed 1500 rpm (25Hz) (Nominal speed)
Impedances (N/m)	118,8 rad/s (18,9 Hz)	120,6 rad/s (19,2 Hz) Forward precession 116,9 rad/s (18,6 Hz) Reverse precession	121,3 rad/s (19,3 Hz) Forward precession 116,2 rad/s (18,5 Hz) Reverse precession
$\begin{bmatrix} 5.10^{15} & 0 \\ 0 & 5\ 10^{15} \end{bmatrix}$		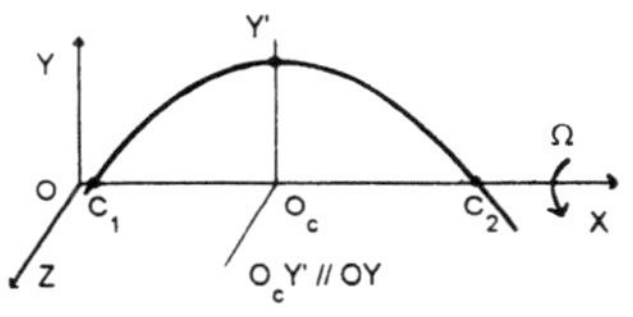	Mode shape in a plane

3. BEARINGS

The natural pulsations and modes of the high pressure rotor and the two De Laval-Jeffcott models are calculated with two kinds of bearings : cylindrical and 3 lobes[1]. The impedance matrix which represent them is described in figure 1 and Annex 2. The stiffness k.. and the damping coefficients C.. depend on Sommerfeld number :

$$S = \frac{\mu \Omega L R}{\pi W}\left(\frac{R}{J}\right)^2$$

with : μ : viscosity (56.10^{-3} Pa.s) R : trunnion radius (200 mm)
Ω : rotational speed (rad/s) W : weight on the bearings (= $2{,}27.10^5$ N)
L : size of bearing (320 mm) J : radial clearance between bearing and trunnion

- For the cylindrical bearing k.. and C.. are given in a CETIM [3] document.

For the 3 lobes bearing, a special software allows one to determine the stiffness and damping coefficients.

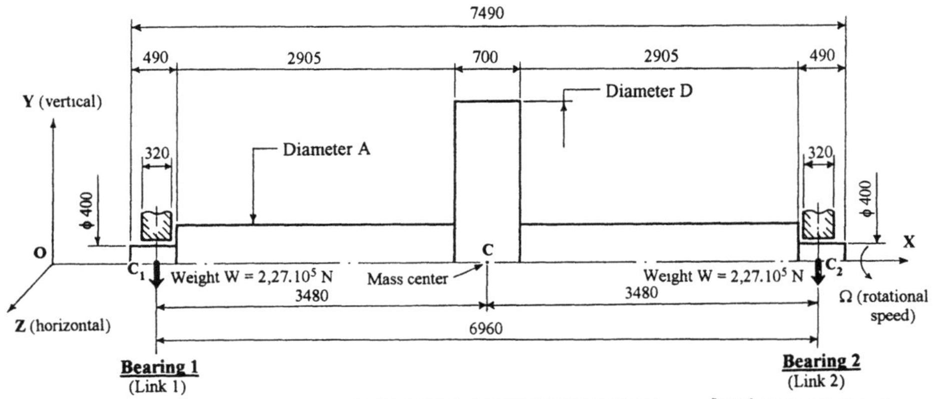

Total mass . . . 46244 kg
Modulus of elasticity . . $2{,}1\ 10^{11}$ Pa
Density . . . 7800 kg/m³
Dynamic viscosity of supplying oil at 50°C. $56\ 10^{-3}$ Pa s

Model	Diameter A	Diameter D
First	817	3384
Second	758	2406

Impedance matrix of displacement at links 1 and 2

$$\begin{bmatrix} k_{yy} + p.C_{yy} & k_{yz} + p.C_{yz} \\ k_{zy} + p.C_{zy} & k_{zz} + p.C_{zz} \end{bmatrix}$$

k.. Stiffness coefficient
C.. Damping coefficient
p. Complex pulsation

Fig. 2 Two models of the high pressure rotor (De Laval - Jeffcott rotor). Geometry and characteristics.

4. FIRST MODEL OF THE HIGH PRESSURE ROTOR

• Figure 2 describes the model - termed "first" - of the high pressure rotor : the trunnions, the distance between centers of the trunnions and the mass center are identical to those of the high pressure rotor.

With the impedance matrixes, where the terms are reduced to direct stiffness equal to 5.10^{15} N/m, the first natural frequency is 18,6 Hz, identical to that of the high pressure rotor ; the geometrical symmetry leads to a first natural frequency independent from the rotating speed ; the gyroscopic moments are nil.

• Annex 2 defines the natural right and left-hand modes of the De Laval-Jeffcott rotor comprising a flexible shaft without mass, a mass disc that is not deflectable, the links represented by impedance matrixes drawn from the stiffness and damping coefficients. The natural modes are expressed in literal terms.

• Tables 2a and 2b define the first natural pulsations and the corresponding natural modes, wich are represented in annex 2. Amplitude ratios and phase differences between modes at the centers of the trunnions (C_1, C_2) and at the center of the disc (C) make comparison possible, (relation [A_2-8] of annex 2).

It should be noted that :

- at 1116 rpm, 1500 rpm and for the first two modes, the disc moves in a radial plane with a forward or reverse precession ; natural pulsations of these modes are different, which is due to the bearing anisotropy,
- amplitude ratios and phase differences at C_1, (C_2) and C are not equal,
- right and left-hand modes are not identical and not in the same plane (formulas of annex 2),
- modes are symmetrical with respect to a radial plane parallel to Oyz passing by O_c.

<table>
<tr><td colspan="5">Table 2a De Laval-Jeffcott rotor First model Natural pulsations Right-hand natural modes ———, left-hand natural modes - - - - - - - -</td></tr>
<tr><td rowspan="2">Bearings
Impedances
$(k_{..} + C_{..}p)\ 10^7$ N/m</td><td colspan="2">Rotational speed : 1116 rpm (18,6 Hz)
(≈ Critical speed)</td><td colspan="2">Rotational speed : 1500 rpm (25 Hz)
(Nominal speed)</td></tr>
<tr><td colspan="2">Natural pulsations - Natural modes</td><td colspan="2">Natural pulsations - Natural modes</td></tr>
<tr><td>Cylindrical
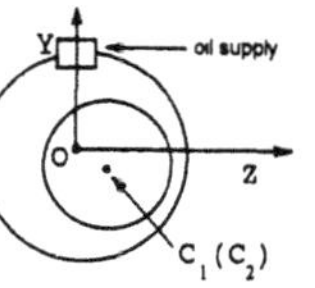
Radial clearance . J = 0,4mm (0,2 %)</td><td>$p = -12{,}90 + j\,98{,}99$ (rad/s)
$p = -2{,}05 + j\,15{,}75$ (Hz)
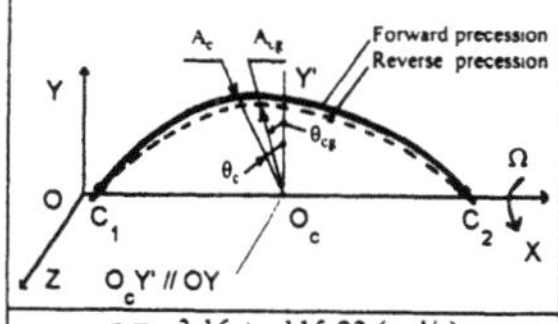</td><td>$\frac{A_1}{A_{1g}} = \frac{A_2}{A_{2g}} = 0{,}92$
$\theta_1 - \theta_{1g} =$
$\theta_2 - \theta_{2g} = 346{,}0°$
$\frac{A_c}{A_{cg}} = 1{,}06$
$\theta_c - \theta_{cg} = 7{,}8°$</td><td>$p = -3{,}97 + j\,98{,}15$ (rad/s)
$p = -0{,}63 + j\,15{,}62$ (Hz)
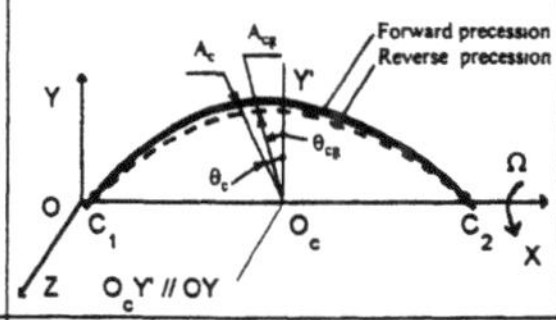</td><td>$\frac{A_1}{A_{1g}} = \frac{A_2}{A_{2g}} = 1{,}09$
$\theta_1 - \theta_{1g} =$
$\theta_2 - \theta_{2g} = 10{,}1°$
$\frac{A_c}{A_{cg}} = 1{,}05$
$\theta_c - \theta_{cg} = 6{,}7°$</td></tr>
<tr><td>At 1116 rpm
$\begin{bmatrix} 96 + 1{,}3p & 25 - 1{,}2p \\ -190 - 1{,}2p & 160 + 3{,}5p \end{bmatrix}$
At 1500 rpm
$\begin{bmatrix} 110 + 1{,}3p & 72 - 0{,}76p \\ -210 - 0{,}76p & 110 + 2{,}3p \end{bmatrix}$
p pulsation</td><td>$p = -3{,}16 + j\,115{,}23$ (rad/s)
$p = -0{,}50 + j\,18{,}34$ (Hz)
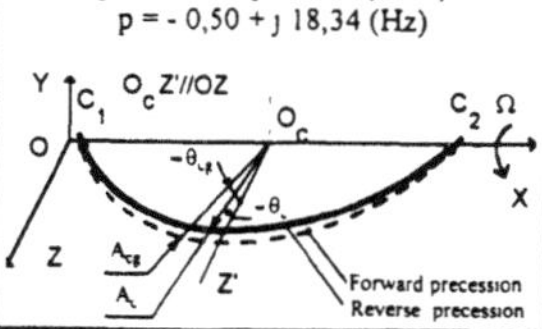</td><td>$\frac{A_1}{A_{1g}} = \frac{A_2}{A_{2g}} = 1{,}19$
$\theta_1 - \theta_{1g} =$
$\theta_2 - \theta_{2g} = 301{,}5°$
$\frac{A_c}{A_{cg}} = 0{,}94$
$\theta_c - \theta_{cg} = 8{,}2°$</td><td>$p = -4{,}15 + j\,114{,}83$ (rad/s)
$p = -0{,}66 + j\,18{,}27$ (Hz)
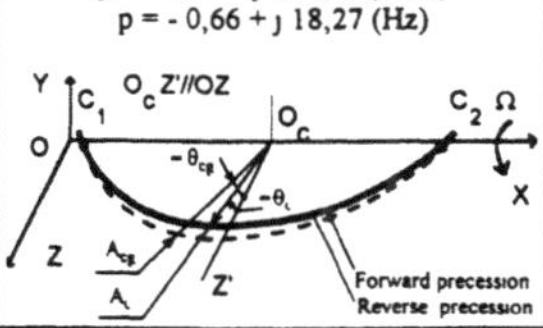</td><td>$\frac{A_1}{A_{1g}} = \frac{A_2}{A_{2g}} = 0{,}98$
$\theta_1 - \theta_{1g} =$
$\theta_2 - \theta_{2g} = 255{,}6°$
$\frac{A_c}{A_{cg}} = 0{,}94$
$\theta_c - \theta_{cg} = 8{,}8°$</td></tr>
</table>

<table>
<tr><td colspan="5">Table 2b De Laval-Jeffcott rotor First model Natural pulsations Right-hand natural modes ———, left-hand natural modes - - - - - - - -</td></tr>
<tr><td rowspan="2">Bearings
Impedances
$(k_{..} + C_{..}p)\ 10^7$ N/m</td><td colspan="2">Rotational speed : 1116 rpm (18,6 Hz)
(≈ Critical speed)</td><td colspan="2">Rotational speed : 1500 rmp (25 Hz)
(Nominal speed)</td></tr>
<tr><td colspan="2">Natural pulsations - Natural modes</td><td colspan="2">Natural pulsations - Natural modes</td></tr>
<tr><td>3 lobes (m * = 0,5)
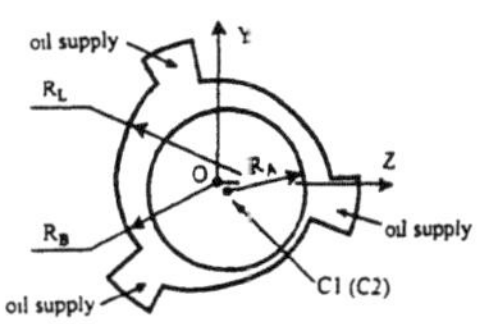
Radial clearance** J=0,4mm (0,2%)</td><td>$p = -6{,}23 + j\,107{,}45$ (rad/s)
$p = -0{,}99 + j\,17{,}1$ (Hz)
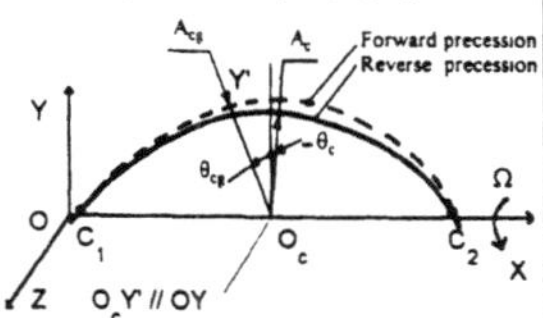</td><td>$\frac{A_1}{A_{1g}} = \frac{A_2}{A_{2g}} = 0{,}99$
$\theta_1 - \theta_{1g} =$
$\theta_2 - \theta_{2g} = 359{,}0°$
$\frac{A_c}{A_{cg}} = 0{,}98$
$\theta_c - \theta_{cg} = 344{,}2°$</td><td>$p = -2{,}98 + j\,106{,}60$ (rad/s)
$p = -0{,}47 + j\,16{,}97$ (Hz)
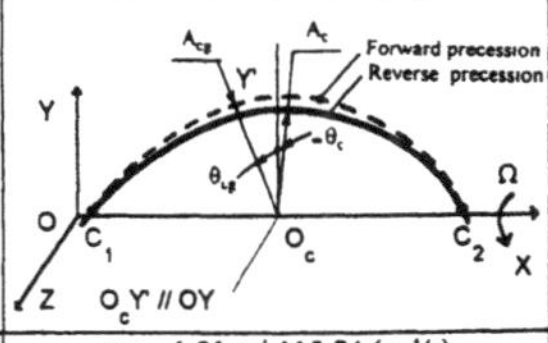</td><td>$\frac{A_1}{A_{1g}} = \frac{A_2}{A_{2g}} = 0{,}96$
$\theta_1 - \theta_{1g} =$
$\theta_2 - \theta_{2g} = 344{,}2°$
$\frac{A_c}{A_{cg}} = 0{,}96$
$\theta_c - \theta_{cg} = 337{,}7°$</td></tr>
<tr><td>At 1115 rpm
$\begin{bmatrix} 65 + 1{,}4p & -230 - 1{,}3p \\ 310 - 1{,}4p & 800 + 9{,}1p \end{bmatrix}$
At 1500 rpm
$\begin{bmatrix} 62 + 1{,}3p & -330 - 1{,}6p \\ 340 - 1{,}7p & 960 + 8{,}3p \end{bmatrix}$
p pulsation</td><td>$p = -1{,}12 + j\,115{,}85$ (rad/s)
$p = -0{,}18 + j\,18{,}44$ (Hz)
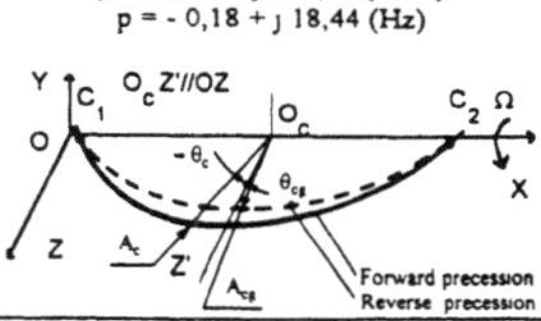</td><td>$\frac{A_1}{A_{1g}} = \frac{A_2}{A_{2g}} = 0{,}93$
$\theta_1 - \theta_{1g} =$
$\theta_2 - \theta_{2g} = 16{,}9°$
$\frac{A_c}{A_{cg}} = 1{,}02$
$\theta_c - \theta_{cg} = 345{,}0°$</td><td>$p = -1{,}01 + j\,115{,}74$ (rad/s)
$p = -0{,}16 + j\,18{,}42$ (Hz)
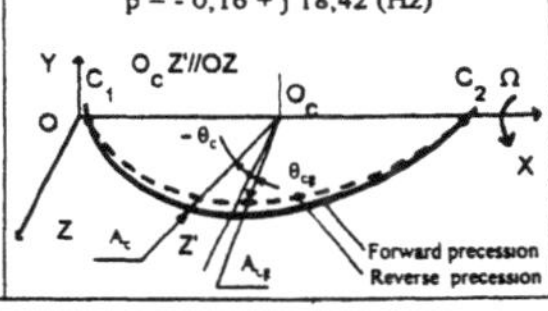</td><td>$\frac{A_1}{A_{1g}} = \frac{A_2}{A_{2g}} = 1{,}0$
$\theta_1 - \theta_{1g} =$
$\theta_2 - \theta_{2g} = 5{,}8°$
$\frac{A_c}{A_{cg}} = 1{,}04$
$\theta_c - \theta_{cg} = 339{,}6°$</td></tr>
<tr><td colspan="5">* $m = (R_L - R_B)/(R_L - R_A)$ ** $J = R_B - R_A$</td></tr>
</table>

5. SECOND MODEL OF THE HIGH PRESSURE ROTOR

• Figure 2 describes the model - termed "second" - of the high pressure rotor , the density of the shaft is equal to $7{,}8.10^3\ kg/m^3$.

• Tables 3a and 3b define the first natural pulsations and the corresponding natural modes. Amplitude ratios and phase differences between the right and the left-hand modes at the centers of the trunnions (C_1, C_2) and at the center of the disc (C), make comparison possible.

Similar conclusions to those of the first model (paragraphe 4), can be expressed.

<table>
<tr><td colspan="5">Table 3a De Laval-Jeffcott rotor Second model Natural pulsations Right-hand natural modes ———, left-hand natural modes - - - - - - - -</td></tr>
<tr><td>Bearings
Impedances
(k.. + C..p) 10⁷ N/m</td><td colspan="2">Rotational speed : 1116 rpm (18,6 Hz)
(≈ Critical speed)
Natural pulsations - Natural modes</td><td colspan="2">Rotational speed : 1500 rpm (25 Hz)
(Nominal speed)
Natural pulsations - Natural modes</td></tr>
<tr><td>Cylindrical
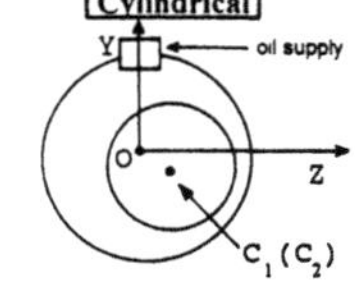
Radial clearance : J = 0,4mm (0,2 %)</td><td>p = - 12,92 + j 101,09 (rad/s)
p = - 2,05 + j 16,09 (Hz)
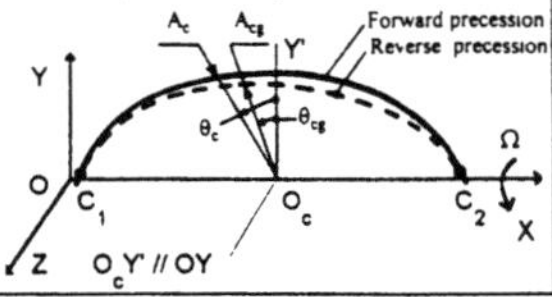</td><td>$\frac{A_1}{A_{1g}} = \frac{A_2}{A_{2g}} = 0,90$
$\theta_1 - \theta_{1g} =$
$\theta_2 - \theta_{2g} = 341,4°$
$\frac{A_c}{A_{cg}} = 1,05$
$\theta_c - \theta_{cg} = 6,6°$</td><td>p = - 4,33 + j 99,63 (rad/s)
p = - 0,69 + j 15,86 (Hz)
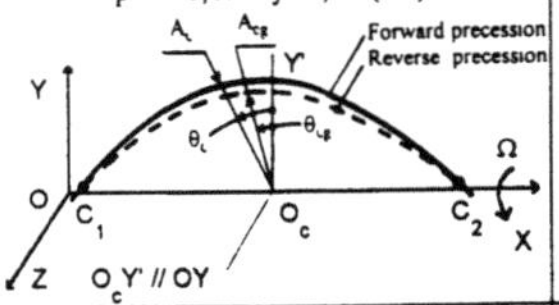</td><td>$\frac{A_1}{A_{1g}} = \frac{A_2}{A_{2g}} = 1,09$
$\theta_1 - \theta_{1g} =$
$\theta_2 - \theta_{2g} = 9,7°$
$\frac{A_c}{A_{cg}} = 1,05$
$\theta_c - \theta_{cg} = 7,2°$</td></tr>
<tr><td>At 1116 rpm
$\begin{bmatrix} 96 + 1,3p & 25 - 1,2p \\ -190 - 1,2p & 160 + 3,5p \end{bmatrix}$
At 1500 rpm
$\begin{bmatrix} 110 + 1,3p & 72 - 0,76p \\ -210 - 0,76p & 110 + 2,3p \end{bmatrix}$
p pulsation</td><td>p = - 2,81 + j 115,53 (rad/s)
p = - 0,45 + j 18,39 (Hz)
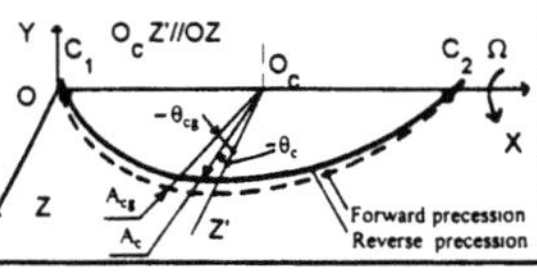</td><td>$\frac{A_1}{A_{1g}} = \frac{A_2}{A_{2g}} = 1,3$
$\theta_1 - \theta_{1g} =$
$\theta_2 - \theta_{2g} = 303,8°$
$\frac{A_c}{A_{cg}} = 0,95$
$\theta_c - \theta_{cg} = 6,9°$</td><td>p = - 3,67 + j 115,04 (rad/s)
p = - 0,58 + j 18,31 (Hz)
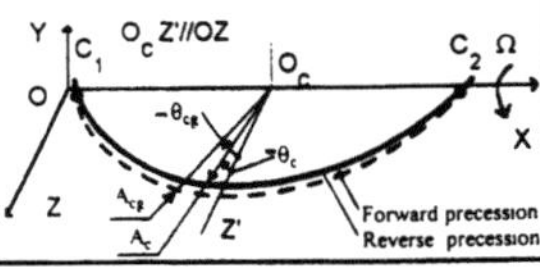</td><td>$\frac{A_1}{A_{1g}} = \frac{A_2}{A_{2g}} = 1,02$
$\theta_1 - \theta_{1g} =$
$\theta_2 - \theta_{2g} = 256,8°$
$\frac{A_c}{A_{cg}} = 0,94$
$\theta_c - \theta_{cg} = 8,6°$</td></tr>
</table>

<table>
<tr><td colspan="5">Table 3b De Laval-Jeffcott rotor Second model Eigenvalues Right-hand natural modes ———, left-hand natural modes - - - - - - - -</td></tr>
<tr><td>Bearings
Impedances
(k.. + C..p) 10⁷ N/m</td><td colspan="2">Rotational speed : 1116 rpm (18,6 Hz)
(≈ Critical speed)
Natural pulsations - Natural modes</td><td colspan="2">Rotational speed : 1500 rpm (25 Hz)
(Nominal speed)
Natural pulsations - Natural modes</td></tr>
<tr><td>3 lobes (m * = 0,5)
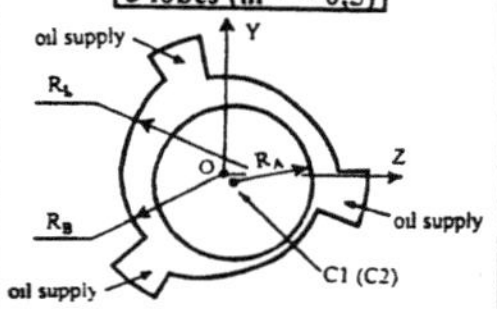
Radial clearance**: J=0,4mm (0,2%)</td><td>p = - 5,85 + j 108,53 (rad/s)
p = - 0,93 + j 17,27 (Hz)
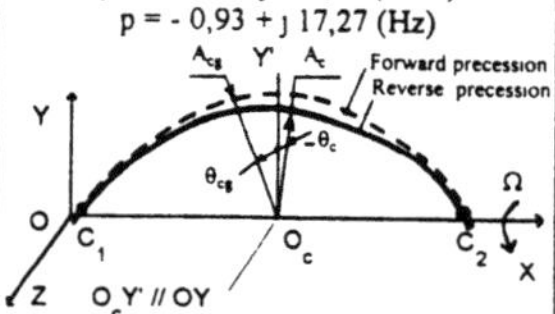</td><td>$\frac{A_1}{A_{1g}} = \frac{A_2}{A_{2g}} = 1,0$
$\theta_1 - \theta_{1g} =$
$\theta_2 - \theta_{2g} = 0,0°$
$\frac{A_c}{A_{cg}} = 0,98$
$\theta_c - \theta_{cg} = 342,8°$</td><td>p = - 2,80 + j 107,53 (rad/s)
p = - 0,45 + j 17,11 (Hz)
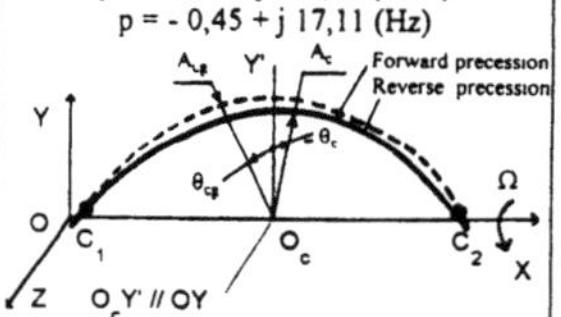</td><td>$\frac{A_1}{A_{1g}} = \frac{A_2}{A_{2g}} = 0,96$
$\theta_1 - \theta_{1g} =$
$\theta_2 - \theta_{2g} = 345,1°$
$\frac{A_c}{A_{cg}} = 0,96$
$\theta_c - \theta_{cg} = 337,2°$</td></tr>
<tr><td>At 1116 rpm
$\begin{bmatrix} 65 + 1,4p & -230 - 1,3p \\ 310 - 1,4p & 800 + 9,1p \end{bmatrix}$
At 1500 rpm
$\begin{bmatrix} 62 + 1,3p & -330 - 1,6p \\ 340 - 1,7p & 960 + 8,3p \end{bmatrix}$
p pulsation</td><td>p = - 1,07 + j 116,27 (rad/s)
p = - 0,17 + j 18,50 (Hz)
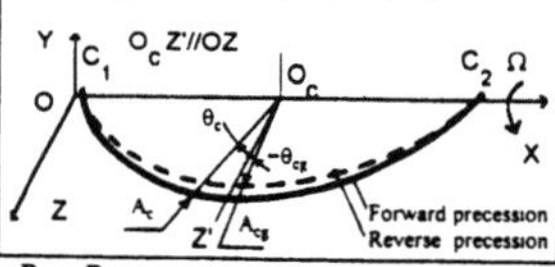</td><td>$\frac{A_1}{A_{1g}} = \frac{A_2}{A_{2g}} = 0,91$
$\theta_1 - \theta_{1g} =$
$\theta_2 - \theta_{2g} = 11,6°$
$\frac{A_c}{A_{cg}} = 1,03$
$\theta_c - \theta_{cg} = 343,4°$</td><td>p = - 1,02 + j 116,23 (rad/s)
p = - 0,16 + j 18,50 (Hz)
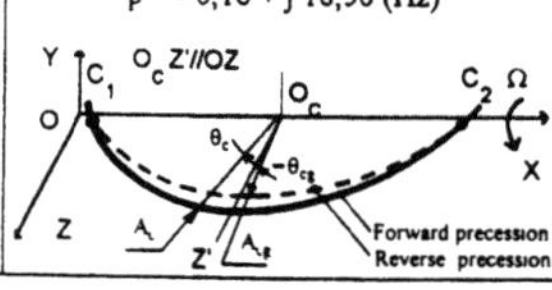</td><td>$\frac{A_1}{A_{1g}} = \frac{A_2}{A_{2g}} = 0,92$
$\theta_1 - \theta_{1g} =$
$\theta_2 - \theta_{2g} = 17,2°$
$\frac{A_c}{A_{cg}} = 1,04$
$\theta_c - \theta_{cg} = 338,3°$</td></tr>
<tr><td colspan="5">* $m = (R_L - R_B)/(R_L - R_A)$ ** $J = R_B - R_A$</td></tr>
</table>

6. THE HIGH PRESSURE ROTOR

• Figure 1 describes the rotor. Tables 4a and 4b define the first natural pulsations and the corresponding natural modes, determined by the use of special software applied to a model composed of 36 elements (see paragraph 2).

Amplitude ratios and phase differences between the right and left-hand modes at the centers of the trunnions (C_1, C_2) and at the center of the rotor (C) make comparison possible. Similar conclusions to those of the second model (paragraphe 5), can be expressed.

Table 4a High pressure rotor - Natural pulsations Right-hand natural modes ———, left-hand natural modes - - - - - - - -

Bearings Impedances $(k + C\,p)\ 10^7$ N/m	Rotational speed : 1116 rpm (18,6 Hz) (≈ Critical speed) — Natural pulsations - Natural modes		Rotational speed : 1500 rpm (25 Hz) (Nominal speed) — Natural pulsations - Natural modes	
Cylindrical 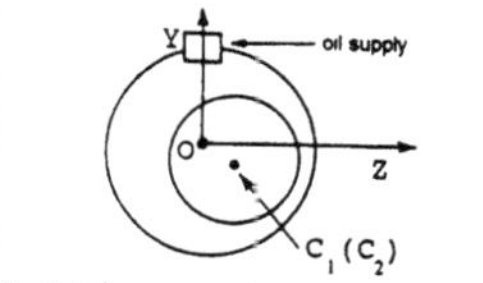 Radial clearance J = 0,4 mm (0,2%)	p = - 11,75 + j 107,69 (rad/s) p = - 1,87 + j 17,14 (Hz) 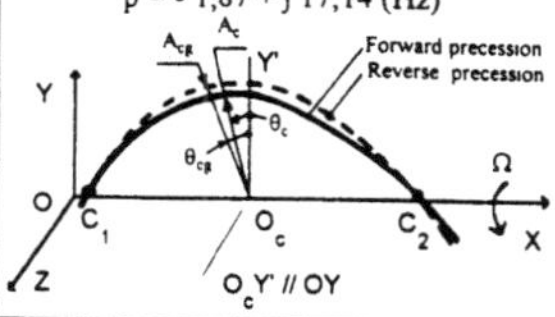	$\frac{A1}{A1g}$ = 0,87 $\theta_1 - \theta_{1g}$ = 337,3° $\frac{Ac}{Acg}$ = 0,98 $\theta_c - \theta_{cg}$ = 357,4° $\frac{A2}{A2g}$ = 0,87 $\theta_2 - \theta_{2g}$ = 335°	p = - 5,56 + j 105,29 (rad/s) p = - 0,88 + j 16,76 (Hz) 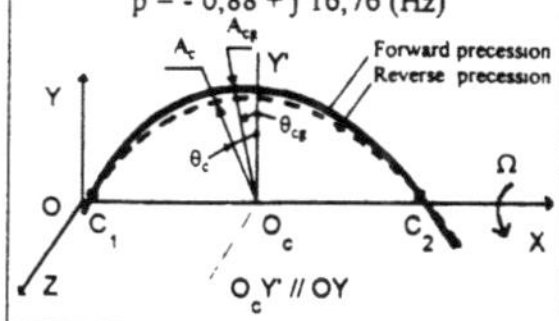	$\frac{A1}{A1g}$ = 0,98 $\theta_1 - \theta_{1g}$ = 349,5° $\frac{Ac}{Acg}$ = 1,07 $\theta_c - \theta_{cg}$ = 7,7° $\frac{A2}{A2g}$ = 0,99 $\theta_2 - \theta_{2g}$ = 348,9°
At 1116 rpm $\begin{bmatrix} 96 + 1,3p & 25 - 1,2p \\ -190 - 1,2p & 160 + 3,5p \end{bmatrix}$ At 1500 rpm $\begin{bmatrix} 110 + 1,3p & 72 - 0,76p \\ -210 - 0,76p & 110 + 2,3p \end{bmatrix}$ p pulsation	p = - 2,06 + j 116,84 (rad/s) p = - 0,33 + j 18,60 (Hz) 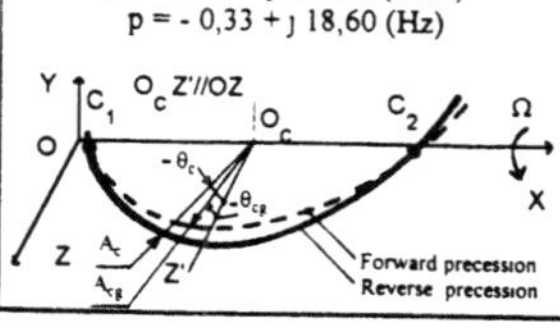	$\frac{A1}{A1g}$ = 1,61 $\theta_1 - \theta_{1g}$ = 327,7° $\frac{Ac}{Acg}$ = 1,04 $\theta_c - \theta_{cg}$ = 355,6° $\frac{A2}{A2g}$ = 1,44 $\theta_2 - \theta_{2g}$ = 309,4°	p = - 2,38 + j 115,57 (rad/s) p = - 0,38 + j 18,39 (Hz) 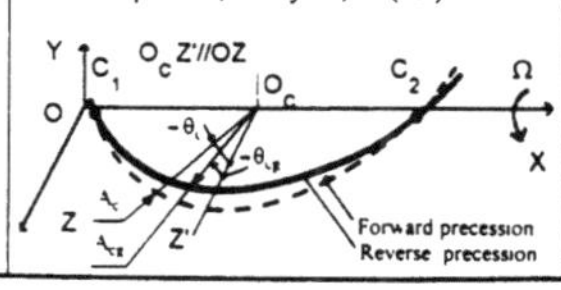	$\frac{A1}{A1g}$ = 2,32 $\theta_1 - \theta_{1g}$ = 321,9° $\frac{Ac}{Acg}$ = 0,96 $\theta_c - \theta_{cg}$ = 4,0° $\frac{A2}{A2g}$ = 2,05 $\theta_2 - \theta_{2g}$ = 305,6°

Table 4b High pressure rotor Eigenvalues Right-hand natural modes ———, left-hand natural modes - - - - - - - -

Bearings Impedances $(k + C\,p)\ 10^7$ N/m	Rotational speed : 1116 rpm (18,6 Hz) (≈ Critical speed) — Natural pulsations - Natural modes		Rotational speed : 1500 rpm (25 Hz) (Nominal speed) — Natural pulsations - Natural modes	
3 lobes (m * = 0,5) 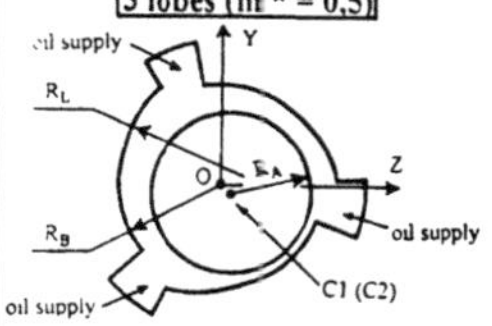 Radial clearance** J=0,4mm (0,2%)	p = - 4,37 + j 111,37 (rad/s) p = - 0,70 + j 17,73 (Hz) 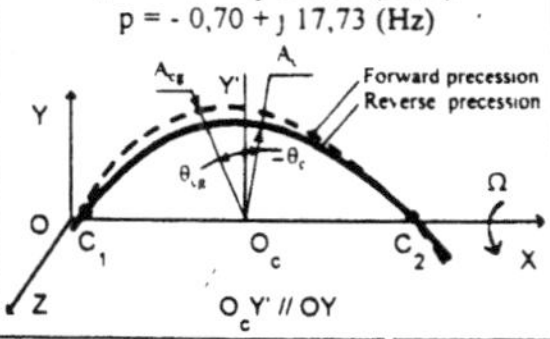	$\frac{A1}{A1g}$ = 1,01 $\theta_1 - \theta_{1g}$ = 1,5° $\frac{Ac}{Acg}$ = 0,96 $\theta_c - \theta_{cg}$ = 333,5° $\frac{A2}{A2g}$ = 1,0 $\theta_2 - \theta_{2g}$ = 1,8°	p = - 1,89 + j 109,85 (rad/s) p = - 0,30 + j 17,48 (Hz) 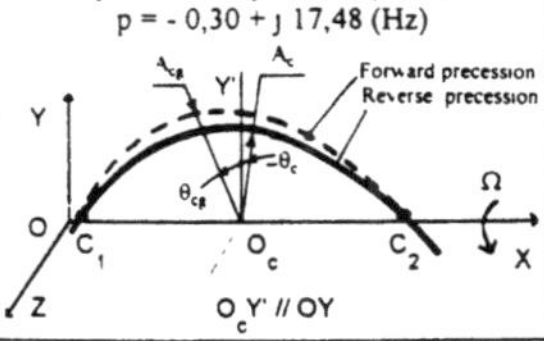	$\frac{A1}{A1g}$ = 0,94 $\theta_1 - \theta_{1g}$ = 347,1° $\frac{Ac}{Acg}$ = 0,96 $\theta_c - \theta_{cg}$ = 335,9° $\frac{A2}{A2g}$ = 0,92 $\theta_2 - \theta_{2g}$ = 347,5°
At 1116 rpm $\begin{bmatrix} 65 + 1,4p & -230 - 1,3p \\ 310 - 1,4p & 800 + 9,1p \end{bmatrix}$ At 1500 rpm $\begin{bmatrix} 62 + 1,3p & -330 - 1,6p \\ 340 - 1,7p & 960 + 8,3p \end{bmatrix}$ p pulsation	p = - 1,37 + j 118,99 (rad/s) p = - 0,22 + j 18,94 (Hz) 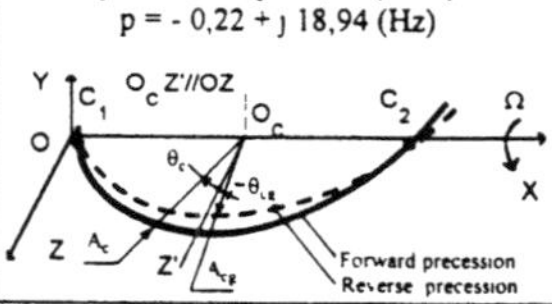	$\frac{A1}{A1g}$ = 0,77 $\theta_1 - \theta_{1g}$ = 338,4° $\frac{Ac}{Acg}$ = 1,05 $\theta_c - \theta_{cg}$ = 332,9° $\frac{A2}{A2g}$ = 0,67 $\theta_2 - \theta_{2g}$ = 151,5°	p = - 1,48 + j 119,47 (rad/s) p = - 0,24 + j 19,01 (Hz) 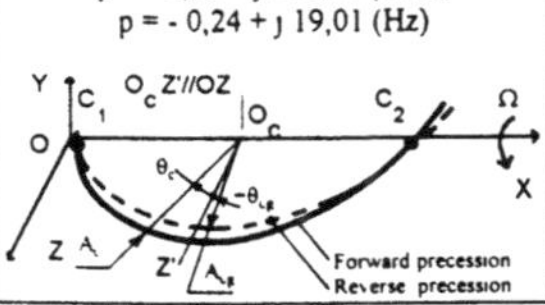	$\frac{A1}{A1g}$ = 1,0 $\theta_1 - \theta_{1g}$ = 351,6° $\frac{Ac}{Acg}$ = 1,05 $\theta_c - \theta_{cg}$ = 335,7° $\frac{A2}{A2g}$ = 0,91 $\theta_2 - \theta_{2g}$ = 345,4°

* $m = (R_L - R_B)/(R_L - R_A)$ ** $J = R_B - R_A$

7. COMPARISON BETWEEN RIGHT AND LEFT-HAND NATURAL MODES

• Table 5, computed for the three models, makes the comparison between the amplitudes and the phases of right and left-hand modes :

- the maximum amplitude ratio value is between 1,19 and 2,32, the minimum ratio value is between 0,67 and 0,92,
- the maximum phase difference is between 359,0° and 345,1° the minimum difference is between 0,0° and 5,8°.

Table 5 Amplitude ratio, phase differences of right-hand and left-hand natural modes

Rotor	Amplitude ratio Max	Amplitude ratio Min	Phase differences ( ) Max	Phase differences ( ) Min	Kind of bearing	Differences * (%) E_k	Differences * (%) E_c	Rotational speed (rpm)	Natural pulsation (rad/s) $p_\cdot = \delta_\cdot + j\omega_\cdot$
First model	1,19 (C_1,C_2)				Cylindrical	153,5	0,0	1116	-3,16 + j 115,23
(De Laval		0,92 (C_1,C_2)			Cylindrical	153,5	0,0	1116	-12,9 + j 98,99
shaft without			359,0 (C_1,C_2)		3 lobes	29,6	7,4	1116	-6,23 + j 107,45
mass)				5,8 (C_1,C_2)	3 lobes	3,0	6,1	1500	-1,01 + j 115,74
Second model	1,30 (C_1,C_2)				Cylindrical	153,5	0,0	1116	-2,81 + j 115,53
(De Laval		0,90 (C_1,C_2)			Cylindrical	153,5	0,0	1116	-12,92 + j 101,09
shaft			345,1 (C_1,C_2)		3 lobes	3,0	6,1	1500	-2,80 + j 107,53
with mass)				0,0 (C_1,C_2)	3 lobes	29,6	7,4	1116	-5,85 + j 108,53
	2,32 (C_1)				Cylindrical	97,9	0,0	1500	-2,38 + j 115,57
High		0,67 (C_2)			3 lobes	29,6	7,4	1116	-1,37 + j 118,99
pressure			357,4 (C)		Cylindrical	153,5	0,0	1116	-11,75 + j 107,69
				1,5 (C_1)	3 lobes	29,6	7,4	1116	-4,37 + j 111,37

* Matrix $\begin{bmatrix} k_{yy}+p.C_{yy} & k_{yz}+p.C_{yz} \\ k_{zy}+p.C_{zy} & k_{zz}+p.C_{zz} \end{bmatrix} \rightarrow$ Differences : $E_k = 100 \dfrac{\left| |k_{yz}| - |k_{zy}| \right|}{\frac{|k_{yz}| + |k_{zy}|}{2}}$; $E_c = 100 \dfrac{\left| |C_{yz}| - |C_{zy}| \right|}{\frac{|C_{yz}| + |C_{zy}|}{2}}$

8. COMPARISON OF THE TWO DE LAVAL MODELS OF THE HIGH PRESSURE ROTOR MODEL

• The quality of the models may be evaluated with the help of natural pulsation ; for example :

first model .... -12,90 + j 98,99
second model .... -12,92 + j 101,09
high pressure rotor -11,75 + j 107,69

Considering the high pressure rotor as reference, with the first model we notice an increase of $|\delta|$ (≈ 10 %) and a decrease of ω (≈ 10 %).

• This quality may be as well evaluated with the help of amplitude ratios and differences of phases ; for example :

First model $\begin{bmatrix} \text{cylindrical bearing} \\ p=-3,97 + j\,98,15 \end{bmatrix}$

$$\frac{A_1}{A_{1g}} = \frac{A_2}{A_{2g}} = 1,09 \qquad \theta_1 - \theta_{1g} = \theta_2 - \theta_{2g} = 10,1°$$

$$\frac{A_c}{A_{cg}} = 1,05 \qquad \theta_c - \theta_{cg} = 6,7°$$

Second model $\begin{bmatrix} \text{cylindrical bearing} \\ p=-4,33 + j\,99,63 \end{bmatrix}$

$$\frac{A_1}{A_{1g}} = \frac{A_2}{A_{2g}} = 1,09 \qquad \theta_1 - \theta_{1g} = \theta_2 - \theta_{2g} = 9,7°$$

$$\frac{A_c}{A_{cg}} = 1,05 \qquad \theta_c - \theta_{cg} = 7,2°$$

High pressure rotor $\begin{bmatrix} \text{cylindrical bearing} \\ p=-5,56+j\,105,29 \end{bmatrix}$

$$\frac{A_1}{A_{1g}} = 0,98, \qquad \theta_1 - \theta_{1g} = 349,5°$$

$$\frac{A_2}{A_{2g}} = 0,99, \qquad \theta_2 - \theta_{2g} = 348,9°$$

$$\frac{A_c}{A_{cg}} = 1,07, \qquad \theta_c - \theta_{cg} = 7,7°$$

Considering the high pressure rotor as reference, with the first model we notice an increase of both the amplitude ratio (≈ 6 %) and the phase difference (≈ 13°).

9. CONCLUSIONS

• For the high pressure rotor equipped with cylindrical and 3 lobes bearings, right and left-hand natural modes are different. With the cylindrical bearing the maximum amplitude ratio is equal to 2,32, the minimum equal to 0,87 ; the maximum phase difference is equal to 357,4 , the minimum equal to 4,0 . With the 3 lobes bearing the maximum amplitude ratio is equal to 1,05, the minimum equal to 0,67 ; the maximum phase difference is equal to 351,6 , the minimum equal to 1,5 .

• Differences between right and left-hand modes depend on the assymetry of impedance matrixes of the links and on the gyroscopic moments.

• Two De Laval rotors help to define the influence of the hypothesis (shapes, shaft mass), which constitute the two models, on the natural pulsations, amplitude ratios and phase differences.

• This study shows the interest of using models in order to determine unbalance responses of rotating machines with the help of modal analysis ; in this way it introduces modal balancing methods in which taking into account only the right-hand modes may leads to false results.

REFERENCES :

1. Ref. 1 : Application of Modal Analysis to the balancing of Rotating Machines R. Bigret - 4th International Modal Analysis Congress (IMAC) - Los Angeles, USA - 1986.

2. Ref. 2 : Modal Testing Theory of Rotor-Bearing Systems Yang-Gyu Jei, Young-ju Kim - Journal of Vibration and Acoustics - April 1993 - Vol. 115.

3. Ref. 3 : Raideurs et Amortissements des Paliers Hydrodynamiques Les Memoires Techniques du CETIM - France - N 32 - December 1977.

Annex 1 - Natural pulsations and natural modes [1] [2]

Within a fixed reference system, the free vibratory behaviour of the model of a rotor-link-structure unity is characterized by :

$$[M]\left\{\frac{d^2u}{dt^2}\right\}+[A_v]\left\{\frac{du}{dt}\right\}+[K_d]\{u\}=\{0\} \qquad [A_1\text{-}1]$$

with :

- $[G]$ gyroscopic moments matrix
- $[K_d] = [K] + [K_i]$
- $[K]$ stiffness matrix
- $[K_i]$ internal damping coupling matrix
- $\{u\}$ linear and angular deflection matrix
- $[A_v] = [A_e] + [A_i] + [G]$
- $[A_e]$ external damping matrix
- $[A_i]$ internal damping matrix
- $[M]$ mass matrix

Solutions of the form $\{u\}=R[\{D\}\exp p.t]$, with R - real part operator and p.- complex pulsation, yield : $[[M]p^2+[Av]p.+[K_d]]\{D\}=\{0\}$. The natural pulsations, $p_k=\delta_k+j\omega_k$, are obtaind by : $\det[[M]p^2+[Av]p.+[K_d]]=0$

The natural modes, associated to pulsations p_k, are defined by : $[[M]p_k^2+[Av]p_k+[K_d]]\{D\}_k=\{0\}$ for the right-hand modes $\{D\}_k$, and by ${}^t\{D\}_{gk}[[M]p_k^2+[Av]p_k+[K_d]]={}^t\{0\}$ which means : $[{}^t[M]p_k^2+{}^t[A_v]p_k+{}^t[K_d]]\ \{D\}_{gk}=\{0\}$ for the left-hand modes $\{D\}_{gk}$, (t → transposition).

The natural modes are defined to a nearest multiplying constant ; a component of $\{D\}_k$ and of $\{D\}_{gk}$ must be chosen as norm. Normalisation may be obtained by two orthogonality relations, between right and left-hand natural modes. When matrixes [M], [Av] and $[K_d]$ are symmetrical, right-hand natural modes are identical to left-hand ones. Left-hand natural modes may be defined with the help of the conjugated pulsation $\overline{p_k}$; so they are conjugated to $\{D\}_{gk}$. Forced regimes are expressed by $\{D\}_k$, $\overline{\{D\}_k}$, $\{D\}_{gk}$, $\overline{\{D\}_{gk}}$.

Annex 2 - De Laval-Jeffcott rotor. Shaft without mass

- Definition

Figure 2 - first model - defines a rotor guided by two identical links. In the fixed reference system Oxyz, the rotor-links section is represented by a model characterized by :

- C the barycenter of the median section and G mass barycenter of the disc,
- a shaft having no mass and finite bending stiffness,
- k_a bending stiffness of the shaft for C_1 et C_2 on rigid swivel supports,
- e excentricity ($e=\overline{CG}$),
- k.. and C.. identical stiffness and damping coefficients on links 1 and 2.

- Formulation : disc in translation

Considering the displacements of C_1 and C_2 as identical, the movement of C and C_1 is expressed by :

$$\begin{bmatrix} m&0&0&0\\0&0&0&0\\0&0&m&0\\0&0&0&0\end{bmatrix}\begin{Bmatrix}\ddot{y}_c\\\ddot{y}_1\\\ddot{z}_c\\\ddot{z}_1\end{Bmatrix}+\begin{bmatrix}0&0&0&0\\0&2C_{yy}&0&2C_{yz}\\0&0&0&0\\0&2C_{zy}&0&2C_{zz}\end{bmatrix}\begin{Bmatrix}\dot{y}_c\\\dot{y}_1\\\dot{z}_c\\\dot{z}_1\end{Bmatrix}+\begin{bmatrix}k_a&-k_a&0&0\\-k_a&k_a+2k_{yy}&0&2k_{yz}\\0&0&k_a&-k_a\\0&2k_{zy}&-k_a&k_a+2k_{zz}\end{bmatrix}\begin{Bmatrix}y_c\\y_1\\z_c\\z_1\end{Bmatrix}=\begin{Bmatrix}me\Omega^2\cos\Omega t\\0\\me\Omega^2\sin\Omega t\\0\end{Bmatrix}\quad [A_2\text{-}1]$$

The two links are represented by an impedance matrix furnished by the ratios of force and displacement Laplace transformation, (p - Laplace symbol, pulsation) :

$$2\begin{bmatrix} k_{yy}+pC_{yy} & k_{yz}+pC_{yz} \\ k_{zy}+pC_{zy} & k_{zz}+pC_{zz} \end{bmatrix} \qquad [A_2\text{-}2]$$

• Natural pulsations

The natural pulsations coresponding modes characterize the free mouvements ruled by the equation $[A_2\text{-}1]$ in which e = 0.

Solutions are of the form : u. = U.exp pt $[A_2\text{-}3]$, with : $u \to y_c, y_1, z_c, z_1$ and $U \to Y_c, Y_1, Z_c, Z_1$. $[A_2\text{-}3]$ broght to $[A_2\text{-}1]$ leads to the equation :

$$\mathrm{Det}\begin{bmatrix} (p/\omega_a)^2+1 & -1 & 0 & 0 \\ -1 & 1+2\frac{k_{yy}}{k_a}+4(p/\omega_a)\xi_{yy} & 0 & 2\frac{k_{yz}}{k_a}+4(p/\omega_a)\xi_{yz} \\ 0 & 0 & (p/\omega_a)^2+1 & -1 \\ 0 & 2\frac{k_{zy}}{k_a}+4(p/\omega_a)\xi_{zy} & -1 & 1+2\frac{k_{zz}}{k_a}+4(p/\omega_a)\xi_{zz} \end{bmatrix}=0 \quad [A_2\text{-}4]$$

The matrix of $[A_2\text{-}4]$ represents the global impedance of the system, with :

$$\omega_a=\sqrt{\frac{k_a}{m}};\ \xi_{yy}=\frac{C_{yy}}{2\sqrt{k_a m}};\ \xi_{zz}=\frac{C_{zz}}{2\sqrt{k_a m}};\ \xi_{yz}=\frac{C_{yz}}{2\sqrt{k_a m}};\ \xi_{zy}=\frac{C_{zy}}{2\sqrt{k_a m}}$$

The six relative natural pulsations $\left[\left(\frac{p}{\omega_a}\right)_{i\pm}=\frac{\delta_i}{\omega_a}\pm j\frac{\omega_i}{\omega_a};\ j^2=-1;\ i=1,2,3\right]$ depend on 8 groups having no dimensions : $k_{..}/k_a$ and $\xi_{..}$.

• Right-hand natural modes

For each of the natural pulsations, the equation $[A_2\text{-}1]$ leads to :

$$\left[I_g\right]_{i\pm}{}^t\left\{Y_c\ Y_1\ Z_c\ Z_1\right\}_{i\pm}=\{0\} \qquad [A_2\text{-}5]$$

where ${}^t\{Y_c\ Y_1\ Z_c\ Z_1\}_{i\pm}$ defines the right-hand natural mode associated to the natural pulsations $(p/\omega_a)_{i\pm}$.

Constituents $\{Y_c, Y_1, Z_c, Z_1\}_{i\pm}$ are defined to the nearest multiplying constant chosen at random, with $Y_c = Y_{co}$ real number representing a norm :

$$\left(\frac{Y_1}{Y_{co}}\right)_{i\pm} = 1 + \left(\frac{p}{\omega_a}\right)_{i\pm}^2$$

$$\left(\frac{Z_1}{Y_{co}}\right)_{i\pm} = \frac{-1}{2}\frac{k_a}{k_{yz}}\left\{\frac{2\left[1+\left(\frac{p}{\omega_a}\right)_{i\pm}^2\right]\left[\frac{k_{yy}}{k_a}+2\xi_{yy}\left(\frac{p}{\omega_a}\right)_{i\pm}\right]+\left(\frac{p}{\omega_a}\right)_{i\pm}^2}{1+2\xi_{yz}\frac{k_a}{k_{yz}}\left(\frac{p}{\omega_a}\right)_{i\pm}}\right\} \qquad [A_2\text{-}6]$$

$$\left(\frac{Z_c}{Y_{co}}\right)_{i\pm} = \frac{-1}{2}\frac{k_a}{k_{yz}}\left\{\frac{2\left[1+\left(\frac{p}{\omega_a}\right)_{i\pm}^2\right]\left[\frac{k_{yy}}{k_a}+2\xi_{yy}\left(\frac{p}{\omega_a}\right)_{i\pm}\right]+\left(\frac{p}{\omega_a}\right)_{i\pm}^2}{\left[1+\left(\frac{p}{\omega_a}\right)_{i\pm}^2\right]\left[1+2\xi_{yz}\frac{k_a}{k_{yz}}\left(\frac{p}{\omega_a}\right)_{i\pm}\right]}\right\}$$

• Left-hand natural modes

For each of the natural pulsations, according to annex 1 :

$${}^t\left[I_g\right]_{i\pm}\left\{Y_{cg} \quad Y_{1g} \quad Z_{cg} \quad Z_{1g}\right\} = \{0\} \qquad [A_2\text{-}7]$$

The comparaison of equations [A_2-5] and [A_2-7] shows that left-hand modes are defined from right-hand modes (relation [A_2-6]) by substituting :

k_{zy} to k_{yz}, ξ_{zy} to ξ_{yz} and Y_{cog} to Y_{co}.

• Representation of natural modes

The mouvements of C and C_1 (C_2) may be represented by rotating vectors defined at the moment t = 0 by fig. A_2-1; natural modes may be represented by a line connecting C_1C (C_2). Left-hand modes are repesented by the subscript g : A_{1g}, A_{cg}, θ_{1g}, θ_{cg}.

• Comparison between right and left-hand natural modes

Considering $Y_{cog} = Y_{co}$, comparisons are possibles with the help of A_1/A_{1g}, A_c/A_{cg}, $\theta_1 - \theta_{1g}$, $\theta_c - \theta_{cg}$.

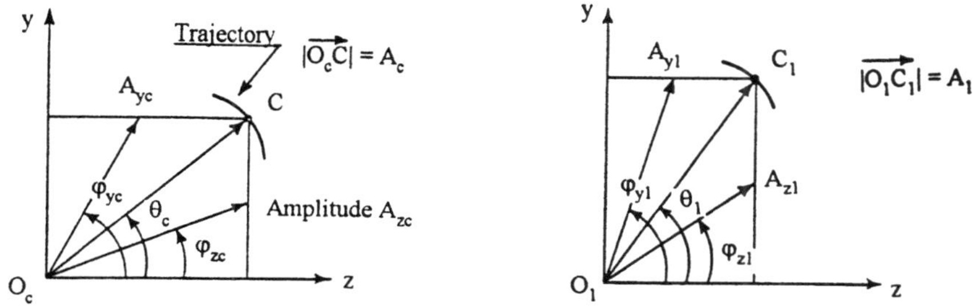

Center of the disc — Center of the trunnion

$$A_{y.} = 2\sqrt{\left[R\left(\frac{Y.}{Y_{co}}\right)\right]^2 + \left[I\left(\frac{Y.}{Y_{co}}\right)\right]^2} \qquad tg\,\varphi_{y.} = R\left(\frac{Y.}{Y_{co}}\right) \Big/ I\left(\frac{Y.}{Y_{co}}\right)$$

$$A_{z.} = 2\sqrt{\left[R\left(\frac{Z.}{Y_{co}}\right)\right]^2 + \left[I\left(\frac{Z.}{Y_{co}}\right)\right]^2} \qquad tg\,\varphi_{z.} = R\left(\frac{Z.}{Y_{co}}\right) \Big/ I\left(\frac{Z.}{Y_{co}}\right) \qquad [A_2\text{-}8]$$

$$A_{.} = \sqrt{A_{y.}^2 + A_{z.}^2} \qquad tg\,\theta_{.} = A_{z.} / A_{y.} \qquad \text{for C: } R\left(\frac{Y_c}{Y_{co}}\right) = 1 \qquad . \to c, 1$$

Forward precession $0 < \varphi_{y.} - \varphi_{z.} < \pi$ Reverse precession $\pi < \varphi_{y.} - \varphi_{z.} < 2\pi$

Fig. A_2-1 Representation of trajectories by rotating vectors. R( ) real operateur unit, I( ) imaginary operator unit

• Remark

For the natural pulsation p_{i+}, modes are defined by :

$${}^t\{1 \quad Y_1/Y_{co} \quad Z_c/Y_{co} \quad Z_1/Y_{co}\}_{i+}; \qquad {}^t\{1 \quad Y_{1g}/Y_{co} \quad Z_{cg}/Y_{co} \quad Z_{1g}/Y_{co}\}_{i+}$$

and for p_{i-} $(= \overline{p_{i+}})$, by: ${}^t\{1 \quad \bar{Y}_1/\bar{Y}_{co} \quad \bar{Z}_c/\bar{Y}_{co} \quad \bar{Z}_1/\bar{Y}_{co}\}_{i-}$; ${}^t\{1 \quad \bar{Y}_{1g}/\bar{Y}_{co} \quad \bar{Z}_{cg}/\bar{Y}_{co} \quad \bar{Z}_{1g}/\bar{Y}_{co}\}_{i-}$

When $k_{zy} = - k_{yz}$ and $\xi_{zy} = - \xi_{yz}$ we have :

$$Y_{co} = Y_{cog}; \quad Z_c / Y_{co} = -Z_{cg} / Y_{co}; \quad Y_1 / Y_{co} = Y_{1g} / Y_{co}; \quad Z_1 / Y_{co} = -Z_{1g} / Y_{co} \quad [A_2\text{-}9]$$

Defining left-hand modes with p_{i-} $(= \overline{p_{i+}})$ [Annex 1], relations $[A_2\text{-}9]$ become :

$$Y_{co} = Y_{cog}; \quad Z_c / Y_{co} = -\overline{Z_{cg}} / Y_{co}; \quad Y_1 / Y_{co} = \overline{Y_{1g}} / Y_{co}; \quad Z_1 / Y_{co} = -\overline{Z_{1g}} / Y_{co}$$

The ratios of modulus A_1/A_{1g}, A_c/A_{cg} are equal to 1, phases differences are equal to π ; precessions of right-hand modes are opposit to precessions of left-hand modes. Gyroscopic moments, wich exist in the high pressure rotor and in the second model by $\pm \Omega I_a p$ (Ω - rotation speed, I_a - axial inertia) leads to the same properties because the assymmetry that they introduce is of the same kind that the one introduced by $k_{..}$ and $\xi_{..}$.

EXPERIMENTAL VERIFICATION OF A NON INTRUSIVE METHOD FOR DETECTING STEAM TURBINE BLADE RESONANCES

C. Cavagnino, G.L. Lapini and M. Zippo
CISE S.p.A., Segrate, Italy

P. Moretti
ENEL S.p.A. DCO-AITC, Rome, Italy

R. Bernante
ENEL S.p.A. DSR-CRTN, Milan, Italy

ABSTRACT

The paper reports the results of an experimental program carried out for establish a method for detecting the resonances that contribute to turbine blade fatigue failures.
The proposed method combines modal analysis and integral order trend evaluation of the vibrations measured by accelerometers on stationary parts; it is expected to be especially useful in situation in which classical experimental approaches have short life or suffer limitations due to high temperature environments.
In order to verify and implement this diagnostic technique, several tests have been carried out on a large scale test facility based on a two stage steam turbine. Strain gauges sensors mounted on twelve blades of the second stage provided a check for the presence of blade resonances, while accelerometers located at different positions on the housing were used to verify the signal attenuation along different propagation paths.
A computer controlled real time system has been used to acquire, analyze and display data.

1. INTRODUCTION

Experience indicates that turbine blade resonances are a serious problem. Vibrations at resonance conditions strongly increase the dynamic stresses on blades and can lead to fatigue failures, with consequent loss of machine integrity.

Several methods for detecting excessive blade vibration levels and resonance conditions have been proposed and are in used, but both the classical strain gauges approach and the more recent optical and blade tip time-of-arrival methods are intrusive and require special transducers [1,2].

The authors' aim was to verify the possibility of detecting resonance conditions of blades using simple instrumentation which does not disturb the turbine operating conditions of a turbine. Moreover, it was considered more important to have the possibility of identifying and monitoring conditions causing resonances, in order to avoid them, rather than to measure blade stresses; for this reason a method that permitted quite long monitoring periods was sought.

The technique described in this work, called BRED (**B**lade **RE**sonance **D**etection), was first proposed by Simmons and Smalley [3,4] and is a non-intrusive method for detecing the presence of blade resonances that contribute to turbine blade fatigue failure. Basically it is based on monitoring and analyzing the vibrations transmitted from the blades to the fixed parts of the machine (the bearings).

As ordinary vibration transducers (typically accelerometers) are used at accessible locations, the technique is well suited for:

- use where no other system would survive;
- emergency situations where measurements are needed quickly;
- cases in which a long term trending is required.

The method combines modal analysis and integral order trend evaluation to identify blade resonances.

The major problem in its application is to recognize true blade resonance conditions from other turbine component resonances which may also transmit vibrations to the accelerometers. Thus additional data are needed to verify that blade resonances do exist and that they can be detected by the sensors. In particular preliminary modal testing is necessary to:

- exactly determine the dynamic characteristics of the blades in terms of modal parameters;
- verify in stationary conditions the good transmissibility of the signal from the blades to the accelerometers.

The subsequent identification of resonances at the expected frequencies is based on the well known fact that the vibrations of a structure reach an amplitude peak and exhibit a phase shift of 180° with respect to the forcing function, on passing through a resonance. This can be seen in a Bode diagram or, preferably, in a Nyquist diagram, where a characteristic circle appears.

During the tests, described below, the evaluation and checking of the natural frequency in dynamic conditions and their true excitability in operation was carried out also using a series of strain-gauges installed on the blades, allowing the correlation of any variation of the vibration signal measured by the accelerometers to a possible blade resonance.

The experimental verification program was performed at a test facility, located in Italy, which includes a steam generation plant and an experimantal two-stages turbine (maximum power, 22 MW). This turbine has two indipendente shafts, running at different speeds, and driving two hydraulic dynamometers (water brakes) (Fig.1).

Nothwithstanding its unusual configuration (it was designed to test new blades) this plant offers extreme flexibility of operating conditions, does not suffer the constraints of a full scale plant, and consequently was chosen to conduct accurate and undisturbed experimental work.

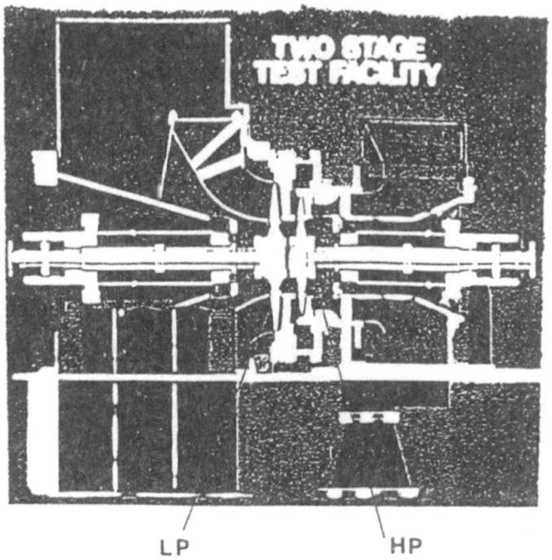

Fig.1 - Schematic trasverse section of the test facility

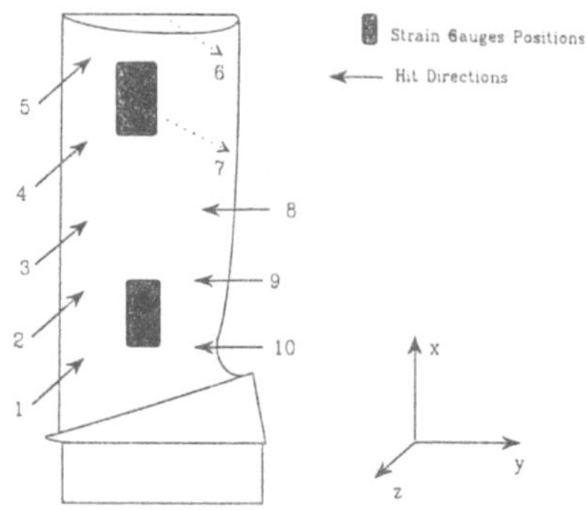

Fig.2 - Sketch of a single blade showing the hit directions used for impact test and strain gauges positions for rotational tests

2. BLADE MODAL TESTING

Modal testing have been carried out during turbine assembly in order to determine natural frequencies and modes of blades mounted on the LP stage of the turbine and also the best location for the strain-gauges, used as a check method for detecting resonance condition.

Impulsive excitation (impact-test) has been used to dynamically characterize the blades; a modal analysis software has allowed determination of the modal parameters [5,6,7].

The whole structure (disk+blades) has been tested in a "free" configuration, suspending the disk from a steel-wire cable.

A sketch of the blade is reported in Fig.2, indicating the impact and the measurement points as well as the strain gauge positions.

Values of the first six natural frequencies for the tested blades have been reported in Table 1: the spread of frequencies along the different blades has a maximum of 6% for the 3rd mode.

BLADE No.	1st mode	2nd mode	3rd mode	4th mode	5th mode	6th mode
1	303	560	667	808	921	1161
2	301	557	646	807	920	1163
3	305	557	650	807	920	1158
5	295	557	643	808	921	1159
16	291	556	630	806	920	1162
31	300	560	643	809	921	1160
46	291	558	629	806	921	1164
61	305	561	668	809	922	1162
76	296	558	646	808	922	1158
91	299	560	644	809	921	1161
106	304	557	645	808	922	1161

Tab. 1 - Experimental values of the natural frequencies of blades [Hz]

Modes 1 (axial mode), 3 and 4 (pitching or X modes) were the easiest to excite, and consequently the most interesting ones to monitor during operating tests.

3. TRANSMISSIBILITY TESTS

Once the turbine was mounted on its pedestals it was possible to perform additional tests, in order to verify the transmissibility of the signals from the source of vibration (the LP blades) to the measurement points (the bearing housings).

The blades were excited using an instrumented hammer and the response simultaneously measured by the strain gauges on the blades and by the accelerometers located on the fixed parts of the turbine.

A total of nine accelerometers were used, four in radial direction on bearing 5 (two mounted directly on the bearing and two on the external housing), four in radial direction on bearing 6, as on bearing 5, and one on bearing 6, but in axial direction.

Typical transfer functions obtained between the applied force and accelerometers on bearings are reported in Fig.3. It is evident thet good transmissibility of the signals to the accelerometers exists, in particular for modes 1, 3 and 5.

The transfer function also shows high coherence levels at the corresponding frequencies. It is interesting that while the frequency of 1st mode picked up by strain-gauges matches well with accelerometers, for the 3rd mode there is a slight shift of the resonance peak, probably due to a damping effect on the propagation path.

In addition, comparison between the amplitudes of the different transfer functions allows an evaluation of the signal attenuation along the whole path. Typically a value of 30 for the attenuation between the vibration signal on the blade and that on bearing 5 (0.4m from the bladed disk) was found, although this depends on the excited blade, while a further attenuation of about 60% was obtained between bearing 5 and bearing 6 (0.8m from the previous one).

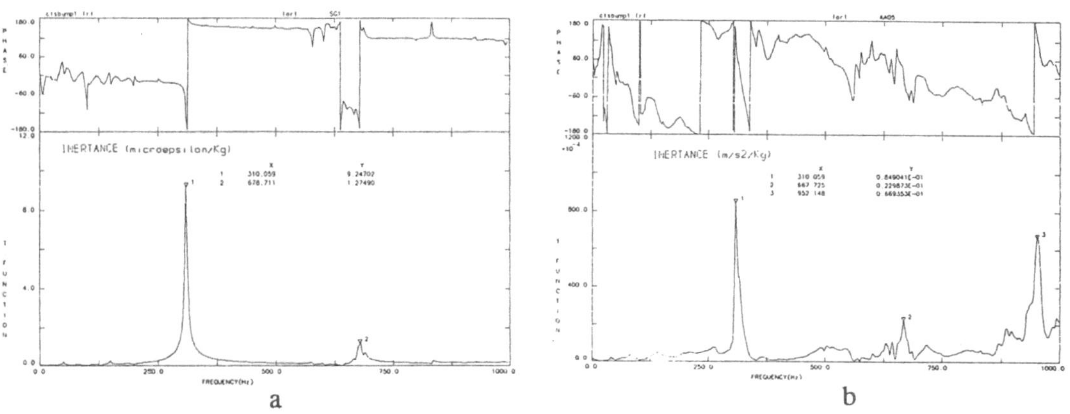

Fig.3 - Transfer functions exhibited during transmissibility tests: a)strain gauge - b)accelerometer

A similar series of tests was carried out with the bearing hydrostatic pumps switched on in order to study the effect of oil film on signal transmissibility. A significant decrease of transmissibility (about 35%) was detected for the first axial mode at 320 Hz, while for higher ones some of the sensors even indicated a slight increase. In any case also in these conditions the signal were well detected by the accelerometers.

In order to evaluate the spread of resonance frequencies, tests have been repeated for 25 blades and a probability density function has been calculated, giving a good correlation with the normal distribution.

4. ROTATIONAL TESTS

Tests in rotation have been carried out both at variable and rated speed; in this second case, in order to change possible resonant conditions, it has been possible to act on the following operating parameters:

- load (by means of brake regulation)
- condenser vacuum
- inlet steam flow
- bearing oil temperature

4.1. Speed Transient Tests

Run up and run down tests in the operating speed range of 3500÷6000 rpm (under

3500 rpm the accelerometer signals were very low and not useable), were performed to identify, from the strain gauge signals, the true resonance conditions of the blades.

A typical waterfall diagram of one of the strain gauges on blades during a rotor acceleration from 3500 rpm to 6000 rpm is reported in Fig.4a.

Some of the harmonic component show remarkable amplification, due to passage through resonances.

In particular there is a definite increment of the vibration components 5 x rev. and 10 x rev. around 3900 rpm, of 4 x rev. and 8 x rev. at about 4900 rpm, and of 7 x rev. at 5700 rpm.

This fact corresponds to the excitation of the 1st blade mode (at about 320 Hz) and of the 3rd blade mode (at about 650 Hz).

From the analysis of all data it is evident that passages through resonances of different blades do not happen at exactly the same speed, because of the blade mistuning underlined in modal testing.

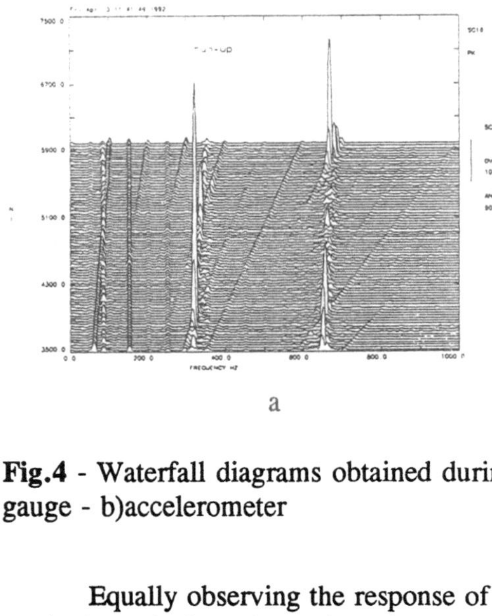

a

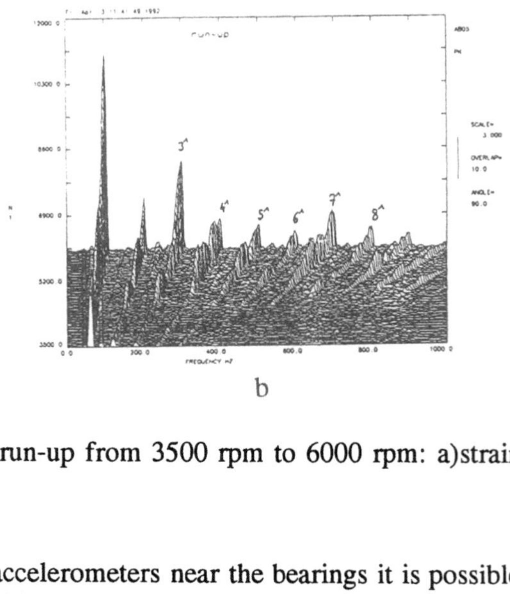

b

Fig.4 - Waterfall diagrams obtained during a run-up from 3500 rpm to 6000 rpm: a)strain gauge - b)accelerometer

Equally observing the response of the accelerometers near the bearings it is possible to obtain typical results such as those presented in Fig.4b, which shows a spectrum cascade for one of the accelerometers mounted on bearing 5.

In this figure the components 1 x rev. and 2 x rev. depend on the general vibrational behaviour of the turbine and are not necessarily related to blade resonances. For all the others components a generalized increment takes place above 4500 rpm, making identification of multiple order excitation more difficult than for strain gauges.

In fact some points must be remarked:

- the transmission of a vibration signal of frequency "f" from a rotating coordinates system to a fixed one produces three components at frequency "f-w", "f", "f+w" (w indicates speed of rotation), where "f-w" and "f+w" are affected by both tangential and radial components of blade vibration while "f" is affected only by the axial one [8];

- the stiffening of the blades with speed, that cause a slight increase of all the resonance frequencies;
- the spread of resonance frequencies (within 10% of resonance), that produce a mistuning effect, as already described [9];
- the accelerometers are also sensitive to blade vibration modes not detected by strain gauges because of their position on the blades.

Owing to all these factors spectrum cascades are not the best way to identify resonances, even if some peaks at constant frequency (modes) are recognizable.

The analysis of data by means of Campbell Diagrams gives better results. Fig.5 shows this kind of graphic for one of the strain gauges: of particular note are the two modes at 320 and 650 Hz, clearly shown by amplitude increments next to the shaft orders.

The same kind of diagram for one of the accelerometers is shown in Fig.6.

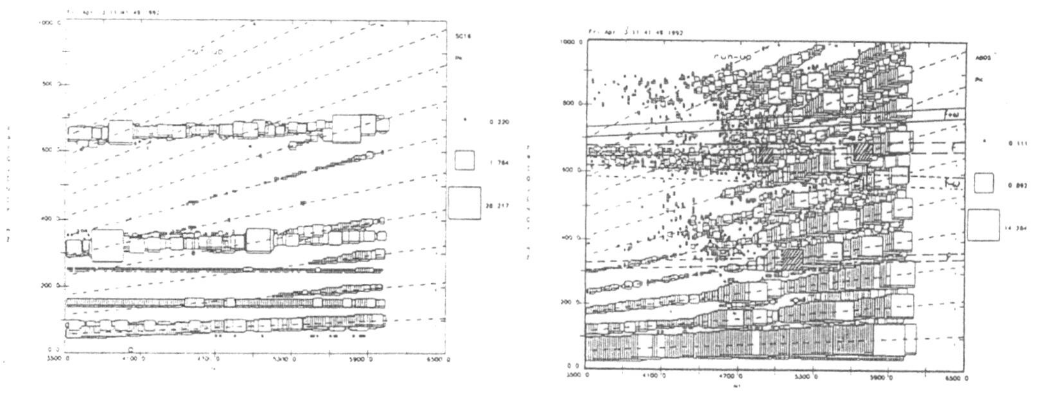

a b

Fig.5 - Campbell diagram obtained during a run-up from 3500 rpm to 6000 rpm: a)strain-gauge - b)accelerometer

The situation here is more complex than that for strain gauges, but in any case it is possible to detect blade resonance crossing. In particular:

- the 4th order shows a passage through the 1st mode at about 5150 rpm (as strain gauges do);
- the 7th and 8th orders show increasing amplitude on the same imaginary line that represent the 3rd mode, at about 5700 rpm and 4800 rpm respectively .
- *finally* the 6th, 7th and 8th orders show, at about 5720 rpm, a contemporary increment

of the amplitude, corresponding to the intersection with the components "f-w", "f" and "f+w" of the 3rd mode at 650 Hz.

Unlike strain-gauges, accelerometers exhibit also some amplitude increments for higher order components highlighting the presence of superior modes.

A better way to identify the presence of real resonances is to examine Bode and Nyquist diagrams. The typical pattern of the more interesting signal harmonics for one of the accelerometer mounted on bearing 5 are plotted on Fig.6.

In particular the Nyquist plots (Fig.6b) show a definite circular pattern for the 4 x rev. and for the 8 x rev. components, with maximum at about 5100 rpm (corresponding to 340 Hz) and 4900 rpm (corresponding to 650 Hz) respectively.

This analysis confirms beyond doubt the presence of resonances. Comparison with plots for the strain-gauges shows an high level of coherence, so that it is possible to identify the first mode at 320 Hz detected by 4th order component, and the third mode at 650 Hz highlighted by 7th and 8th orders.

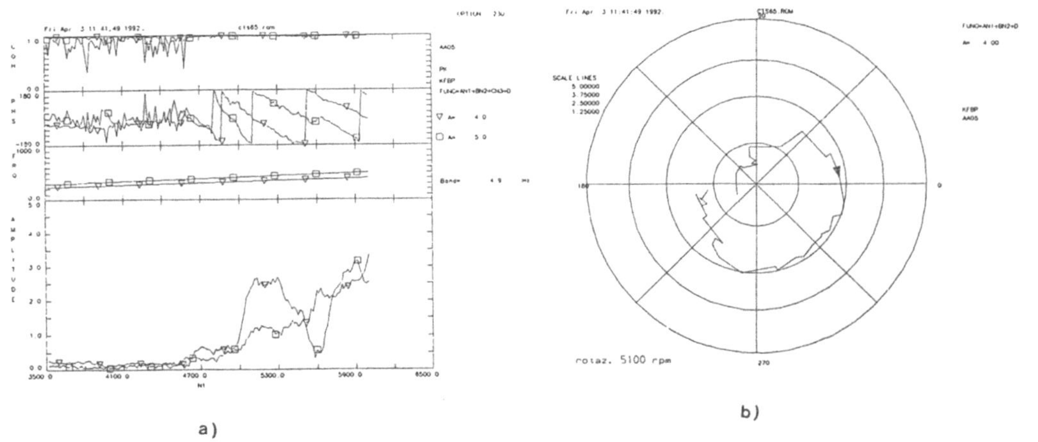

Fig.6 - Analysis of some of the components exhibited by one accelerometer on bearing 5: a) Bode diagram for the 4xRev and 5xRev - b) Nyquist diagram for the 4xRev

4.2. Rated speed tests

The purpose of a test cycle at rated speed, was to simulate the effect of varying some plant operating parameters which on a full scale plant can drive blades in or out of resonance or disturb measurements.

A secondary aim of these tests was to find more correlations between the accelerometers signals and the blade vibrations.

The most significative results were obtained by varying condenser vacuum and earing oil temperature.

A vacuum variation causes a change of pressure drop across LP stage, and an increase r decrease of the static and dynamic forces acting over the blade, while an oil temperature erturbation causes a change of lubricant viscosity with consequent variation of ransmissibility.

These tests were carried out both at and out of blade resonance condition in order to onfirm that the measured accelerometer signal variations were related to blade vibrations.

Since at resonance a significant number of blades vibrated during the whole test acuum variation produced large effects in the vibratory response measured by the ccelerometers.

Fig.7 collects the main results related to a vacuum variation cycle carried out with he LP turbine at a speed of 5120 rpm, corresponding to a resonance of the 1st and 3rd nodes which are excited by the 4th and the 8th harmonics.

As it can be seen, these components show significant variations as the cycle rogresses, with acceptable correlation with vacuum and strain-gauges signals.

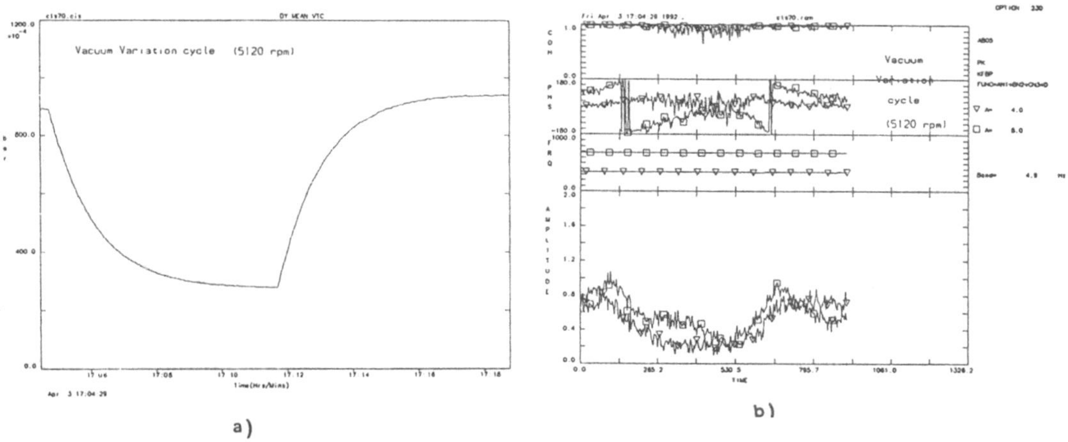

Fig.7 - Evaluation of blede vibration at resonance condition during a vacuum variation cycle: a) Vacuum variation cycle - b) Changes of 4xRev and 8xRev on the accelerometer

A similar test, performed at 4750 rpm in which none of the relevant modes were significantly excited, showed an absence of correlation between vacuum variation and 4th and 8th harmonics trend.

Another interesting test conducted at the rated speed involved the variation of the lubricant oil temperature.

The main purpose of this testwas to verify the change of the transmissibility of the vibration signals from the rotating parts to fixed ones while varying the phisics characteristics of the lubricant hence illustrating the importance of keeping this parameter constant during resonance tests.

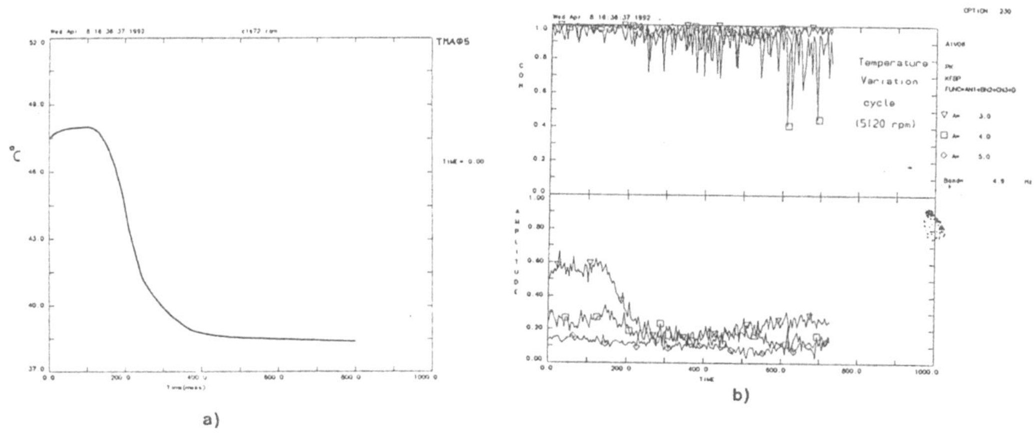

Fig.8 - Evaluation of blade vibration at resonance condition during an oil temperature variation cycle: a) temperature variation - b) Changes of 3xRev, 4xRev and 5xrev components on the accelerometer on bearing 6

From Fig.8 it is possible to see how a decrease in oil temperature (increase of viscosity) causes a clear and well correlated variation in vibration levels measured by the accelerometers (shown here for 3rd, 4th and 5th harmonics).

5. CONCLUSIONS

The possibility to detect the existence of turbine blade resonances through the signals recorded by accelerometers placed on machine bearings has been demostrated in a test cycle carried out on an experimental two stages steam turbine.

Provided the blade resonance frequencies are known (experimentally measured by impact tests) and there is a good signal transmissibility between the blades and the bearings, blade resonances are clearly visible and can be distinguished from other structural resonances.

The BRED method is fairly easy to apply, requires simple sensors and instumentatione and is non-intrusive. Unlike other methodos BRED is not able to measure the dynamic stresses on turbine blades, but can only determine if blade resonances are present or not. In any case it promises to be particularly useful for machines on which recurrent fatigue ruptures occur.

The machine behaviour can be followed without problem for long periods and the operating conditions that induce or increase blade vibrations can be determined. This is very useful, from a plant management point of view, because by avoiding the more damaging operating conditions, failures can be prevented or at least blade life can be increased.

ACKNOWLEDGEMENTS

The authors would like to express their thanks to Mr. Harold R. Simmons of Southwest Research Institute for his useful advice in preparation and execution of the test program.

REFERENCES

1. Donato V., Bannister R.L., De Martini J.F., "Measuring Blade Vibration of Large Low Pressure Steam Turbines", Power Engineering, March 1991
2. Mase M., Rozelle P.F., "Blade Vibration Monitoring Technology", EPRI Steam and Combustion Turbine Blading Conference, Orlando, Jan.1992
3. Simmons, H.R. "Non Intrusive Detection of Turbine Blade Resonance" presented at Third EPRI Conference on Incipient Failure Detection in Power Plant, Philadelphia, Dec.1987
4. Simmons, H.R. "Blade Resonance Analysis" Southwest Research Institute Document n.04-3112-210, Sept.1990
5. Prohl, M.A. "A Method for Calculating Vibration Frequency and Stress of a Banded Group of Turbine Buckets2 ASME Paper 56-A-116
6. Halvorsen, W.G. and Brown, D.L. "Impulse Technique for Structural Frequency Response Testing" Journal of Sound & Vibration, Vol.96(3), 1984
7. Ewing, D.J. "Modal Testing: Theory and Practice", Research Study Press, 1984
8. Tanaka, S. and oth. "Monitoring of Blade Vibration Through Detecting of Bearing Pedestal Vibration" ASME Paper n.83-JPGC-Pwr-36
9. Ewing, D.J. and Han, Z.S. "Resonant Vibration Levels of a Mistuned Bladed Disk" presented at The Ninth Biennial Conference on Mechanical Vibration and Noise of the Design and Production Engineering Technical Conference, Dearborn, Michigan, Sept.1983

NON-CONTACT BLADE VIBRATION INFORMATION SYSTEM BeSSI

M. Gloger and M. Jung
Siemens AG, Mülheim, Germany

0. Abstract

By the use of two electric sensors in the casing above the rotating blades' tip the vibrational deflection of all blades of a wheel can be measured. The system, well proved for temporary measurements in model turbines as well as in steam and gas turbines, is now developed for the uninterrupted observation of the vibrational behavior of all last stage rotating blades of high power steam turbines. Operational conditions leading to life-reducing stresses will thus be made obvious and can then be avoided. If mavoidable the remaining service life of each blade is determined continuously and a warning is given before consumption of all fatique life, i.e. before starting of a crack.

1. Introduction

Free standing rotating blades in steam LP turbines and gas turbines are designed at Siemens with regard to their aerodynamic and mechanical characteristics with the aid of advanced computer programs. Siemens' 45 years of experience in designing free standing blades and the use of in-process model tests are valuable assets for establishing the requisite outline conditions for these designs. New types of blades are subjected to extreme loads in a model turbine to ensure that they can withstand dynamic stresses not accessible in design calculations. In the event of limit loadings or extreme operating conditions, the dynamic stresses placed on newly developed last-stage blading is measured for all possible operating conditions on its first use in a full-scale turbine [1].

Consequently, the blades are tested under all currently-known operating conditions to ensure that they are not subjected to conditions which may reduce the service-life of the blades. Design modifications and confirmatory analysis also ensure that the blades cannot be excited under resonance to excess vibration amplitudes by unusual stresses which have recently been identified, e.g. by 100 Hz torsional vibration induced by high-voltage side line short-circuits. It can therefore be postulated that, based on current knowledge, all newly developed last-stage blades will not be subjected to life-reducing stresses at any time.

Installation of a blade vibration information system is being proposed by Siemens/KWU to cover any possible cases of increased stress, e.g. caused by alteration of the operating mode, even though this is improbable on the basis of present knowledge.

As the characteristics of the operating condition which can cause increased stresses for last-stage blading is not known - this occurring, for example, for a short period of time but with intense magnitude - data from all last-stage blades must be continuously collected and checked. These data are then discarded in the event of normal operational stresses.

For stresses somewhat higher than normal, the blade loading data are registered along with characteristic turbine operating data to allow the cause for the increased loadings to be determined at a later date.

For stresses even greater than those just described, but which are not blade life-reducing, data is registered as previously described and a warning is simultaneously issued to the operating personnel. This warning signals that the trend towards a hazardous operating condition is being counteracted by intervention in the operating mode (e.g. changing of condenser pressure or output).

The overall objective of the non-contact blade vibration information system is the timely detection of increased stress on last-stage blading so as to exit the

operating condition which caused the increased stress before this leads to any reduction in service life.
In the event that the blades are subjected to life-reducing stresses before a hazardous condition can be avoided, the remaining service life of each blade is determined continuously and an alarm is then issued for each blade at an appropriate safety margin before consumption of all fatigue life.

2. Functional Description

A procedure which has been applied for many years is that of alternating-stress measurement on one or more rotating blades using strain gauges. For utility turbines the signals from these gauges must be transmitted by radio from the rotating shaft. Along with the advantage of obtaining continuous blade vibration signals, the disadvantages of this type of data transmission must also be accepted, e.g. data from only a few blades can be acquired, system installation is very elaborate and system service life is very limited.
Siemens has developed a measuring method for examining vibration behavior of all the free standing blades of a gas or steam turbine blade wheel in a model or prototype turbine which does not require application of any equipment of the rotating blades. It is based on an initial idea of Raby [2] to measure the vibrational deflection of the blade tips and; similar to a proposal of Roth [3], two sensors, located in the casing above the rotating blades, measure, by means of a 100 MHz-clocked counter, the time required by each blade to cover the distance between the two sensors. If the time difference for a non-vibrating blade is subtracted from the time difference obtained for a vibrating blade, the vibration displacement of the blade tip in the circumferential direction is obtained, provided that speed, sensor spacing and the distance between blade tips are taken into account. If blade vibration is not synchronous with speed, or a multiple thereof, the maximum positive or negative amplitude is obtained after several revolutions. Amplitude is likewise measured when a resonance point is passed through during coasting up or down the turbine.
Allocation of individual signals to each blade is accomplished by the phase sensor, which sends a pulse signal from the shaft.
If the time signal of a blade is analyzed in relation to the frequency, the alias frequencies for the true natural vibrations of the blades are obtained, as the vibration signal is only scanned one time for each revolution. If the approximate position of the natural frequencies in relation to the harmonic frequencies for running speed is known, the natural frequencies for each individual blade can be precisely determined from the frequency analyses. By definition, only the vibration modes which lead to noticeable blade tip

amplitude can provide measurable signals. In general these are the first to third or fourth natural modes of a free standing blade.
Allocation of maximum alternating stress of the blade to deflection amplitude for the blade tip is achieved by calculating the characteristic modes and the associated alternating stress distributions.

3. Setup of System for Acquiring Information on Blade Vibration of Several Blade Wheels of a Turbine

Since 1991, a device developed on the basis of SIEMENS MMC modules and signal processors has been in use for blade vibration monitoring on the last stage of the LP2 turbine in the Unterweser nuclear power plant.
On the basis of this device, a new concept has been developed by Siemens for continuous vibration measurement of all the blades of up to six blade wheels with up to 100 blades each. Acquisition of vibration data is performed in three processing stages.

3.1 Analog Pre-processing Unit

The start and stop signals provided by the two analog sensors for each separate LP turbine flow are linked to an amplifier unit by a short coaxial cable. In this amplifier unit a digital sampled signal of 2.4 μs is obtained from the amplified analog signal with the aid of a trigger threshold. In addition, a shaft marking signal is amplified and processed (Fig. 1).
For startup support the analog processing circuit has a counter circuit which compares the number of start and stop pulses per revolution with a reference number that can be input via a DIL switch panel. The result is shown on an LED display.
The complete unit is integrated into a closed wall-mounted case. The mechanical design of the unit is such that it can be installed directly on the outside of the turbine casing near the sensors and can operate under the associated extreme environmental and temperature conditions. Input of the conditioned signals to the digital pre-processing unit up to 100 meters away is implemented with electrical isolation by using optical fibers.

3.2 Digital Signal Pre-processing

The signals for an LP turbine that are provided by the analog processing unit are linked to a signal processing unit in the computer cabinet (Fig. 1). This signal processing unit comprises a powerful signal processor for each LP turbine which processes the start and stop signals in real-time. With the aid of a 100 MHz counter the start and stop signals from the 2 separate LP turbine flows are converted to time differentials whose resolution is 10 nsec and then converted to vibration displacement amplitudes for the blade tips using the circumferential velocity. An integrated monitoring logic circuit ensures that when erroneous signals are received from the turbine the entire measurement is marked as invalid and is replaced by the last previous correct series of measurements.
The data records of up to 100 revolutions of both turbine blade rows are stored in the main memory of the signal processor. Three trigger criteria compare these data records with configurable vibration limits. When one of these criteria is violated, transfer of the data record to the downstream analysis and memory computer is triggered.
If one or more optional analog modules are inserted, up to 8 analog channels can be read in step with shaft rotation by each module. The number of scans for each revolution can be configured.

3.3 Data Transmission Medium

A fiber-optic modem is connected to the IEC bus of the digital signal pre-processing unit. The parallel data are converted bidirectionally via single-mode fibers. All of the data of up to 3 parallel signal pre-processing units are transferred to the workstation via this fiber-optic interface.

3.4 Analysis and Storage Computer

Vibration data is transferred via the fiber-optic link by the signal processing unit only when triggered. During normal operation, no data is transmitted through the fiber-optic link to the analysis and storage computer. However, as data transmission is designed for continuous storage, continual recording of measured vibration data is also possible for a trigger threshold of 0.
The data at the bidirectional fiber-optic interface is reconverted to IEC bus specification in the analysis and storage computer by means of a second fiber-optic modem. An advanced high-power workstation with adequate memory has been selected for use as the analysis and storage computer.

An outline of the overall system is shown in Figure 1. Data stored after triggering can be transferred by modem to the laboratories for Engineering Testing and Mechanics in Mülheim/Ruhr for in-depth analysis.

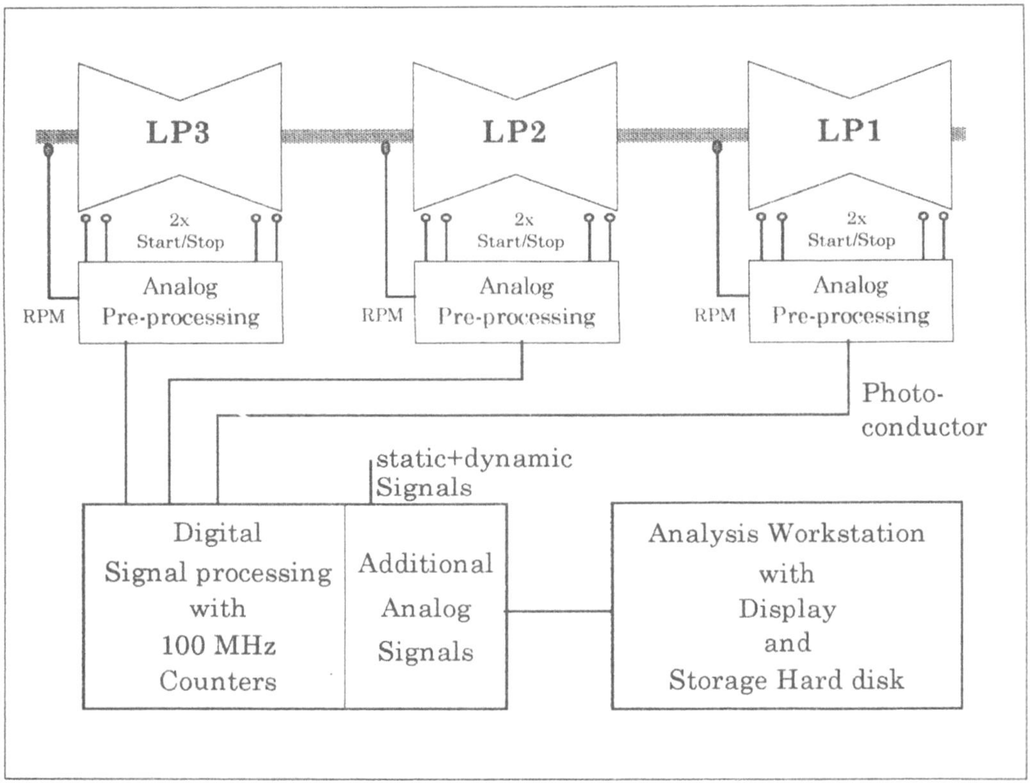

Fig. 1: Survey of system components

4. System Triggering and Signaling

Three trigger criteria with two trigger thresholds are planned for implementation:

Threshold 1: Blade vibration signals exceed the amplitude which corresponds to approximately **33% of the fatigue limit**, b (e.g. 5) number of times within a (e.g. 100) number of revolutions.

Result: The vibration signals for all the blades of all the blade wheels are stored in the memory for approximately 1 minute, including 20 seconds pre-event history, together with the characteristic analog variables. An appropriate message is issued to the control room personnel.

Threshold 2: Approximately 75% of the fatigue limit has been exceeded.

Result: Same as for threshold 1. A warning is issued to control room personnel signaling that the cause of increased blade vibration must be eliminated, for example by changing condenser pressure or output. Analysis of vibration signals and associated analog variables should be implemented without delay.

Criterion 1: Thresholds are applied to each individual blade of all the blade wheels (Appendix 1).

Criterion 2: Thresholds are applied to ½ the difference between the maximum positive and negative amplitude of an arbitrary blade of a blade wheel for one revolution so as to detect system vibrations which are synchronous with speed or which are not caused by speed. This criterion must be applied over one complete revolution, offset by 180°.

Criterion 3: Blade vibration signals are recorded along with the analog variables for approximately 1 minute, regardless of the momentary magnitude of amplitude, at regular intervals, e.g. every ½ hour.

Using a menu, the maximum amplitudes of all the blades of one stage can be displayed with a time window of approximately 1 minute during which one or more blades exceeded a trigger threshold (Appendix 2).
In the event that the fatigue limit of one or more blades is exceeded during a suddenly occurring event, the relative consumption of service life is calculated for each blade from the stress magnitude and the duration of stress (Miner sum, see Section 5). On consumption of 75% of the service life of a blade, a warning is issued to the control room personnel (Appendix 3).
A trend display can be generated at any time from the ½-hour values of the associated statistical and dynamic analog variables (Appendix 4).

4.1 Miner Sum Calculation

Figure 2 shows the calculation process for a Miner sum in the form of a flow chart. The following paragraphs explain more precisely the computational algorithm used for calculating the Miner sum.

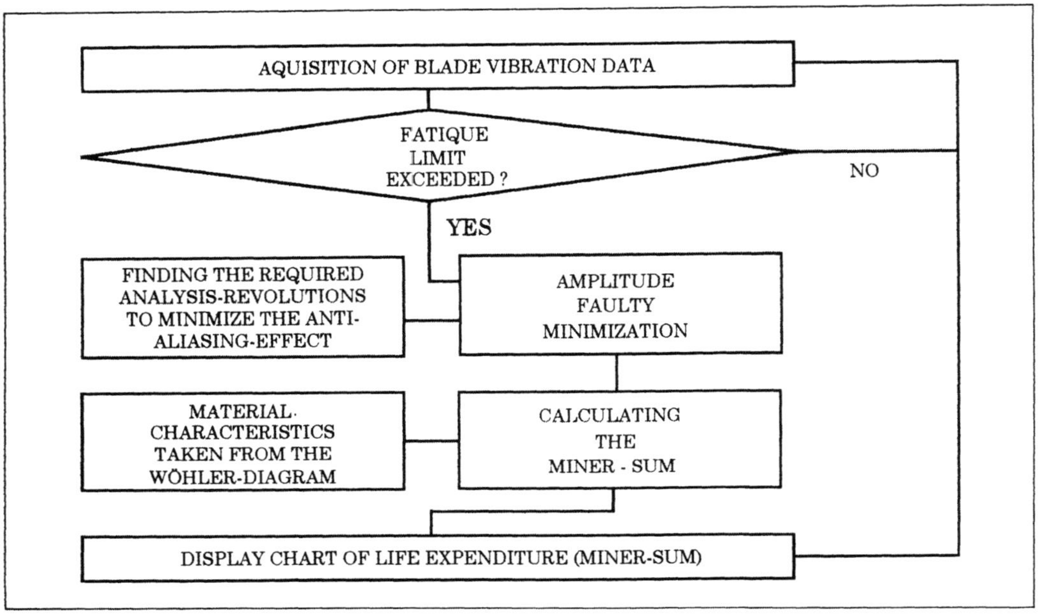

Fig. 2: Miner-Sum-Calculation

Since a continuous assignment of experienced stress (blade deflection) to a permissible load-cycle number is not meaningful, **A_i** classes are used. Each class is assigned a permissible load-cycle number **N_{perm}**.

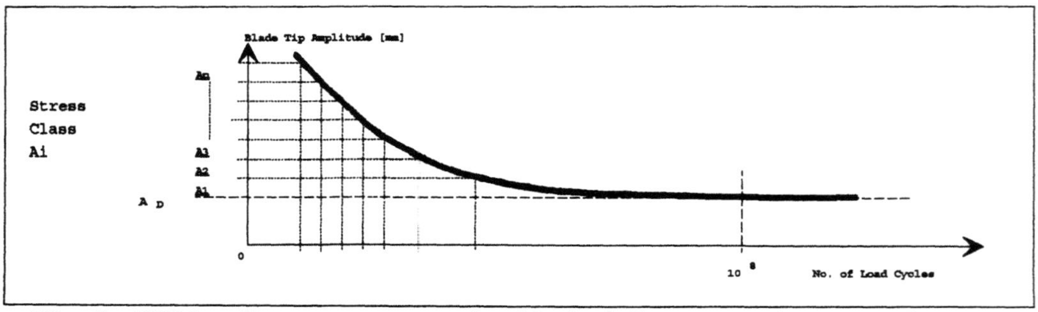

Fig. 3: Wöhler diagram converted to blade deflection in the circumferential direction.

The amplitude **a** = ½(Y_{max} - Y_{min}), obtained previously for a time period **dT**, is assigned to a corresponding **Ai** class. Earlier tests have shown that the error resulting from an inadequate scanning duration is less than 1% for a scanning time period **dT** = **1s**. By simply multiplying **dT** by the blade frequency **f**, the service life-reducing load cycle number N_{act} is obtained.
Now that all of the variables required for calculating the Miner sum have been determined, the Miner sum **M** can be calculated.

$$M = M + \frac{N_{act}}{N_{perm}}$$

5. Results of Measurements

Using measurements taken at Rye House, the potential uses of the system are presented in the following. The scope of this report does not, however, allow for thorough elucidation of all of the measurement results and types of presentations.
At Combinated Cycle Power Plant Rye House with three gas turbines and one steam turbine with a direct air cooled condensing system the vibrations of the last stage blades (10 m²) had been measured during initial plant start-up. The aim of these measurements was to demonstrate the reliability of the blades even at low load operating conditions at elevated back pressure. These conditions might occur under extraordinary circumstances operating in a direct air cooled system.
Appendix 5 shows the diagram of the vibration signals of blades #1 to #20. At different times higher blade amplitudes are clearly visible. An in depth evaluation unveils the vibration modes concered and their amplitude and stress response.
Figure 4 shows the unfiltered time signal and the spectrum of blade No. 6 under normal operating conditions. The maximum amplitude is about 0.7 mm. Mainly the first vibration mode (alias frequency about 5.2 Hz) and to a smaller degree the second one (alias frequency about 21.0 Hz) are concerned.

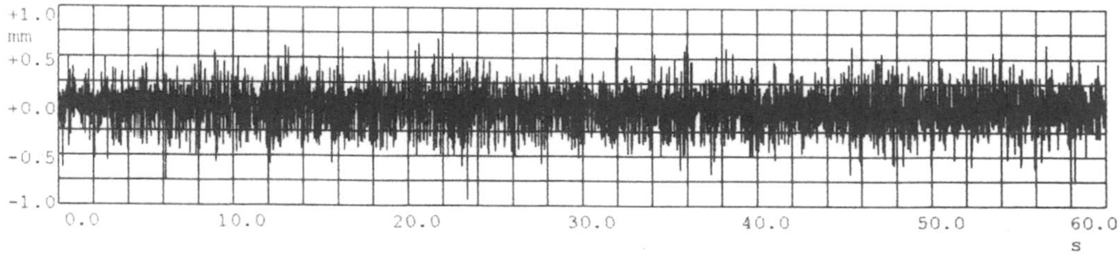

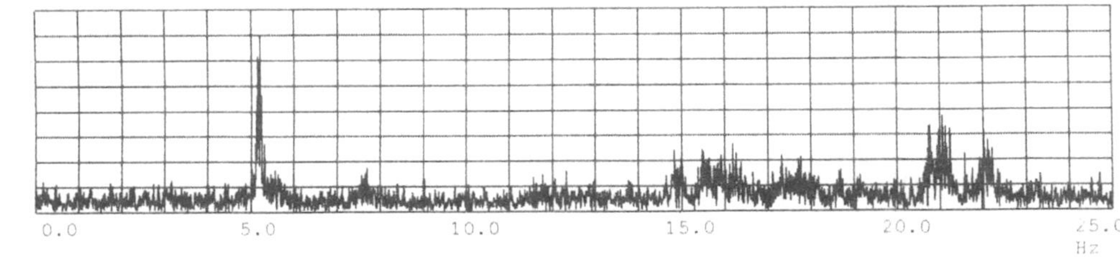

<u>Fig. 4</u>: Unfiltered time signal and alias frequency spectrum of blade No. 6 at normal operating conditions (T=3400 s, see Appendix 5)

At elevated back pressure of 270 mbar, this condition was achieved by stopping operation of a number of ventilators of the air cooled condensor, the flow number φ is reduced to a value of 0.05, characterizing windage conditions for the last stage blade row with the typical stochastic vibrational excitation of the blades. As you can see from Fig. 5 mainly the second mode is excited.

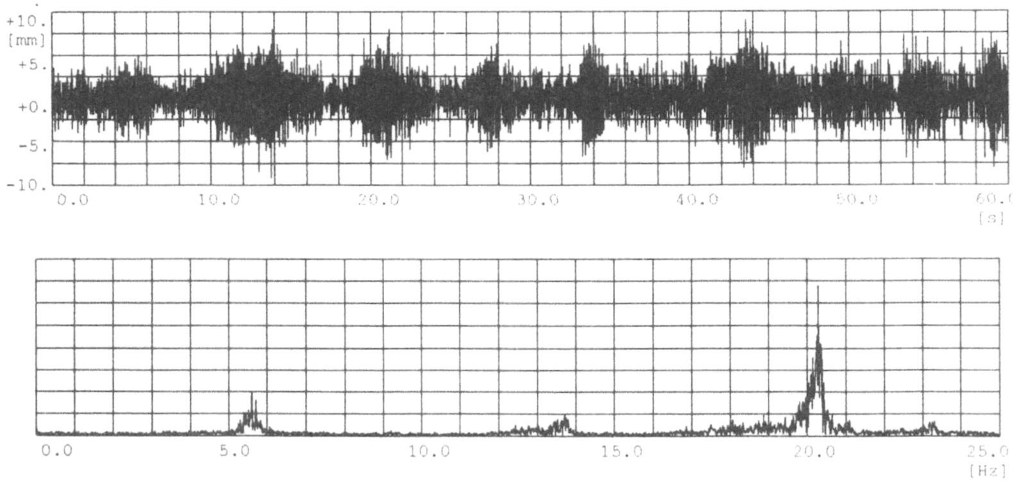

<u>Fig. 5</u>: Unfiltered time signal and alias frequency spectrum of blade No. 6 at windage condition at high back pressure (T=2400 s, see Appendix 5)

The maximum stress at the blade surface is shown in Fig. 6 for the 1st mode and in Fig. 7 for the 2nd mode.

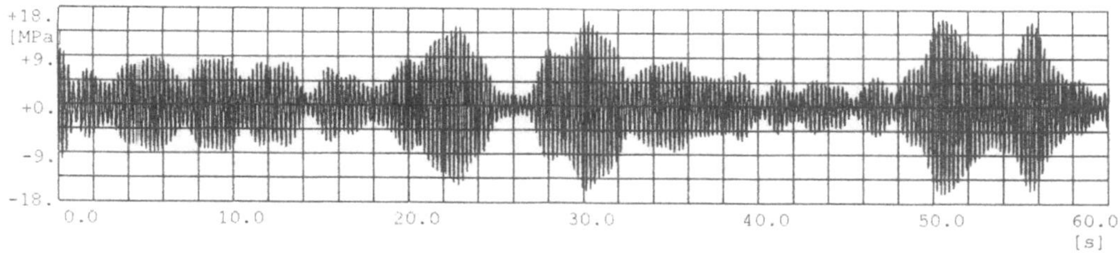

Fig. 6: Bandpass filtered time signal (5 to 6 Hz) of blade No. 6 1st mode blade vibration (T=2400 s, see Appendix 5)

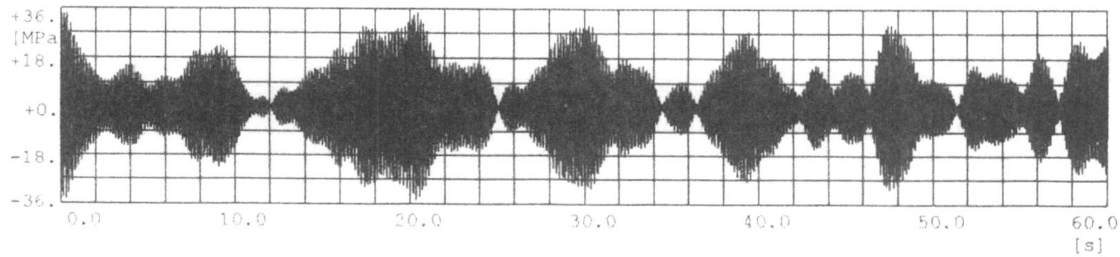

Fig. 7: Bandpass filtered time signal (19.5 to 20.5 Hz) of blade No. 6 2nd mode blade vibration (T=2400 s, see Appendix 5)

Fig. 8 displays the amplitudes of the unfiltered signals of all blades at the same time intervall and Fig. 9 shows the prevailing alias frequency at which these amplitudes occur.

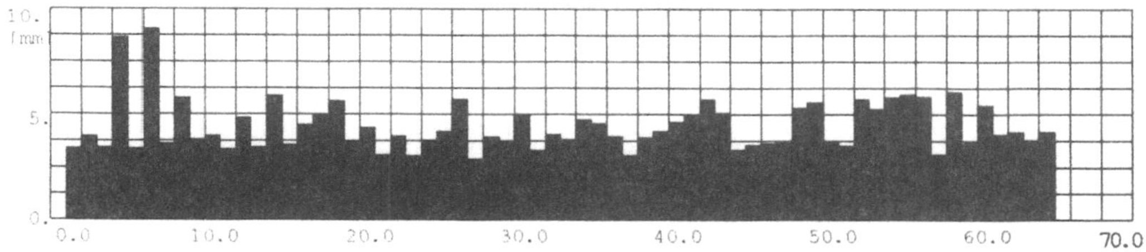

Fig. 8: Vibration amplitudes of all blades at T=2400 s (see Appendix 5)

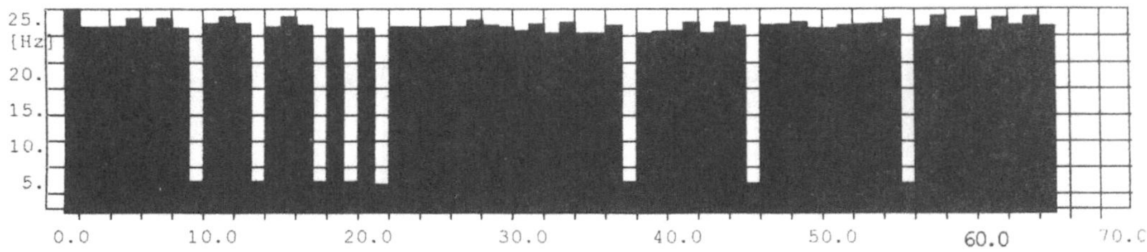

Fig. 9: Prevailing alias frequency of all blades at T=2400 s (see Appendix 5)

6. Bibliography

[1] M.Gloger, K.H. Neumann, D. Bergmann, M. Termühlen: Advanced LP Turbine Blading - A Reliable and Highly Efficient Design Presented at the 1992 IJPGC, Atlanta, Georgia

[2] H.Raby: Rotor Blade Vibration Observed from the Casing. Pres. at Conf. on Methods of Transmitting Signals from Rotating Plant, Leatherhead, June 1970.

[3] H.Roth: Vibrational Measurements on Turbomachine Rotor Blades with Optical Probes. Pres. at the ASME Joint Fluids Engineering Gas Turbine Conf., New Orleans, March 1980.

[4] M. Gloger: Berührungslose Schaufelschwingungsmeßtechnik. VGB-Fachtagung Dampfturbinen und Dampfturbinenbetrieb 1990 Nürnberg, Essen, Nov./Dec. 1990

Appendix 1

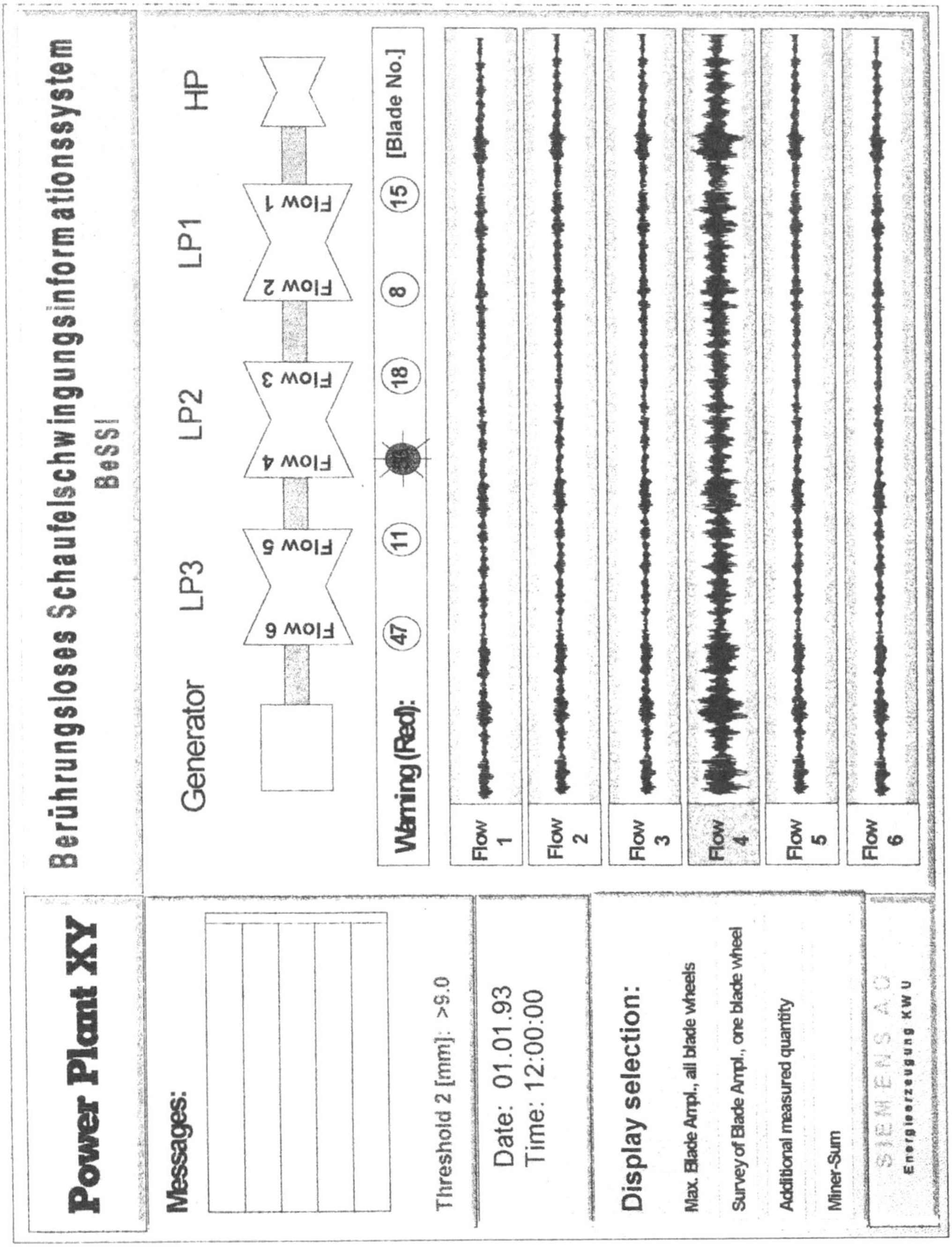
Power Plant XY
Berührungsloses Schaufelschwingungsinformationssystem
BeSSI
Generator
LP3
LP2
LP1
HP
Flow 6
Flow 5
Flow 4
Flow 3
Flow 2
Flow 1
Warning (Red):
47
11
18
8
15
[Blade No.]
Flow 1
Flow 2
Flow 3
Flow 4
Flow 5
Flow 6
Messages:
Threshold 2 [mm]: >9.0
Date: 01.01.93
Time: 12:00:00
Display selection:
Max. Blade Ampl., all blade wheels
Survey of Blade Ampl., one blade wheel
Additional measured quantity
Miner-Sum
SIEMENS AG
Energieerzeugung KWU

Appendix 2

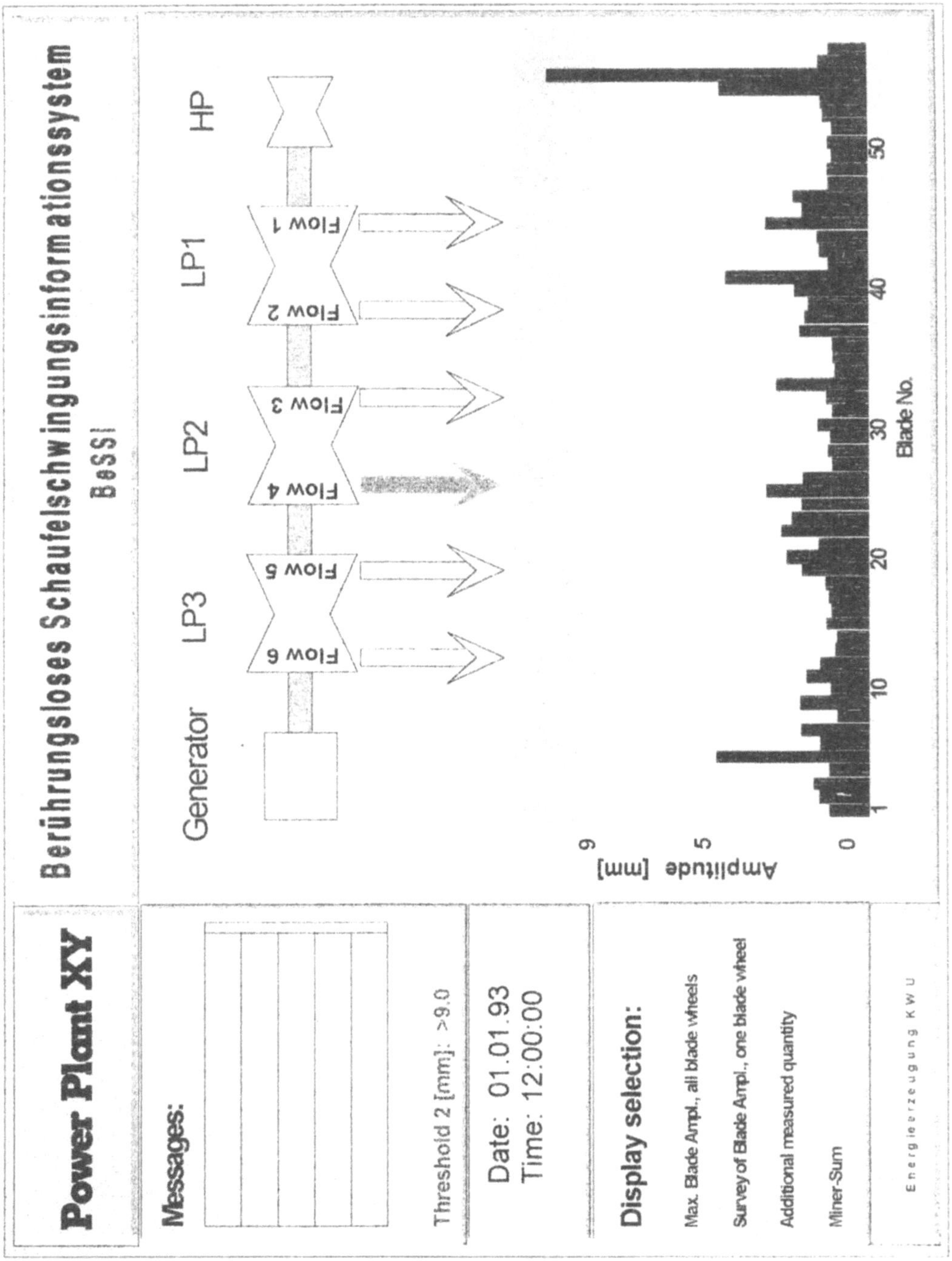
Power Plant XY
Berührungsloses Schaufelschwingungsinformationssystem
BeSSI
Messages:
Threshold 2 [mm]: >9.0
Date: 01.01.93
Time: 12:00:00
Display selection:
Max. Blade Ampl., all blade wheels
Survey of Blade Ampl., one blade wheel
Additional measured quantity
Miner-Sum
Energieerzeugung KWU
Generator
LP3
LP2
LP1
HP
Flow 1
Flow 2
Flow 3
Flow 4
Flow 5
Flow 6
Amplitude [mm]
Blade No.

Appendix 3

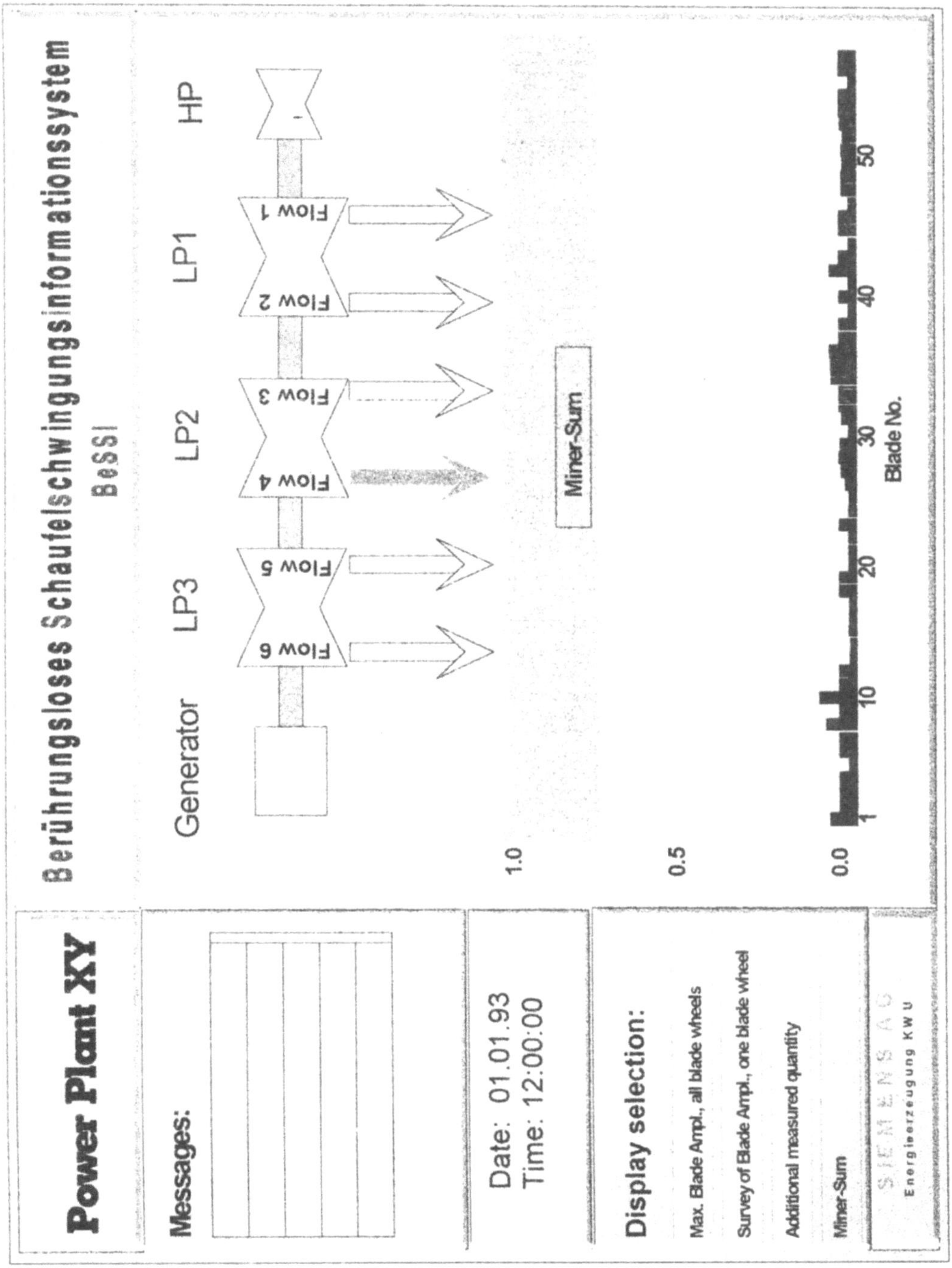

Power Plant XY
Berührungsloses Schaufelschwingungsinformationssystem
BeSSI
Generator
LP3
LP2
LP1
HP
Flow 6
Flow 5
Flow 4
Flow 3
Flow 2
Flow 1
Miner-Sum
1.0
0.5
0.0
1
10
20
30
40
50
Blade No.
Messages:
Date: 01.01.93
Time: 12:00:00
Display selection:
Max. Blade Ampl., all blade wheels
Survey of Blade Ampl., one blade wheel
Additional measured quantity
Miner-Sum
SIEMENS AG
Energieerzeugung KWU

Appendix 4

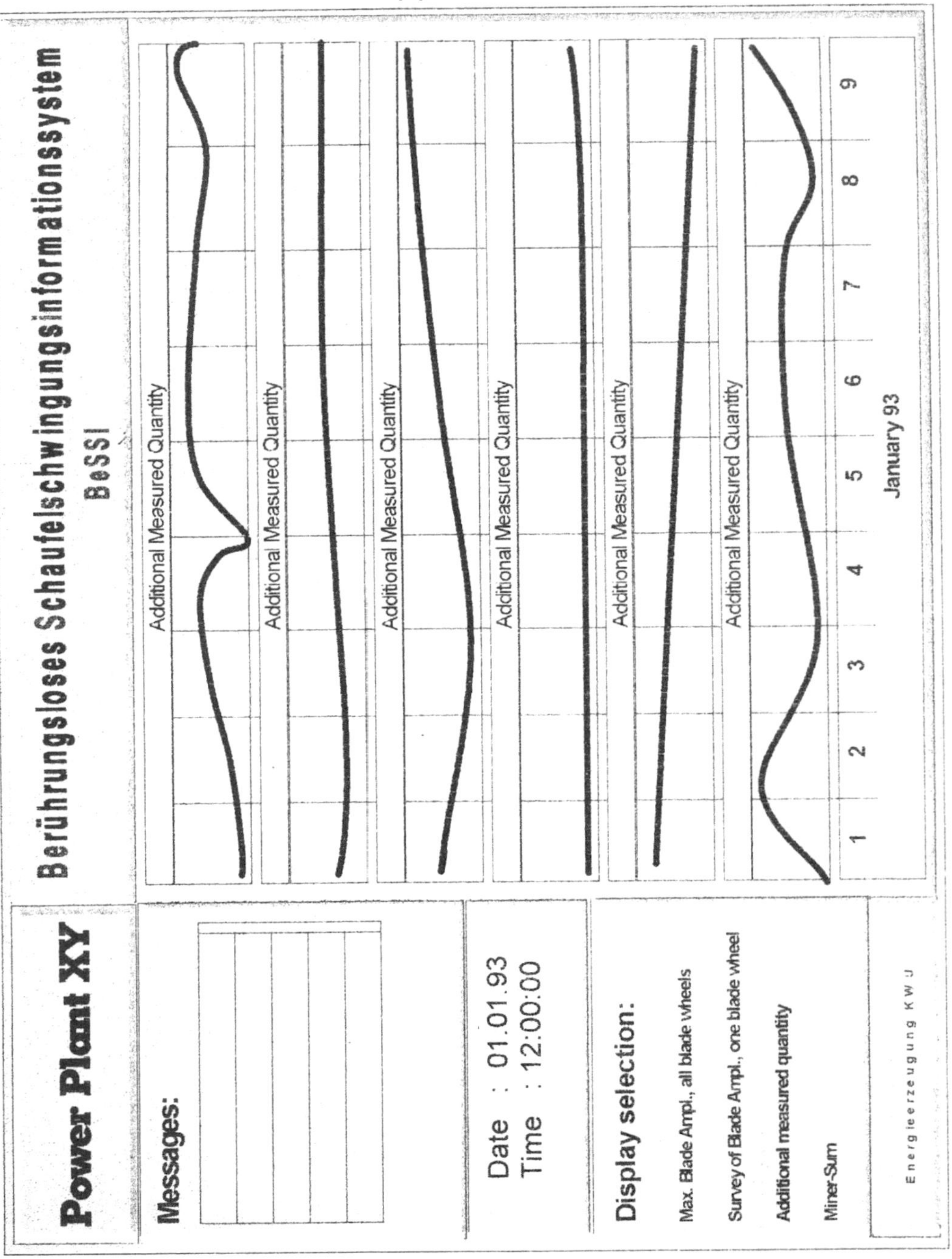
Berührungsloses Schaufelschwingungsinformationssystem
BeSSI
Additional Measured Quantity
Additional Measured Quantity
Additional Measured Quantity
Additional Measured Quantity
Additional Measured Quantity
Additional Measured Quantity
1 2 3 4 5 6 7 8 9
January 93
Power Plant XY
Messages:
Date : 01.01.93
Time : 12:00:00
Display selection:
Max. Blade Ampl., all blade wheels
Survey of Blade Ampl., one blade wheel
Additional measured quantity
Miner-Sum
Energieerzeugung KWU

Appendix 5

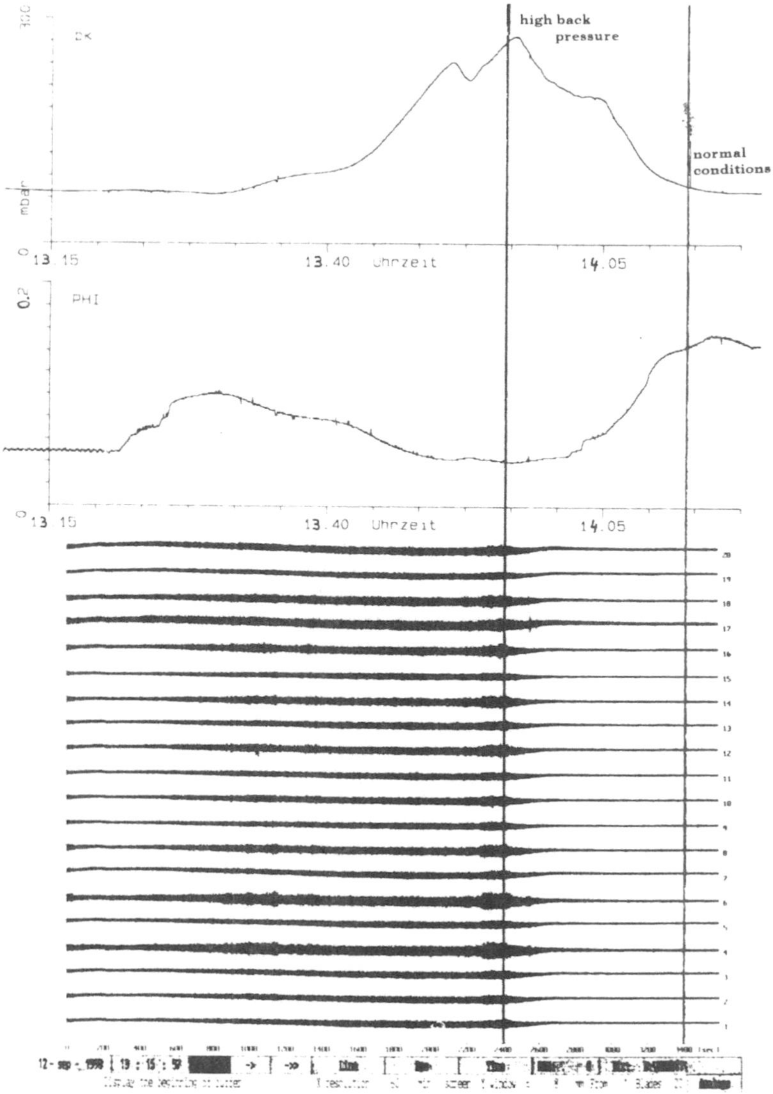
high back pressure
normal conditions
mbar
13.15
13.40 Uhrzeit
14.05
0.2
PHI
0
13.15
13.40 Uhrzeit
14.05

EXPERIMENTAL INVESTIGATION OF TURBINE BLADES IN SERVICE

O. Danek, Fr. Vanek and J. Cibulka
Academy of Sciences of Czech Republic, Prague, Czech Republic

ABSTRACT

The article deals with the investigation of the rotor blade vibrations in the last low pressure stage of the steam turbine. The first part of this article is devoted to the functional analyse of the mathematical model and in the second part some results of the measurements on the blades in the service are presented.

1.INTRODUCTION

The rotor blades, long and thin shaped, from the last stages of the low pressure turbine part belong to the most strained parts in the machinery. They work under extreme conditions such as the fields of strong centripetal accelerations and external dynamic forces from a wet steam flow. Their tension, bending and torsion stresses have in addition to high static also dynamic components, whose influences are dominant in resonances and especially in the case of self-excited vibrations.

Since the resonant domains are determined by the properties of the blades, they can be eliminated by the suitable aerodynamic design. In nominal service the proper designed blades behave as asymptotically stable systems with admissible both amplitudes and strains and with sufficient life-times. For such blades the increase of the displacement and strain amplitudes can be only caused by aeroelastic interactions between rotating vibrating blades and a flowing steam what can occur in extreme service conditions. However, for every blading admissible service conditions exist, and therefore the aim of the investigation is to determine these conditions from both the mathematical model and measurements on blades.

2. MATHEMATICAL MODEL

For the first eigenmodes with less numbers of nodal lines the blades can be considered as the thin and distorted non-prismatic beam with the length L stressing both by the bending in two rectangular planes xy and xz and torsion around the axis x. The relation between the forced displacement $\mathbf{v}(x,t)$ and excitation force $\mathbf{f}(x,t)$ in each point $x \in \langle 0, L \rangle$ and in any time $t > 0$ can be expressed by the well-known general equation :

$$\mathbf{Kv}(x,t) + \mathbf{B}\dot{\mathbf{v}}(x,t) + \mathbf{M}\ddot{\mathbf{v}}(x,t) = \mathbf{f}(x,t)\,, \tag{1.1}$$

where $\mathbf{v}$ and $\mathbf{f}$ are three dimensional vectors and $\mathbf{K}$, $\mathbf{B}$, $\mathbf{M}$ generally full matrices of the third order of linear differential operators (up to the 4th order for bending and to the 2nd order for twisting) with respect to the coordinates x. These operators have real coefficients, which are independent on the time and can be considered in the form

$$\begin{aligned} \mathbf{K} &= \mathbf{K}_0 + \varpi^2\mathbf{K}_1 + c^2\mathbf{K}_2, \\ \mathbf{B} &= \mathbf{B}_0 + c\mathbf{B}_2, \\ \mathbf{M} &= \mathbf{M}_0, \end{aligned} \tag{1.2}$$

where ω is a constant angular frequency of the rotation and c is an "average" velocity of steam flow. The operators $\mathbf{K}_0$, $\mathbf{B}_0$, $\mathbf{M}_0$; $\mathbf{K}_1$ correspond to the blade structure and are self-adjoin and positive-definite. Then the blades behave as an asymptotically stable dynamic system. The destabilizing aeroelastic couplings are represented by the operators $\mathbf{K}_2$, $\mathbf{B}_2$:

$$\mathbf{K}_2 = \mathbf{K}_{2_s} + \mathbf{K}_{2_{ns}}, \quad \mathbf{B}_2 = \mathbf{B}_{2_s} + \mathbf{B}_{2_{ns}}.$$

These operators are generally non self-adjoin, but their self-adjoin parts are dominant ($\|\mathbf{K}_{2_{ns}}\| << \|\mathbf{K}_{2_s}\|$, $\|\mathbf{B}_{2_{ns}}\| << \|\mathbf{B}_{2_s}\|$) and are either directly negative-definite or they become negative definite with limit values of service parameters.

According to (1.2), for small velocities c the operators **K** and **B** are positive-definite. These operators become singular by critical speeds c_1 and c_2 :

$$c_1^2 = \frac{(\mathbf{v}, (\mathbf{K}_0 + \varpi^2 \mathbf{K}_1)\mathbf{v})}{(\mathbf{v}, \mathbf{K}_2 \mathbf{v})}, \ /^*$$

$$c_2 = \frac{(\mathbf{v}, \mathbf{B}_2 \mathbf{v})}{(\mathbf{v}, \mathbf{B}_0 \mathbf{v})}, \qquad (1.3)$$

which cause the lost of the static rigidity and the undamping of the system, respectively. For the greater velocities than c_1 or c_2 these operators are negative-definite. Let the operators $\mathbf{K}_0$, $\mathbf{B}_0$ and $\mathbf{M}_0$ in (1.2) are self-adjoin, **K** positive and $\mathbf{K}_2$, $\mathbf{B}_2$ negative-definite. Then for $\mathbf{f}(x,t) = \mathbf{0}$ and $\mathbf{v}(x,t) = \mathbf{v}(x)e^{st}$ the (1.1) leads to the eigenvalue equation

$$(\mathbf{K} + s\mathbf{B} + s^2\mathbf{M}) = \mathbf{0}, \qquad (1.4)$$

which is identical with an adjoin eigenvalue equation. In this case only one set of eigenvectors $\mathbf{v}(x)$ corresponds to the spectrum of generally complex eigenvalues s.

Let the quantities s and $\mathbf{v}(x)$ are known and the scalar product with the function $\mathbf{v}(x)$ is introduced into (1.4). We get a scalar quadratic equation for parameter s :

$$ms^2 + bs + k = 0 \qquad (1.5)$$

/* The brackets denote the scalar product with generally complex vectors in the whole definition domain, e.g. $(\mathbf{v}, \mathbf{B}\mathbf{v}) = \int_0^l {}^H\mathbf{v}(x)\mathbf{B}(x)\mathbf{v}(x)dx$, where $\mathbf{v}(x)$ represents a hermitian conjugation to the vector $\mathbf{v}(x)$.

with the roots

$$s = \frac{(-b \pm \sqrt{b^2 - 4km})}{2m}.$$

For (1.5) the next statement holds:

if $c_1 > c < c_2$ then $k = (\mathbf{v},\mathbf{Kv}) > 0$; $b = (\mathbf{v},\mathbf{Bv}) > 0$; $m = (\mathbf{v},\mathbf{Mv}) > 0$;
if $c_1 < c > c_2$ then $k = (\mathbf{v},\mathbf{Kv}) < 0$; $b = (\mathbf{v},\mathbf{Bv}) < 0$; $m = (\mathbf{v},\mathbf{Mv}) > 0$.

The blade is asymptotically stable (Re $s < 0$) only when $b - 4km < 0 < b$. In all other cases it is either on the limit of the stability or unstable ($0 <$ Re s). If in the latter case $\mathbf{K}_2$ has nonzero non self-adjoin part ($\mathbf{K}_{2_{ns}} \neq 0$), the decisive role for the stability of the system, described by (1.4) and vibrated with angular frequency Ω, plays the definiteness of the corresponding hermitian operator $\mathbf{Z} = \Omega\mathbf{B}_s + i\,\mathbf{K}_{ns}$ /* [1], where $\mathbf{B}_s = \mathbf{B}_0 + c\mathbf{B}_{2s}$ and $\mathbf{K}_{ns} = c^2\,\mathbf{K}_{2ns}$. Assuming $\|\mathbf{K}_{2_{ns}}\| << \|\mathbf{K}_s\|$ the non self-adjoin part $\mathbf{K}_{2ns}$ causes only a small change of c_1.

The operators $\mathbf{K}_0$, $\mathbf{B}_0$, $\mathbf{M}_0$ can be verified by the identification evaluation of the forced harmonic vibration measured in the sufficiently wide frequency range. The twist of blades caused by the centrifugal forces corresponding to $\mathbf{K}_1$ can be evaluated from the measurement on the test equipment Campbell which works up to "centrifugal revolutions" in a vacuum. The assumptions for deriving $\mathbf{K}_2$, $\mathbf{B}_2$ could be verified from the differences between the "service measurements" and the "Campbell measurements". However, this resarch is in addition to a certain degree of a speculativeness very time and financial consuming.

2. THE EXPERIMENTAL RESEARCH

Since the full identification of the operators **K**, **B**, **M** of the mathematical model (1) is a very complicated issue we aimed at a more simple problem, namely, at the limitation of the admissible service. From the measurements of service parameters we can

/* Non self-adjoin part $\mathbf{B}_{2ns}$ of the operator $\mathbf{B}_2$ has no influence on the stability. It has only together with $\mathbf{K}_{2ns}$ influence on the eigenvalue problem, because (4) is distinct from the adjoin equation $(\mathbf{K} + s\mathbf{B} + s^2\mathbf{M})^*\,\mathbf{w}(x) = \mathbf{0}$ and the two sets of eigenvectors $\mathbf{v}(x)$, $\mathbf{w}(x)$ correspond to the one set of eigenvalues. If $\|\mathbf{K}_{2_{ns}}\| << \|\mathbf{K}_2\|$ and $\|\mathbf{B}_{2_{ns}}\| << \|\mathbf{B}_2\|$, $\mathbf{B}_{2_{ns}}$ and $\mathbf{K}_{2_{ns}}$ evoke only small changes of eigenvalues and eigenvectors.

evaluate a mass flow rate, pressure, temperature and flow velocity of a steam at an input and a pressure of steam at an output or in a condenser. For detecting frequency spectrum of the excitation we can also measure the time pressure distributions in front of and behind the blades. Under a service the strains and displacements of blades can be picked up at their foot and free ends, respectively. In our case we chose the first alternative and for the strain measurements of blades we used the radio telemetric equipment. This equipment (Fig.1) for the measuring of stresses in rotors was developed in our Institute (IT AS ČR), manufactured in a collaboration with the Škoda Works and the Research and Experimental Institute of Škoda Plzeň, and was tested on machines manufactured in Škoda Plzeň.

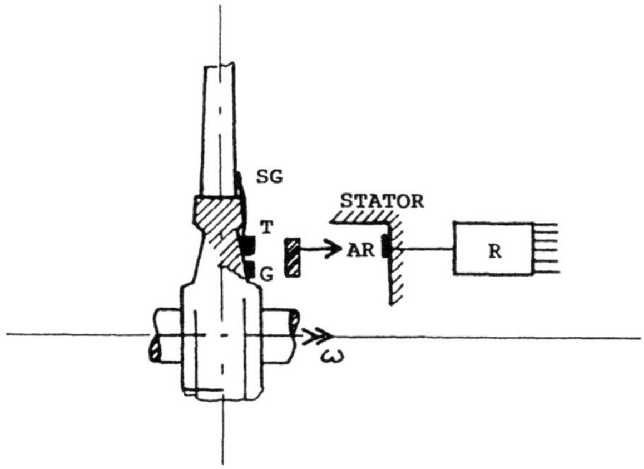

Fig.1 The set up of the radio telemetric equipment on the circulating wheel

The strains of vibrating blades are measured by strain gauges. The signal from strain gauges is amplified by the alternating amplifier and transmitted by the antenna to the other antenna of the receiver. The very small transmitter with the antenna having the dimensions 25x25x4*mm* is fixed on the hub of the wheel. The antenna of the receiver is on the stator of the turbine. The receiver is in a safety distance from the machine.The rotor equipment part is charged from its own generator. The signal from the receiver was observed on the oscilloscope screen, recorded by the measuring tape recorder and evaluated by the two channel analyzer. The quality of these records can be seen in Fig.2, where the comparison between the radio telemetric measurement and the measurement obtained by the slip rings Hottinger SK 12 is shown.

The radio telemetric equipment was tested on the steam turbine 110 MW for the evaluation of the functions $f = f(\psi)$ between resonant frequencies f and revolution frequency ψ in the range $\psi \in 27 \div 54$ Hz. The strains were measured during time interval

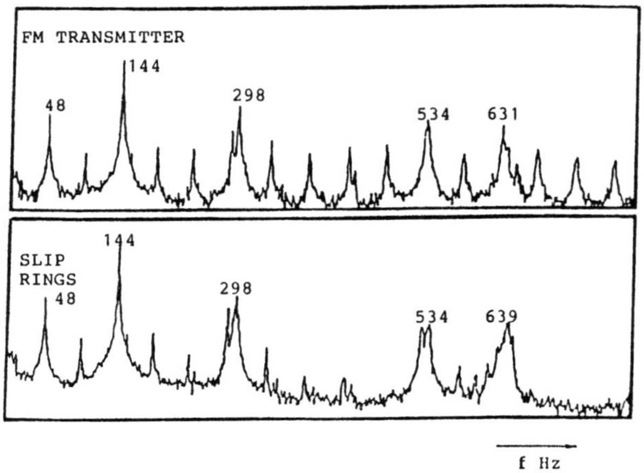

Fig.2 The frequency spectra of the blade vibration obtained by means of the radio telemetric (above) and slip ring (bellow) measurements.

interval 0÷1000ms. These records for two blades which were turned one to the other by 180° and rotated with frequency ψ = 27Hz are presented in Fig.3. It can be seen from this figure that the behaviors of blades distinguish even in a case of their careful selection. These differences are results of a non rotational symmetry of the steam excitation.

The records, above described, were obtained for five different frequencies ψ. The resonant frequencies f evaluated from the harmonic analysis of real time signals are presented in Tab.1.

ψ /Hz/	27	30	32	40	54
f /Hz/	163	164	165	168	175

Tab.1

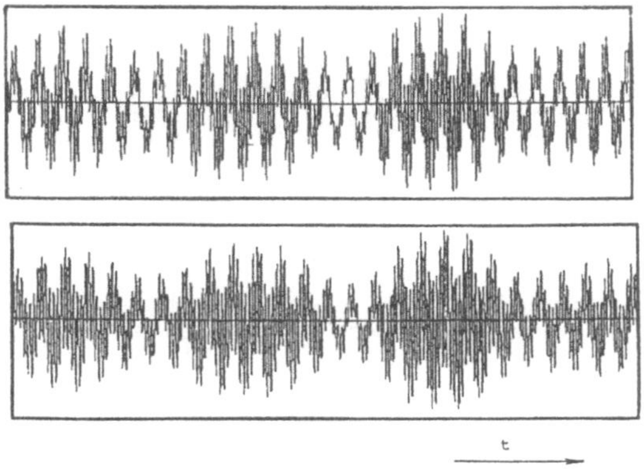

Fig.3 The amplitudes of strain versus a time for two blades turned one to the other by 180° .

As an example the next two records are in Fig.4. The first one represents the instantaneous strain in time interval 0÷500 s and the second corresponds to the amplitude-frequency function in the frequency range $\langle 0,400\rangle$ Hz. The first peak of the latter diagram belongs to the revolution frequency ψ = 54Hz a and it is very likely caused by the unbalanced mass of the rotor; the next two peaks belong to resonant frequencies 175 and 296 Hz.

The coefficients of the function $f(\psi)$, deduced from (1.5) for $\mathbf{K}_2 = \mathbf{0} = \mathbf{B}_2$ [/*] and $f_0 = \frac{|\mathrm{Im}\, s_0|}{2\pi}$ in the form

$$f^2 = f_0^2 + a\varphi^2,$$

where f_0 is the resonant frequency of the non rotating blade, were evaluated by the regression of the data from Tab.1 : $a = 1{,}833, f_0 = 159$ Hz.

The resonant frequency f_0 of the non rotating blade was also measured. We got $f_0 =$ 155Hz, and the small difference (-4Hz) was caused by improper embedding in a case of non rotating blade.

/* The steam flow drove only the turbine without the generator. In this case the small quantity of steam entered in the turbine. Therefore the aerodynamic coupling was weak and the operators $\mathbf{K}_2$,$\mathbf{B}_2$ were negligible.

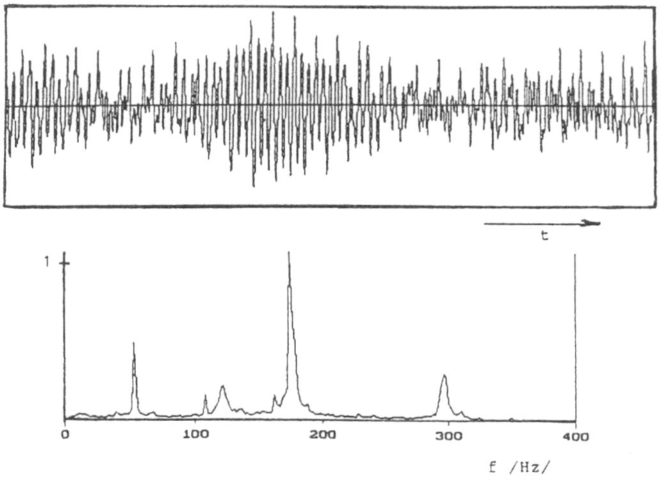

Fig.4 The strain amplitude of the vibrating blade versus a time (above) with the corresponding power spectral density (bellow).

On the other turbine the strains of blades in steam flow were measured by the radio telemetric equipment. At the same time the pressure signals in front of and behind the blades were picked up for detecting the pressure fluctuations. In Fig.5 there are presented the example of this pressure signal behind the blade and the spectrum of the blade strain during this measurement.

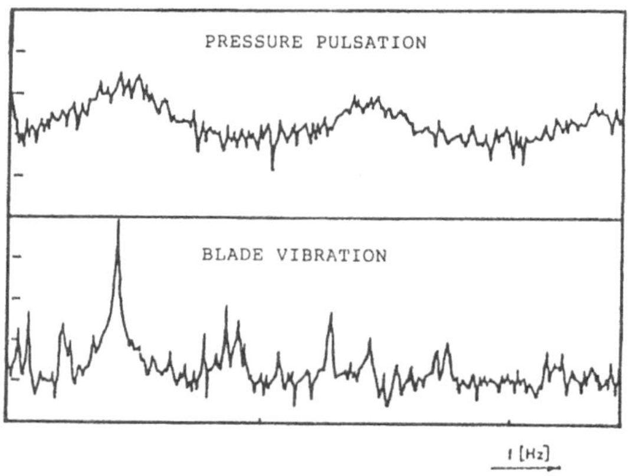

Fig.5 The signal of the pressure behind the blade and the spectrum of blade strain (bellow) during the pressure measurement.

From the last measurement on the turbine 110MW it should be mentioned that the existing radio telemetric equipment has not had a sufficient life. Due to the stream of wet seal steam the disconnection between the strain gauge and the transmitter arose. Therefore our nowadays effort is to increase the life of equipment at least for the registration of the strains in all range of possible service states by improving the shading of the wire connection against the erosion. In futue we plan to complete the measurement of strains with the measurement of displacements as from the simultaneous measurements of the strains and displacements the transfer function "strain - displacement" could be evaluated for more likelihood diagnostics.

ACKNOWLEDGMENT

This theoretical research and evaluation of the measurements was done both under the grant No.27-61-02 funded by the Czech Academy Grant Agency and under the grant reg. No.101/93/1013 of Czech Republic Grant Agency. This source of support is gratefully acknowledged.

REFERENCES

1.Daněk O.: Modal analysis of non conservative continua. Textbook for Short Course on Modal Analysis, Institute of Thermomechanics AS CR, Prague 1993, in Czech.

2.Kopřiva V.: The Measurements of the Blades on the Test Turbine for Aerodynamical Test, Strojirenstvi, 7(1966), p.19-23, in Czech.

3.Roth M.: Vibration Measurement on Turbomachine Rotor Blades with Optical Probes, Int. Fluids Eng. Gas Turbine Conf., New Orleans, Publ. by ASME, New York, p.215-224, March 10-13,1980.

4.VGB report: Non contact Blade Vibration Measurement Technique. VGB Conference Steam Turbines, Nurnberg, 1990.

5.Vaněk F.,Pelc.F.,Cibulka J: Radio Telemetry for Strain Measurements on Rotating Machine Parts, Proceeding 30th International Conference on Experimental Stress Analysis, Czech Technical University, p. 309-313, Prague 1992.

THE HAPPENING POSSIBILITIES OF MAIN VIBRATION FAULTS COMING FROM MORE THAN 160 EXAMPLE CASES IN RMISPP (ROTATING MACHINES IN STEAM POWER PLANTS)

Xu Min
Shanghai Jiao Tong University, Shanghai, China

Chu-Zhen Xu, Jin-Yuan Shi and Heng-Tao Zhang
Shanghai Power Equipment Research Institute, Shanghai, China

ABSTRACTS

In this paper, authors intend to provide "A Real Picture" in RMISPP, "What Kind of Main Vibration Fault Will Happen?" according to the statistics and analysis of more than 160 example cases in the passing two decades by means of 8 statistic – analysis tables, just for giving readers an integrated consideration when they carry out their design work about the on-line diagnostic system, the arrangement of transducers or monitors, the focus on the critical condition and the real time date processing, etc. as the original basis of monitoring and diagnosis.

* This work is supported by
SPC 8 – 5 research project

1. Introduction

It is quite difficult for a senior engineer to prove surely for a certain prototype machine in a power plant that some vibration faults will certainly happen or some vibration faullts will not happen. Because of the many factors, such as: the reliabilities of all elements, every parts of manufacturig process, the quality of installation, the circumstance of machine foundation, the level of maintenance, and so on, even though the machine by using the same drawings, coming from the same manufactory, keeping the same quality of installaton, and so on and so forth. Theoratical inference is one thing, but the real happening is another. A great amount of statistic figures and many cases of a wide statistics give a more strong basis for us to control the real natures of machine faults and to make a more objective decision.

2. The backgroud of statistics

The data sources of statistic cases here come from several ways, such as 20 periodicals in English, 30 periodicals in Chinese, 3 books in Chinese, 6 proceedings and international computer index as well.

The time duration of statistic cases covers mainly in 70s and 80s, just a few in 90s.

All the statistic cases are picked up from the cases of vibration faults, excluding those which are nothing to do with vibration faults directly, such as the faults of blade break caused by material defect, the faults of main shaft caused by initial bending, the faults of rotor corrosion crack, the faults of electric insulating material destroy, the faults of stator winding overheated, etc..

There are more than 160 example cases of vibration faults collected incompletely in this paper with explanation of tables as follows.

3. The main explanatory tables of statistics

A. Table No.1 Several Remarkable Events of Machine Unit Serious Damage.

B. Table No.2 The Natrues of vibration Faults.

C. Table No.3 The Damage Formats of Vibration Faults.

D. Table No.4 The Happening Location of Vibration Faults.

E. Table No.5 The Damage Events Caused by Torsional Vibration.

F. Table No.6 The Low Frequency Steam Oscillation.

G. Table No.7 The Fault Events Caused by Oil Film Oscillation.

H. Table No.8 The Fault Events Caused by Friction Vibration.

Several Remarkable Events of Machine Unit Serious Damage

Table No.1

	Power plant	Power (MW)	Date of Event	Cause	Date of Production
(1)	MU. No.3, South Sea (Japan)	600	June 5, 72	Vibration (Friction v.)	April, 72
(2)	MU. No. 6, Bosvl. (France)	600	July 22, 77	Vibration (Unbalance v.)	March, 73
(3)	MU. No. 2, D. T (China)	200	Oct. 29, 85	Vibration (OFO+TV)	Dec., 84
(4)	MU. No. 1, Nuc. Ener. (Taiwan)	900	July 7, 85	Vibration (Torsional v.)	Begin., 83
(5)	MU. No. 5, Q. L. (China)	200	Feb. 12, 88	Vibration (OFO+Fric. v.)	Febru., 6
(6)	MU. No. 2, Gallatin (USA)	225	June 19, 74	Quality of Material	—
(7)	Nuc. Ener., X Isl. (USA)	—	March, 79	Operation Mistake	—
(8)	MU. No. 1, Rowe (USA)	185	Feb. 4, 80	Corrosion Crack	Nov., 60
(9)	Irsching (Germany)	330	Dec. 31. 87	Forging Defect	1972
(10)	Soviet Union	300	1984	Vibration (Torsional v.)	—
(11)	Nuc. Ener., Chel. (Soviet Union)	—	April, 86	Vibration	—

(MU=Machine Unit, OFO=Oil Film Oscillation, TV=Torsinal Vibration.)

The Natures of Vibration Faults

Table No.2

	Fault Nature Classfication	Case Number	%
(1)	Torsional Vibration	36	21.8%
(2)	Steam Oscillation	28	17.0%
(3)	Oil Film Oscillation	23	14.8%
(4)	Foundation Vibration	20	12.1%
(5)	Unbalance Vibration	16	9.7%
(6)	Friction Vibration	12	7.3%
(7)	Vibration Caused by Standard Height Level Change of SCL	8	4.9%
(8)	Shaft Series Vibration	7	4.2%
(9)	Electromagnetic Vibration	6	3.6%
(10)	Shaft Vibration Caused by Bad Coupling Connection	4	2.4%
(11)	Beat Vibration	2	1.2%
(12)	Unclear Vibration	3	1.8%
	Total=	165	100.0%

Notes: Steam oscillation (2), oil film oscillation (3), friction vibration (6) are self-excited vibration. Others are forced vibration. Forced vibration case number 101 (61.2%). Self-excited vibration case numder 64 (38.8%).

The Damage Formats of Vibration Faults

Table No.3

Damage Format	Case Number	%
(A) Steam Turbine Damage	5	6.0%
(1) Steam Turbine Rotor Crack	30	35.7%
(2) Coupling Screw Damage	10	11.9%
(3) Steam Turbine Rotor Break	9	10.7%
(4) Steam Turbine Blade Break	8	9.5%
(5) Steam Turbine Disc Break	7	8.3%
(6) Bearing Damage	7	8.3%
(7) Steam Turbine Shaft Bending	4	4.8%
(8) Gland Friction Damage	3	3.6%
(9) Coupling Gear Connecter Damage	1	1.2%
Amount to	84	100.0%
(B) Electric Generator Damage	2	5.0%
(1) Electric Generator Rotor Iron Core Damage	8	20.2%
(2) Electric Generator Rotor Break	7	17.5%
(3) Protactive Ring Break	6	15.0%
(4) Coupling Screw Damage	5	12.5%
(5) Bearing Damage	5	12.5%
(6) Coupling Gear Connecter	3	7.5%
(7) Spring Ring Damage	3	7.5%
(8) Electric Generator Rotor Groove Wedge Burn-out	1	2.5%
Amount to	40	100.0%

The Happening Locations of Vibration Faults

Table No.4

Happening Location	Case Number	%
(1) Steam Turbine	81	49.1%
(2) Electric Generator	49	29.7%
(3) Magnetic Exciter	15	9.1%
(4) Foundation	14	8.5%
(5) Machine Unit	6	3.6%
Total=	165	100.0%

The Damage Events Caused by Torsional Vibration

Table No.5

Country	Power Plant	Power (MW)	Date of Event	Cause and Damage Level
(1) USA	M#2, Mohavc	790	Dec., 70	29. 5Hz Subsynchro-Harmo Vibration. Gen. Sha. Break
	M#6, Cleco Coughlin	300	1979	120Hz 2x rpm Harmonic v. Blade Break
(2) S. U.	—	500	—	33.7, 33.9 Subsynchro – H. Vibration, Shaft Break
	—	800	—	24 Hz Harmonic v., Exciter system, Shaft Break, G – E
(3) U. K.	Middle Area	500	1975	Electric Current Shock Generator Crack
	Several MUs	—	—	Misparallel
(4) Japan	Stone – Rever Island	1.65	1980	Electric Network Trouble Shaft Break
	Sakilo Steel Factory	—	—	2x rpm Vibration Generator Rotor Break
(5) Canada	Nuc. Ener. MU	500	Mar., 83	Electo – Liquid Governing Vibration
(6) Germany	N. P. P.	630	Oct., 83	Misparallel Generator Rotor Crack
(7) China	M#2, Y. M. PP	300	Mar., 85	One Phase Charge Generator Rotor Crack.
	M#6, M. D. PP	200	Feb., 85	Quik Control Testing, Turbine Rotor Screw Br.
	M#8, Q. H. PP	200	Oct., 86	Quik Control Testing, Generator Screw Break

The Low Frquency Steam Oscillation

Table No.6 (a)

Power Plant	Power (MW)	Survey
(1) M#1, N. L. (SU)	300	No.1, 2, 3 bearings appear low freq. v., when loading more than 180 MW
(2) M#1, K. D. S. (SU)	300	26 — 27 low frequency oscillation, when loading up to 220 MW
(3) M#7, K. D. S. (SU)	300	26 — 27 Hz low frequency oscillation when loading 290 MW
(4) B. B. C. (Germany)	300	24 — 30 Hz low frequency serious osc. when loading 240 MW (H – P rotor critical frequency = 28 Hz)

The Low Frquency Steam Oscillation

Table No.6 (b)

Power Plant	Power (MW)	Survey
(5) S. M. C. (Germany)	400	Low frequency vibration, f=H−P rot. first critical freq. Amplitude =0.2—0.3 mm
(6) O. N. F. (Germany)	400	Low frequency vibration, f=H−P rot. first critical freq. 32 Hz
(7) M#2 J. Z. (China)	200	26—28 Hz serious vibration, when loading 150 MW. But no vibration when loading down to 110 MW

The Fault Events Caused by Oil Film Oscillation

Table No.7

Power Plant	Power (MW)	Survey
(1) M#12 H. D.	300	n=2500−2600 rpm OFO Fo=15 Hz (1974)
(2) M#1, Y. M.	300	n=3000 rpm OFO Generator bearing (1978)
(3) M#6, X. Z.	200	n=3200 rpm OFO Generator bearing No.6 (1986)
(4) M#2, T. L.	200	n=3000 rpm OFO Fo=18.8 Hz (1985)

The Fault Events Caused by Friction Vibration

Table No.8

Power Plant	Power (MW)	Date of Event	Survey
(1) M#3, Sou. Sea (JAPAN)	600	June 5, 72	Non−linear unsteady fric. vibration, unit destroied
(2) M#1, D. K. (China)	320	Oct. 10, 78	Expand limited, diametrical gap too small, v. increase
(3) M#4, X. S. (China)	125	April, 80	Isolator/blade-disc fric. isolator damage, shaft bending.
(4) M#9, M. H. (China)	125	March, 81	Blades friction, rotor shaft bending

4. The Causal Analyses of Some Main Vibration

4.1 The Causal Analyses of Torsional Vibration

The torsional vibration problem now is the key problem in handling and solving the monitoring and diagnosis technical problems of a huge power transmission network connected with many high power machine units.

The investigation of shaft torsional vibration essentially is to make sure the interaction and the disturbance between the huge transmission network and the many machine units.

The network disturbances are involved in:

(a) Electric faults – single phase short circuit fault, two or three phase short circuit fault, etc;

(b) Switch operation faults – quick switching – off, double swiching – on, twice switching – off, etc.;

(c) Abnormal working condition – losesynchro operation, series capacitance compensation system, quick close steam valve, mistake parallel, quick unloading, ect.

When network disturbances happen, a step torque or a transient torque with wide frequency range shall act on the generator rotor and rotating shaft over. The network – machine (rotationg shaft) couple vibration exists in three types of critical harmonic vibrations. They are subsynchro harmonic vibration, supersynchro harmonic vibration and oscilatory torque shock.

To adjust the critical resonant frequencies, to increase the structure damping of shaft system, to limit the disturbing forces and to set some protective devices are necessary.

4.2 The Causal Analyses of Low Frequency Vibration

Bearing oil film oscillation and gap steam oscillation belong to the category of low frequency vibration. Most of these low frequency vibrations are nothing to do with the mass unbalance of shaft system rotors, but something to do with the steam flow condition and lublicant oil film dynamic condition of bearing – rotor – steam flow system.

Many cases in statistics of oil oscillation and steam oscillation show that the happening and expanding of these vibrations very often lead to a catastrophe event.

The oil film oscillation is closely concerned with oil pressure, oil temperature, whirling frequency and bearing structrue, and derictly reflects on shaft speed. Correctly to desgin the bearing structure, reasonablely to control the bearing operation parameters, accurately to calcurate the shaft steady speed and the spare level of it are quite necessary.

The gap steam oscillation is involved in the density of steam flow, the gravity quantity of steam flow, steam passage – way structure, etc.. The high pressure turbine rotor is quite sensitive to the steam oscillation. Several cases in statistics of steam oscillatin tell us that when the loading of steam power in high pressure turbine is decreased, in other words, in lower loading operation or lowering power parameters operation, the phenomena of low frequency vibration will soon release correspondingly.

4.3 The Causal Analyses of Foundation Vibration, Unbalance Vibration and Friction Vibration.

From the statistic cases, the happening possibility of the foundation vibration is mostly connected with couple vibration between the shaft disturbing frequency and the natural frequency of froundation frame structure, unstraight of shfaft line or the change of standard level of bearing height, unit base distortion caused by the differances of temperature, etc.. In the stages of designing, establishing, mounting and operating maintenance, more attention to above problems must be paid.

Mainly, there are three kinds of unbalance vibrations involving in shaft rotors. They are mass unbalance of turbine rotor, generator rotor and exciter rotor, heat unbalance and unstable unbalance. The causes of them are seperated from each other. So, the handling of them must be one by one carefully.

The friction vibration, or say touching-friction vibration, is mostly caused by certain conditions, such as: the dry friction between journal and bearing metal, the touching-friction of running rotor and isolating plate, etc. The original causes of these cases are the changes of rotor bearing led to the lowering of rotor critical speed, the increase of rotor bending, the increase of isolating plate distortion, etc. The features of this friction vibration have those of nonlinear, unstable and quick happening phenomena. Some mechanism of these vibrations are still unclear.

5. Conclusion

The statistics from more than 160 example cases in order to understanding the happening possibilities of main vibration faults is not so complete yet, particulary just a few cases up to date, but a brief integrated idea or "A Real Picture" of main vibration faults happened in RMISPP could be fairly taken.

It is difficult to describe the full story for a special case of vibration fault one after another in this short paper, but you could get an order of the happening possibilities in statistics in the area of vibration faults, i.e. the first is torsional vibration, the second is steam oscillation, the third is oil film oscillation, the fourth is foundation vibration, etc., meanwhile you could get the key points of the causal analyses of some major vibration faults in RMISPP.

REFERENCES

[1] Chu-Zhen Xu, Jin-Yuan Shi, Others,"Analyses of Important Events of Shaft Faults in High

Power Plant Units, Aborad and Domestic", Research Report, Shanghai Power Epuipment Research Institute, Oct, 1989.
[2] Wei-Xin Shi, "The Vibration of Steam Power Generator Unit", Book, By Water Conservancy and Electric Power Press, 1991.
[3] Zheng-Ya He, Zhao-Yi Wang, "The Selected Editor of Diagnosis Cases of Machinery Faults", Book, Xian Jiao Tong University Press, 1991.
[4] Proceedings of IMMDC '89, '90, '90.
[5] Proceedings of NC-M&EFD '86, '88, '91.

THE VIBROVIEW SYSTEM
A MODERN CONCEPT OF CONDITION-BASED VIBRATION MONITORING OF BIG TURBOGENERATORS

F. Herz

ABB Power Generation Ltd., Birr, Switzerland

1.Abstract.

The investigation of the mechanical vibrations is an excellent method to indicate the current condition of a machine, whereby the vibrations must not be necessarily at high levels. Even within the normal intensity ranges , they are an indicator of the machine condition because they describe the dynamic forces acting on the considered system.(machine frame, foundation, building, etc.). This paper deals with vibration monitoring at large turbogenerators.
At first,general aspects of monitoring are covered, diagnostic strategies and impacts on the plant operation are discussed.Then the impacts of a modern monitoring system to the maintenance strategie of a power plant are described and discussed.
This is followed by the configuration of a modern monitoring system. As an example the description of our condition-based monitoring system "VibroView" is taken, which will be presently implemented in our 1000 - MW nuclear plant "Leibstadt". An explanation of our expert system "ViDiamon" is finally given, which is either a stand alone - System or an option to the "VibroView" - System..The "ViDiamon" - System is a software tool ,in order to diagnose the most common faults of turbine or generator out of the vibration signature. This tool had been also developped by engineers of ABB. Presently 15 fault cases are implemented (such as shaft cracks, loose stator frame, pedestal resonance etc.), but more of them will be added, according to our experience.

2.General aspects of vibration monitoring.

The following list of common machine faults (Fig.1) shows a selection of possible dedecting parameters.

Parameter / Machine Fault	Tempe-rature	Pres-sure	Flow	Oil analysis	Vibra-tion
Out of Balance					×
Misalignment/ Bent Shaft	×				×
Damaged Rolling Elem. Bearings	×			×	×
Damaged Journal Bearings	×	×	×	×	×
Damaged or worn Gears				×	×
Mechanical Looseness					×

Fig.1

It is obvious, that the vibration behaviour is the most sensitive parameter ,covering the entire fault spectrum. Oil analysis or lube oil temperature measurement for example, are only covering a fraction of the fault possibilities.So there is a demand for the continous registration of vibration levels,this leads to a monitoring system.
The vibration monitoring in today`s power plants is mostly limited to the recording and announcement of vibration values (amplitudes). This applies to almost every type of machinery. (gas turbines, steam turbines, hydro units).If predetermined limits are exeeded, and alarm - or trip - signal is triggered.
This philosophy ensures a sufficient protection in the most cases, but it does not permit an analysis of vibration events, diagnosis of the machine condition, or predictive maintenance.
A modern monitoring system surpasses these basic functions and permits a continous recognition of machine conditions with an associated evaluation enabling diagnostic statements.
The basic aim of a modern monitoring system is the increase of the economic efficiency of a plant by means of a condition based maintenance planning.(Fig.2).

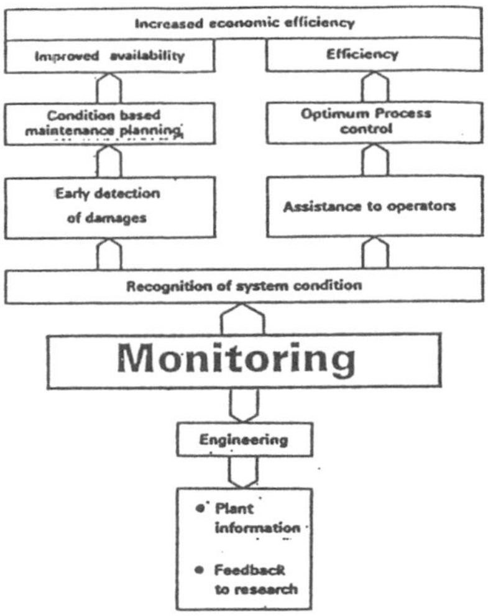

Fig.2

In order to recognize changements of conditions vibrations have to be registered continously far below of the limit value and also evaluated.(Fig.3).

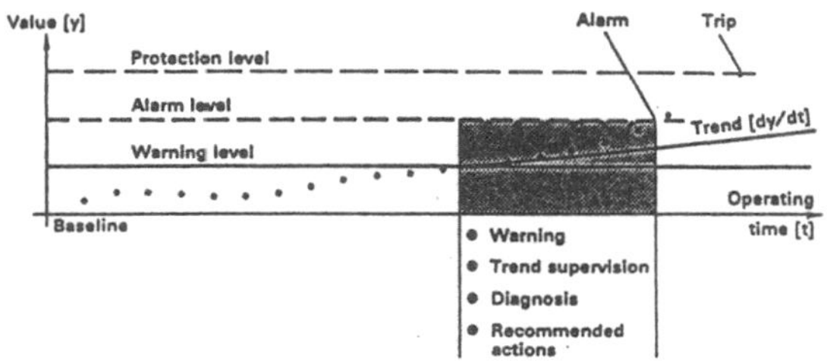

Fig.3

A modern maintenance strategy does not require revisions based on a fixed time scedule, the overhaul planning will be sceduled according to the necessity in regard to the condition of the monitored machine (Fig.4).

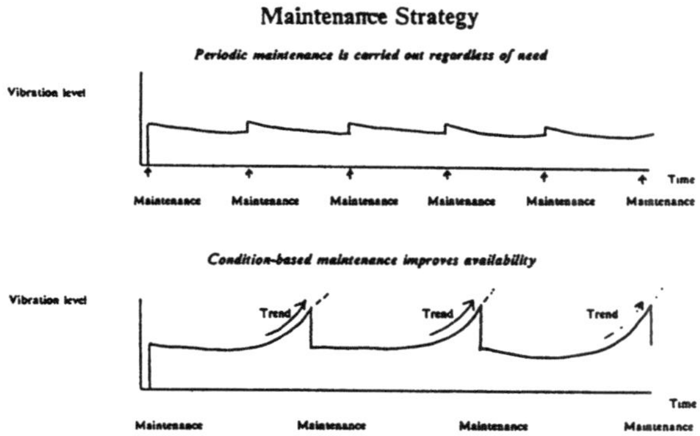

Fig.4

Modern monitoring systems are also going a step further than monitoring the absolute trend values.When there is a "normal" increase of the vibration level by known parameter influences (excitation current, active load etc.)the deterioration (increase of increase,Fig.5) is monitored. A malfunction is then indicated, when the known (or predicted) increase is superimposed by something unknown.

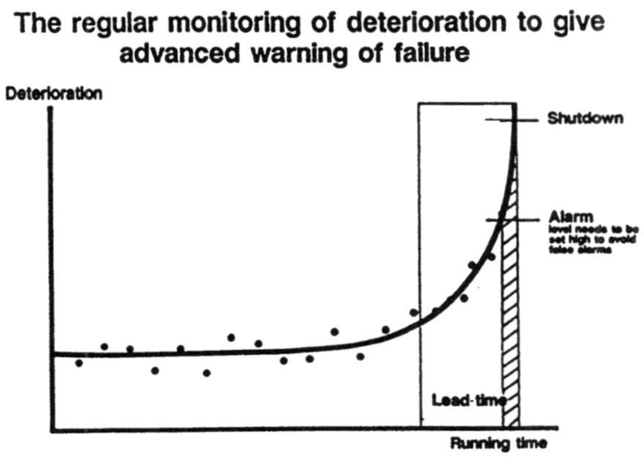

Fig.5

This can be done for the overall vibrations as well as for spectral components (Fig.6).

Change Trend Instead of Absolute Trend

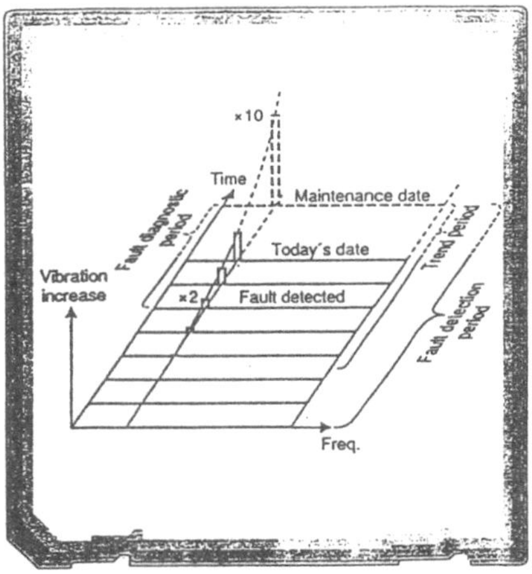

Fig.6

In this case, it is not enough to have a level alarm, considering only the overall value, the alarm philosophy has to be selective in regard to the various frequency components, as shown in Fig.7

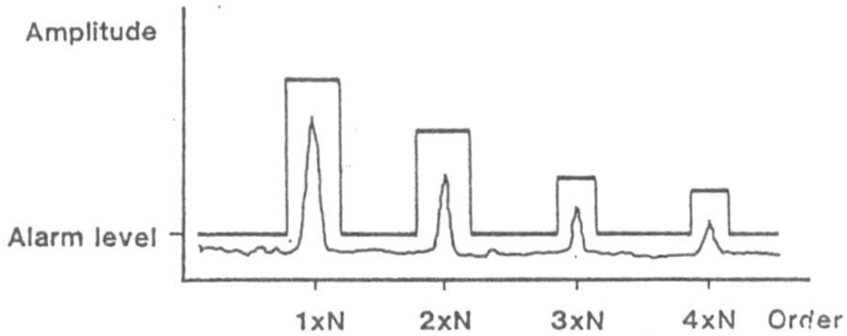

Fig.7

A main question also is : how high is the probability to detect a fault, or how capeable is a monitoring system ?
THe following Fig.8 shows a paper issued at an American Power Conference.

CAPABILITIES OF A VIBRATION DIAGNOSTICS AND MAINTENANCE PREDICTION SYSTEM*

Fault	Detectibility Using Vibration	Can Source Be Located?	Can Time to Failure Be Predicted?
Unbalance	100	75	75
Misalignment	90	90	65
Instability	100	75	75
Damaged Bearings	75	75	50
Elec. Air Gap Misalignment	80	80	50
Seal Rubs	60	30	25
Loose Rotor Parts	70	50	50
Gear Damage	80	80	30
Pump Cavitation	50	100	20
Ecc. Blades or Nozzles	80	80	30
Bent Rotors	80	80	60
Structural Resonance	60	30	20
Pedestal Bolts Loose	70	60	30
Piping Misalignment	30	20	20
Cracked Rotors	50	50	20
Blade Fouling	30	25	25
Thrust Bearing Damage	70	60	40
Torsional Resonance	65	45	40
Hydraulic Pulsations	70	50	30

*Numbers are the percentage of the number of instrumented machines where detecting, locating and predicting time to failure of a fault is possible.

Fig.8

Here the percentage of detectability, locateability and predictability of machine faults by the vibration behaviour is shown. According to this investigation, a monitoring system has quite a good chance of predicting failures.

3.The VibroView - system.

ABB has a long tradition in computerized vibration measuring techniques.
Since over 10 years ASEA-Stal designs and builds vibration monitoring systems (TVM 10 - TVM 300) and BBC introduced the first computer aided data aqcuisition system (CADA) 1971 at the comissioning of "Cumberland".
Out of this vast experience the VibroView system had been developed at ABB-Stal in Finspong under the partizipation of two software engineers from ABB-Birr.

3.1.General aspects of the system.

The VibroView-system is built specially for bigger power plants and can handle up to 16 independently rotating shafts in its basic version.It is based on the most recent technology. We gave great importance to standards, portability, modularity and network capability.
The former generations of vibration monitoring systems had been designed as "closed systems",where the hardware components had not been interchangeable and the software could only be operated on one particular system.The fast development of the computer technology makes it necessary to built an design "open systems" where the software becomes independent of the hardware,the components are exchangeable and the communication between different computers is possible.
The following principles had been realized in the new system:

-Modularity:
It is simple, not only to add, but also to remove harware or software components.

-Portability:
The software is independent from the hardware configuration, it is "portable" from one type of workstation to the other.
The system is runable from different workstations.

-Costumized functionality:
The system can be customized in regard to its size and in regard to the depht of analysis and evaluation to the needs of the customer and the requirements of the plant.

-Long lifetime expectation and standardisation:
The application of common standards, an "open" system architecture ensures a longlasting comatibility of hard-and software components.

Besides of the official standardization organisations like ISO or ANSI there are a number of cooperations of large manufacturers which are active in standardizing open computer systems.The ABB VibroView is based on one of these systems called OSF/Motif.
The following external standards had been applied:
For communication:

- -Ethernet - TCP/IP
- -RS 232
- -GPIB

For the software:

- -Unix(Ultrix32)
- -Ansi C
- -X-Windows
- -OSF/Motif
- -SQL Database
- -Postscript

3.2.The main components of the ABB VibroView

Vibro View is built with the following parts:

The **front end FE** acquires the raw signals from almost any type of analog instrumentation, extract the desired measurements and stores the data into a shared memory.

The **workstation MMC** (man machine communication) catches the data from the front end , checks alarm exeedence, stores the data into a data base and, if required, shows the desired diagrams on the monitor.

A **communication system** (Ethernet,RS 232) controls the dataflow between the several components.
Figure 9 shows the setup of the system.

System layout

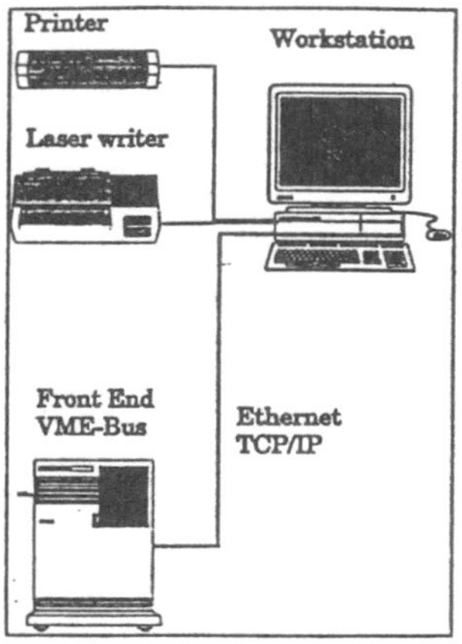

Fig.9

A larger system can contain more than one unit of each kind.One MMC unit can be located in the control room, another MMC unit can be installed at the head office. Several FE's can be positioned at the different machine sites.The minimal system consists of one FE and one MMC.

3.2.1.The MMC - computer.

The MMC-computer is a capeable standard workstation with a fast processor and a satisfactory grafical user surface. The software is independent from the hardware type,the hardware manufacturer can be chosen freely.(e.g.DEC,Sun,IBM Risc,HP/Apollo).
The MMC computer is the interface between the user and the system.From here the user can chose and activate all informations and functions, most of the data will be displayed grafically on the screen.
The man machine communication is based on the grafical user envioroment "X-Windows", which is comparable to "Windows" of Microsoft.The keybord as a communication device is almost completely replaced by a mouse,whereby the user activates the desired functions by moving a pointer and clicking on an icon or symbol.The following Fig.10 shows a typical screen display of the system:

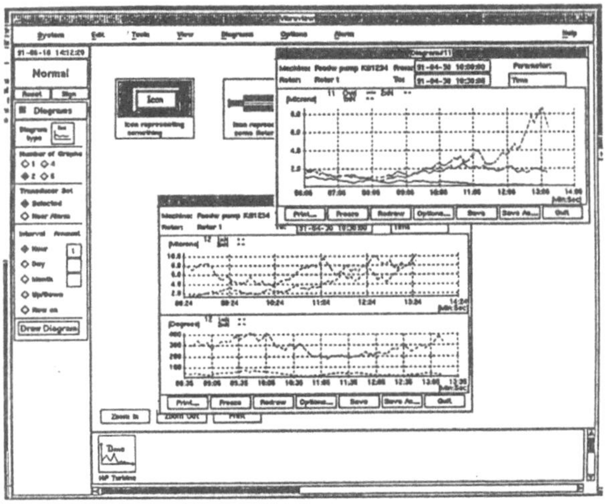

Fig.10

In the upper part there are the pull down menues in order to control the system.A selection of the most essential functions is displayed in the vertical window called "toolbox".The status display just above the toolbox indicates the date .time,and alarm status.The large main window in the center of the display is the "working area" of the user, where he can move,zoom or overlay the different displays.The lower part of the window contains minimized important graphs as icons in a "stand by"- mode, which can be brought to full size at any time.There is also a message box in order to enable the system giving messages to the user.

The MMC computer also takes care of the data storage on a secondary mass storage.
On the MMC- computer also the alarm handling is carried out.Every alarm will be displayed on the screen,listed on a printer resp.indicated on a potentialfree relais or put trough the communication ports to other workstations.
The MMC-computer governs the dataflow from the FE-computer.After having recieved these data basically the following sequenze will be performed:

- Check of the alarm limits,if the alarmstatus had been reached,triggering of the alarm.
- Display of this information on the screen or another peripheral unit.
- Storing of the data on a secundary mass storage unit.
- Transfering the data to the governing control system of the plant as an input for other computers.

3.2.2.The FE(Front End)-data acquisition system.

The front end unit FE is an independent computer with a fast microprozessor (Motorola 68040 + Coprocessor68882),which is primarily designed for acquiring data.A real time multitasking operating system is used,in order to perform evaluation, acquisition and communication simultaneously.The FE is equipped with a VME-bus system in order to accept standard hardware and it is configured with :

128 AC channel inputs for vibrations
128 DC channel inputs for parameters

16 Key phasor inputs for speeds of different shafts
There are different AD-converters for vibration - (AC) and parameter signals (DC).
The FE unit has four basic functions:

- Data acquisition,A to D conversion,scaling and intermediate storage.
- Fast Fourier Transform of data by means of the digital signal processor.
- Receiving and storage of the information coming from the MMC-computer (measuring sequenze,sampling frequency,number of averages,sensitivity etc.).
- Transfer of the measured and scaled data to the MMC-computer.

3.2.3.The communication system.

There are different interfaces provided for communication:

- Ethernet and TCP/IP LAN between MMC-computers and other workstations resp.between the MMC-computer and FE-computer.
- RS 232 between MMC-computer and other computers via modem.

3.3.Functions of the VibroView system.

The system is capeable of measuring vibrations as AC-signals,machine parameters as DC-signals and speeds as impulses per revolution.There are basically three categories of system functions:

- measuring and analysing
- alarm indication and message
- evaluation

There is an additional option of adding a diagnosis software module (ViDiamon) which is capeable of pinpointing the machine fault, using the information of the VibroView (15 fault cases are included).
If desired also a balance software module can be added,in order to establish influence coefficient data bases of large machines with a big number of balancing planes and more than one balancing speed.This software module uses the influence coefficient method and the least mean square approximation.
Both modules provide an automatic data transer from the VibroView.

3.3.1.Measuring and analysing.

Vibration measurements can be performed by time based or speed based sampling.
The time based sampling uses frequencies between 100 and 20000 Hz, the speed based sampling is carried out by orders of the speed (16x,32x,64x).
The following measurements are possible:

- Fast Fourier Transform FFT
- RMS from 0 to max.20 kHz
- Harmonic components: 1,2,3,4x,speed,amplitude+phase
- Overall level
- Orbit of fundamental and harmonics

3.3.2.Alarm functions

There are two alarm niveaus: WARNING and ALARM. A slightly delaying "hysteresis funtion" around the triggering value avoids "flickering" and unnecessary alarm indication.
The alarm functions are divided into PRIMARY and SECONDARY alarms.The primary alarms will be checked every 10 seconds, the secondary alarms are checked every 5 minutes by the system.
The most important primary alarms are:

- Niveau alarm,amplitudes of overall niveau at 1,2,3,4x of speed and at processparameters.
- Circle alarm at 1 and 2x of speed for the vector. The following Fig.11 shows the principle of the circle alarm.This alarm type is sensitive for vibration events which are characterized by amplitude and phase changes(e.g.lateral cracks,thermal unbalances,rotor rubs etc.)

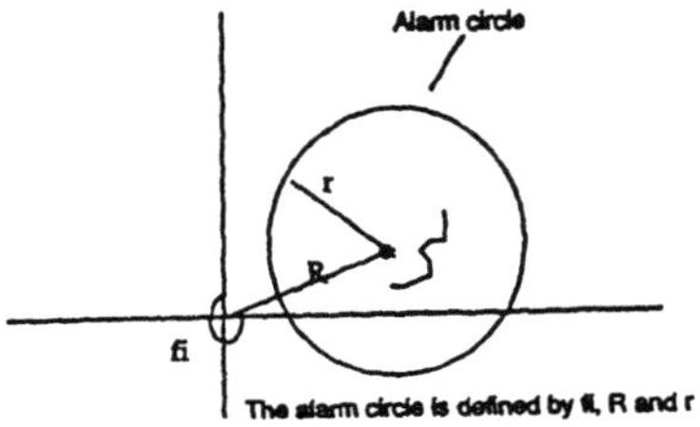

Fig.11

The most important secondary alarms are the different trend alarms.The trends of the vibration vectors will be calculated out of the already measured values by means of a linear regression.There are basically three types of trend alarms:

- fast trend which take into account the measured values of the last hour.
- slow trend which take into account the measured values of the last week.
- jump alarm which is an indication of a very fast(seconds or less) vibration changement. A jump e.g. occurs at the loss of a turbine blade or if a bearing failure occurs.

The different alarms are not only indicated at the screen of the MMC-computer,they will also appear,outside of the monitoring system,at relais-outputs (which are integrated in the system) . In order to set the different alarms, the system provides software tools,whereby statistical and empirical methods are applied.

3.3.3.Evaluation.

The interpretation of measuring results and therefore a fast approach to a possible problem is very much dependant from the possibilities to document the measured results.The VibroView system

has a great variety of diagrams,which can be configured directly on the screen, regardless wether they are consisting of rotors,transducers or components.The scaling according to the size can be done automatically or by hand.Reference curves or alarm niveus resp.values can be easily integrated into the graphs.
Most of the diagrams can be shown on the screen in a static or a continously updated mode.
The following diagram types are available:

- Bode diagram,amplitude+phase or process parameters will be shown as a function of time,speed,or other process parameters.
- Polar diagram,displays fundamental or harmonic components.
- Cascade diagram,also for fundamental and harmonic components.
- Trend diagram,displays trends of transducers or components.
- Spectrum diagram.
- Spectral map (over time or speed),a number of spectra at sequential times or speeds(max.50).
- Orbit diagram,the movement of the shaft center causing the vibrations.
- Diagrams configured by the user.

3.3.4.Diagnosis functions.

By means of the sophisticated alarm philosophy and the various alarm functions a sort of rough "prediagnosis" is already possible,according to the kind of alarm which triggered.
As already mentioned, it is possible to integrate (as an option) the diognose module "ViDiamon" as an expert system ,into the VibroView.
If an alarm occurs at the VibroView,the ViDiamon becomes automatically active, providing an automatic datatransfer from the VibroView.
In order to perform a diagnose ,ViDiamon uses vibration data,information about actions during overhaul,information related to fault cases occured in the past,process data,general data about the machine and manual(keyed in-) information.
Mostly there are given a certain number of fault possibilities,in the squence of their probability.The user will also get an explanation what information the diagnose was based upon.
ViDiamon uses two different knowledge sources,at first a general data base,which contains the experience of our vibration experts,regarding the different vibration phenomena.Secondly,there is a dedicated knowledge base, which contains elder fault cases occured at this machine or at those of the same type.
ViDiamon uses state of the art technology as well as common standards: UNIX as the operating system,X-WINDOWS as the graphical user enviorement,C as the programming language and an SQL relational database.
ViDiamon was developed by ABB-Stal in Finspong with a very close cooperation of the three largest nuclear power plants of Sweden:
BARSEBÄK,RINGHALS and OSKARSHAMN.
At the present time there are 15 fault cases implemented,but this list will be continously updated.
The following Fig.12 shows the schematic layout of the ViDiamon system.

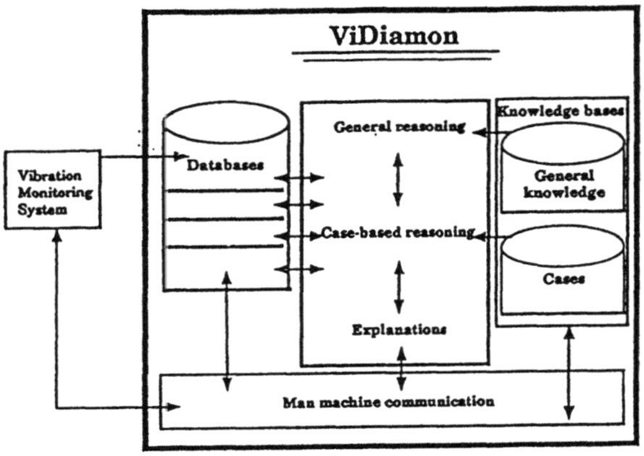

Fig.12

4.Conlusion.

The VibroView is a modern vibration monitoring system,which can handle up to 16 shafts simultaniously in its basic version.The aim of the vibration monitoring is to get higher reliability due to more information about the mechanical operating behaviour of the monitored machine.
This goal will be achived by recording vibrations and process parameter values of important elements of the unit with a modern measuring equipment.
Comparison to ealier measurements (trend) enables the operator to become familiar with a operating history of the machine and is taken as a basis to predict the vibration of the coming time period.According to this extensive information about the operating behaviour , the condition of the machine will be judged.This knowledge enables the customer to reduce the outage rate and to perform an efficient maintenace planning.

THE DIAGNOSTIC TECHNOLOGIES IN POWER PLANTS IN JAPAN

M. Tanaka
The University of Tokyo, Tokyo, Japan

Abstract
A state-of-the-art report is presented on vibration diagnostic & operation support systems for electric power plants in Japan. Such systems emerged from proof test phases in Japan late 80's and are now in practical use in some power plants. This paper presents a short history, basic and advanced functions, system configurations and newly emerging problems of the systems.

1. Introduction

Modern power plants use various rotating machinery of high performance, such as steam turbines, gas turbines, generators, pumps and fans. These machines have inherently numerous vibration problems. Power plants have been haunted by threats of unexpected shutdowns due to vibration problems of rotating machinery.

Operating conditions were continuously heightening severity, threatening the service reliability of power plants in spite of the achievements in the field of rotordynamics research. This accelerated the developement and installation of vibration monitoring and alarm systems which are now conventional standard equipment for power plants. The systems constantly monitor vibration levels of various parts of machines and give an alarm when a preset limit of level is exceeded. If the limit is low enough, the systems can buy precious time for operators to take corrective measures in order to avoid forced shutdown, and could serve to maintain reliability levels.

Vibration amplitude level is not only a preceding index of service reliability of rotating machinery but an influential factor to decide the reliability itself. This paper outlines the current vibration diagnostic and operation support systems for rotating machinery used in power plants in Japan.

2. Alarm Guidance Systems and Vibration Diagnostic & Operation Support Systems

With the increase in shaft speed and flow rate of machines, and also with the scale-up of power plants, monitoring and alarm systems were regarded as insufficient to keep high reliability levels. When an alarm window started blinking on the panel in the central control room, operators had to page through operation manuals and follow previously provided action procedure corresponding to the blinking window. Normally they assumed the cause of what triggered off the alarm, considering all the factors after checking the relevant status of the plant, and decided which corrective measures to take. However the relationship of cause and effect in a power plant is not always simple and clear enough to choose a corrective measure easily, and even experienced operators took not a little time to check all the necessary data and reach a conclusive decision. Suppose multiple alarm windows blink one after another, human brains can not process the vast amount of incoming information at one time and may make a serious mistake. As a matter of fact, more than 200 or 400 alarm windows for one unit of turboset are found on control panels in modern power plants.

For this reason, alarm guidance systems were developed in late 80's. Such a system of the first generation could only show operators the preiviously provided list of actions in operation manuals on a CRT display, reducing the work load of operators' paging manuals. The systems of further generations can identify, automatically or sometimes partly with the assistance of operators, the cause of the alarm and show an appropriate measure to take on a CRT display, which can reduce the work load of operators significantly. When multiple alarms are given at one time, the system sorts out priorities according to the rules in the system so that operators can tackle the most urgent alarm efficiently and smoothly.

Some vibrations grow to a large amplitude so rapidly that the warning is often too late for operators to avoid a forced shutdown. A lower warning limit of vibration level may lead to the increase in false alarms. Consequently, with the alarm level unchanged, new computer-based systems were developed to detect abnormal vibrations at very early stages, identify the causes and tell operators what corrective measures to take. These are vibration diagnostic & operation support systems which are now in practical use to maintain high reliability of power plants.

3. Practical Use at Power Plants

The new systems are emerging from proof test phases, and 8 systems are now in practical use for 14 units at 6 power stations in Japan, to the best knowledge of the author. One system sometimes covers two or three units at one time, and monitors the vibrations of not only main steam or gas turbines together with generators but also auxiliary rotating machinery such as boiler feed pumps, BFP turbines, condenser pumps and circulation pumps, and also boiler equipments such as forced draft fan and gas mixing fan.

The electric power companies in Japan are constantly under intensive public pressure to meet strong demands for stable power supply, and they make every effort to keep the high reliability levels of their power plants. Recently there have emerged some problems which may deteriorate the plant reliability if left alone.

The recent power demand in Japan shows a wide variation during 24 hours in a day, particularly in summer, and the bottom load is only 40 % of the peak load. To meet this demand pattern, thermal power plants (oil-, coal- and LNG-fired) have to be operated in the daily startup & shutdown mode or weekly startup & shutdown mode. The operation of these modes are very hard on the rotating machinery, which is widely accepted as heightening threats to the operational reliability.

Furthermore full automated operation systems have been long used in power plants, and most of recent machinery can be operated free from vibration troubles for long. Consequently operators can have little opportunities to master the skills to fight against vibration troubles with their bare hands.

Consequently the vibration diagnostic & operation support systems are regarded as a new hope to keep highly reliable operation with less manpower of less experience of vibration troubles. Operators are expected to be able to handle the troubles expertly and easily with the assistance of the system.

Most of newly constructed thermal power plants are likely to be armed with the system.

4. Functions and Configurations of the Systems

The current systems have the following functions:

(1) Detect vibration troubles at very early stages before the alarm system is triggered and inform operators of the occurrence.
(2) Send signals to alarm systems when the vibration levels exceed preset limits.
(3) Diagnose the troubles, identify the vibrations and establish the causes.
(4) Show operators the results of the dignosis and possible corrective measures to take, or suggest some investigation.
(5) Store the original signals, processed data, and the process and results of the diagnosis so that operators can consult any time.

Figure 1 shows two examples of hardware configuration of the vibration diagnostic & operation support sytem. Modern power plants have unit control computers for automated operation. Normally the vibration diagnostic & operation support systems do not use the computers to avoid possible interference with conventional control system, but have their own minicomputers or engineering work stations to carry out their jobs. These computers are linked with conventional data highways for unit control systems. Some of the diagnostic & support systems use two computers, each of which is allotted for each of the two different categories of the jobs. The first one is collecting, processing vibration signals and detecting abnormal vibrations, and the second one is diagnosis and guidance.

Large-size colour CRT displays are used independently to interface operators with the systems. Ordinary peripheral hardwares like hard disk systems, printers and plotters are incorporated in the system. Signal conditioners such as filters, integrators and amplifiers and preprocessors like FFT spectrum analyzers are needed between transducers and the computers.

Seismic velocity transducers have been preferably used to measure the vibration of steam turbine rotors by riding on the rotating shaft because of the strong points of low impedance signals and no need of power supply, but non-contact displacement sensors of eddy-current type are gradually increasing in use, particularly for gas turbine rotors, perhaps because of higher reliability resulting from no moving parts inside.

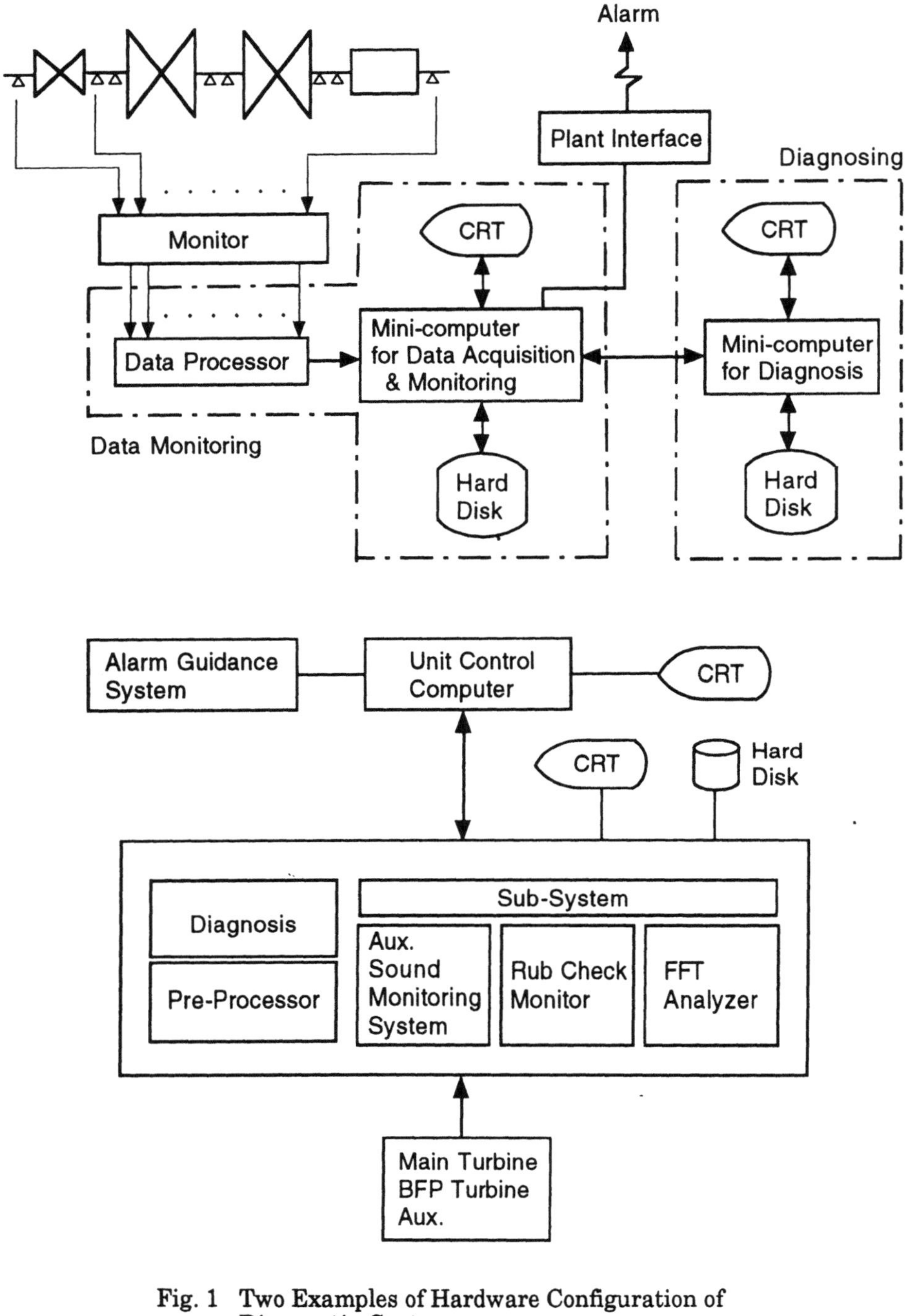

Fig. 1 Two Examples of Hardware Configuration of Diagnostic System

Recent turbines have labyrinth seals of smaller clearances than before to get better turbine efficiency, which tends to increase the risk of rubbing of rotors against stationary parts, particularly at the startup stage. DSS-mode operations enhance the risk because a fast startup may lead to anisotropic dilation of rotors and casings. Once rubbing starts, local thermal dilation of rotor enhances the rubbing, eventually resulting in high vibration levels. Piezoelectric accelerometers are normally used to detect rubbing at the seals by sensing sonic waves of the frequency range around several kHz. AE sensors are sometimes used for the same task by detecting supersonic waves (some tens to several hundreds kHz) from the rubbung part. Advanced signal processing technologies are used to distinguish rubbing from steam noises.

Operators usually patrol the machine site and make visual and aural inspection periodically. For aural inspection sound sticks are often used like stethoscopes. But it needs skills to diagnose the machine by means of the sticks, and operators have to go out of the comfortable control room. Consequently piezoelectric acceleration sensors are installed to collect sonic waves from rolling element bearings or other parts of mainly auxiliary rotating machinery. Collected data are transferred by cable to the central control room, and operators can select each of the channels from various positions and make on-line ispection by listening to the sound through a speaker system in the room. Furthermore an computer-based system is in use to monitor the levels of the sounds automatically.

Some data other than vibration signals are needed to carry out diagnosis of abnormal vibrations, and a lot of sensors are used to collect signals of generator output, phase angle reference of rotor, pressure and temperature of main steam, temperature of gland steam, condenser vacuum, elongation, elongation difference, valve opening angle, journal eccentricity, attitude angle of journal and so on. In total, more than 200 or 300 data per unit are transferred to the computer for monitoring and diagnosis.

5. Vibrations of Rotating Machinery and Detection of Abnormal Vibrations

Vibrations of rotating machinery can be characterised and also categorised by the frequency. The first category represents synchronous, forced vibration of rotor such as unbalance vibration and vibration due to rotor bow. This vibration is caused by loss of rotational parts, coupling problems, bearing problems and bent shaft. A large-scale turbine generator set has to pass several critical speeds at which the vibration amplitudes reach peak values, until its rated speed is reached.

The second category represents subsynchronous, self-excited vibrations such as very dangerous oil whip due to oil film bearings and steam whirl noticeable for light weight rotors of high pressure turbines.

The third one denotes supersynchronous vibrations such as observed in the cases of cracked rotors and rubs.

These vibration categories based on rotordynamics should be modified to fit in with vibration diagnosis for power plants. Figure 2 shows an example of two stage tree-like categorization in practical use. Abnormal vibrations falls into eight sections denoting different abnormal phenomena at the first stage. At the second stage each of the eight phenomena is connected with several causing factors. The tree in Fig. 2 has 33 factors at the second stage.

Figure 3 shows another tree which has seven items at the first stage and 59 factors

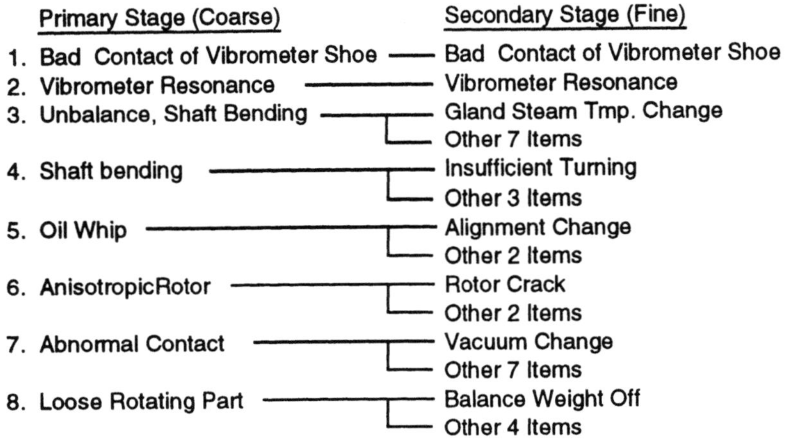

Fig. 2 Two Stage Categorized Decision Tree of Abnormal Vibrations (No.1)

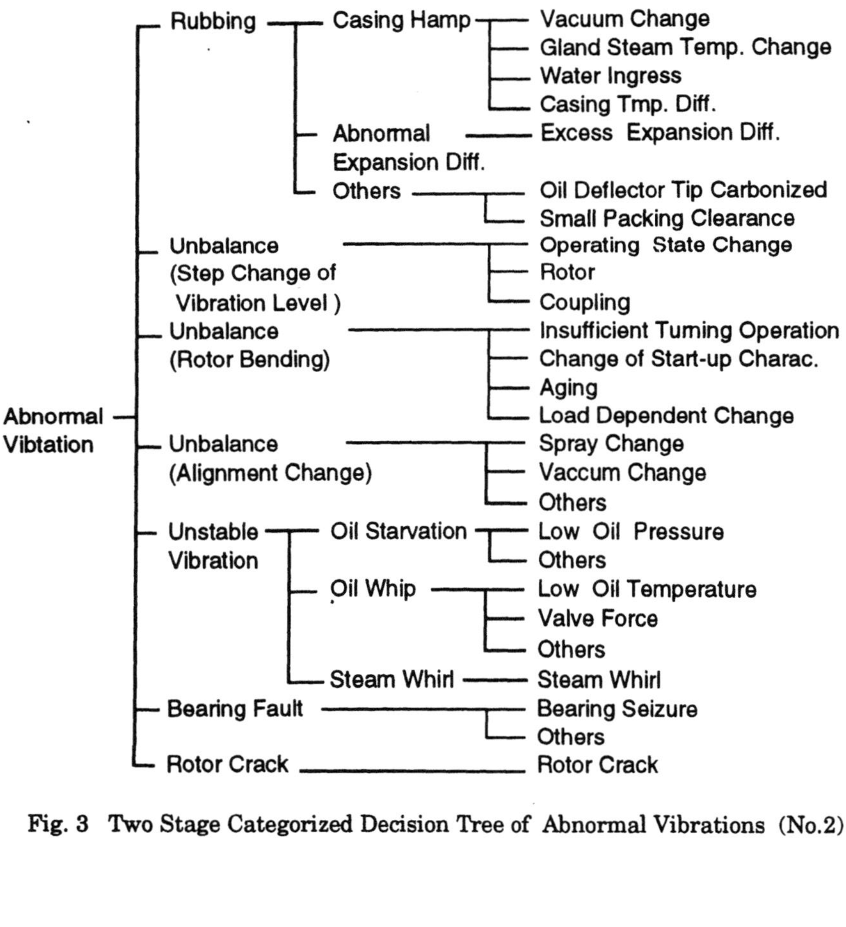

Fig. 3 Two Stage Categorized Decision Tree of Abnormal Vibrations (No.2)

at the second stage. These tree-like knowledge structures are prepared from not only rotordynamics theories but various knowledge of design engineers, operators and maintenance engineers and also FTAs (Fault Tree Analysis) for the accidents or incidents experienced in power plants.

The system monitors the following unprocessed and processed data from vibration sensors.

(a) overall vibration amplitude
(b) vibration amplitude change rate with time
(c) stepwise change of vibration amplitude
(d) vibration amplitude distribution in the frequency domain
(e) change of vibration vector
(f) synchronous component of sound amplitude, higher-order moment or envelope of sound waveform

The system compares the data with each of predetermined limits, and when one of the data is found to have exceeded the correponding limit value, the system starts automatic diagnosing to identify the abnormal phenomenon, establish its cause and select corrective measures or inspection guidance. The results are given on a CRT display for operators to understand easily and quickly.

Formerly alarm signal systems monitored only vibration levels, but the new diagnostic & support systems monitors various processed data to improve the ability of detecting and diagnosing abnormal vibration at an very early stage.

The time interval of monitoring usually varies with shaft speed. In the case of rated speed operation at 3000 rpm or 3600 rpm, the interval is one second, for example. Another system repeats every 15 seconds at rated speed. At start-up or coast-down, the interval is every ten or twenty revolutions of rotor, or every 50 or 100 rpm.

The collected and processed data are reduced through multi-stage gates. For example, the two-hour file stores all the data for the last two hours, and the 24-hour file consists of the data extracted every one minute. The one-week file has the data of every 15 minute interval, and the one-month file stores the data of every hour interval. However all the data for the respective two hours before and after the system starts diagnosing are stored without being reduced.

The selection of the limits values is most important but most unclear factor for the system. No theoretical or practical evidence can be given for a particular value of limit. It seems those values are determined separately for each plant from operational experiences. However the limit must be well lower than the alarm level.

Figure 4 shows a typical variation of overall vibration amplitude of a rotor with shaft speed. The rotor-bearing system has two criticals before the rated speed. Figure 5 shows an example of two different settings of the limit for each speed region with the vibration characteristics in Fig.4 being considered.

Figure 6 shows another setting of the limit in the frequency domain for a rotor rotating at a speed.

6. Diagnosis and Inference Strategy

Figures 2 & 3 shows that various abnormal vibrations can happen to rotating machinery in power plants and also that those vibrations originate from multiple causes. It means that an observed phenomenon does not always correspond to a unique cause.

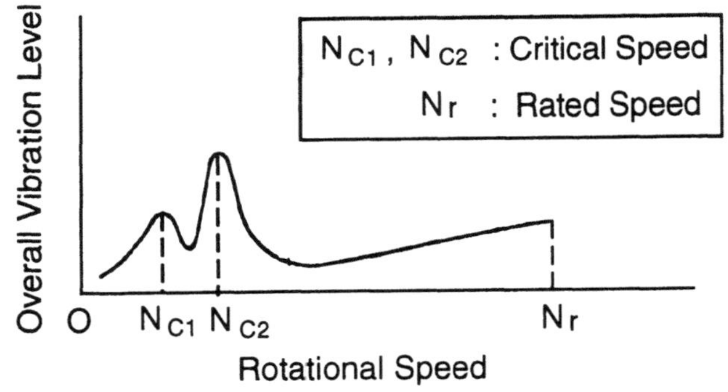

Fig. 4 Amplitude Variation with Speed

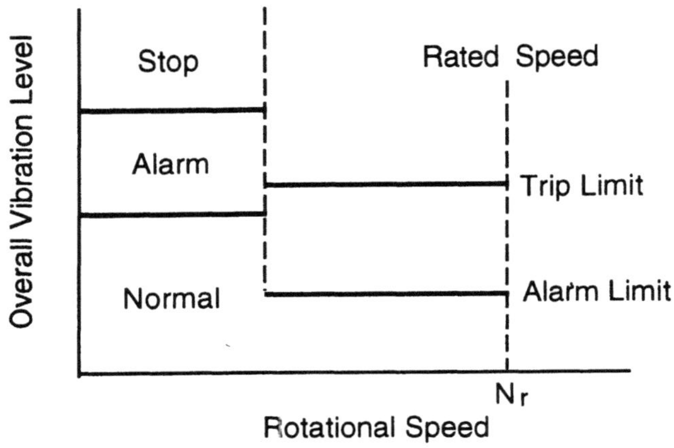

Fig. 5 Change of Vibration Level Limit with Speed

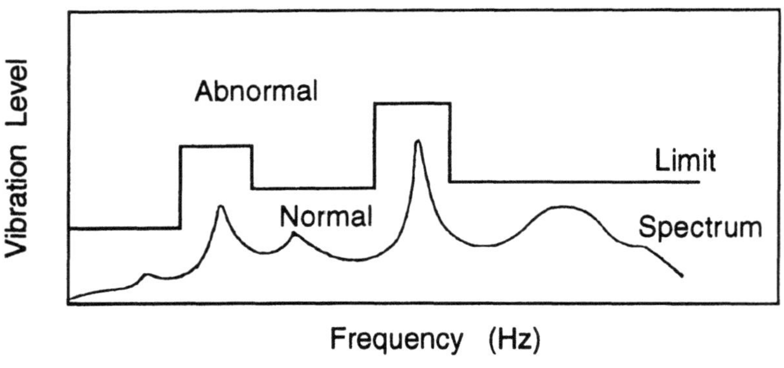

Fig. 6 Limit of Vibration Level in Frequency Domain

Consequently even skilled operators take a long time to identify the abnormal vibration and reach a correct cause of the vibration because various scenarios must be screened by checking the data against each other.

Computer-based expert systems, if carefully designed, can expedite vibration diagnosis significantly. An inference engine in the system searches a knowledge base for rules which agree with given conditions. Most diagnostic expert systems use production rules for knowledge base, and more than 5,000 rules are needed to diagnose vibration troubles of rotating machinery in power plants. Consequently even high speed digital computers consume much time to check all the rules without tactics, and may be outpaced by the growing vibration.

A solution to this problem is dividing the diagnosis into two stages, that is, the fast one which identifies the abnormal vibration which triggered off the diagnostic system by choosing one or two in less than ten, and the detailed one which takes time to establish the causes of the vibration.

Figure 7 shows a matrix-formed decision table which enables real-time diagnosis of the first stage. Each abnormal phenomenon arranged on the left end column corresponds to that in Fig.3, and the items of the status of the machinery or the plant are arranged on the top transverse line. Each value of the certainty factor in each slot shows how closely connected the two elements (a phenomenon and a stauts) are under the current situation, and is calculated from a given function for the current situation of the status item. The certainty factor for an abnormal vibration is calculated by summing up each value on the same transverse line, and is showed in the right end slot. The abnormal phenomenon with the highest value or the second highest value of certainty factor can be reasonably assumed to have happened to the machine. If the matrix does not consist of large number of elements, the calculation of the certainty factor takes a very short time of less than one second, and a real-time diagnosis can be provided. This method uses forward inference which is better for fast diagnosis. The functions which give the values of certainty factor are the key to this method. Sometimes fuzzy membership functions are introduced to give the values of certainty factors.

Vibration	a	b	c	d	e	f	g	Certainty Factor
1. Rubbing								
2. Unbalance(StepChange)								
3. Unbalance (Bending)								
4. Unbalance (Alignment)								
5. Unstable Vibration								
6. Bearing Probrem								
7. Cracked Rotor								

Status
a : Load b : Bearing Position c : Vibration Level d : Vibration Level Trend
e : Stepwise Change of Vibration f : Frequency g : Vibration Vector Change

Fig. 7 Decision Table for Real-Time Diagnosis

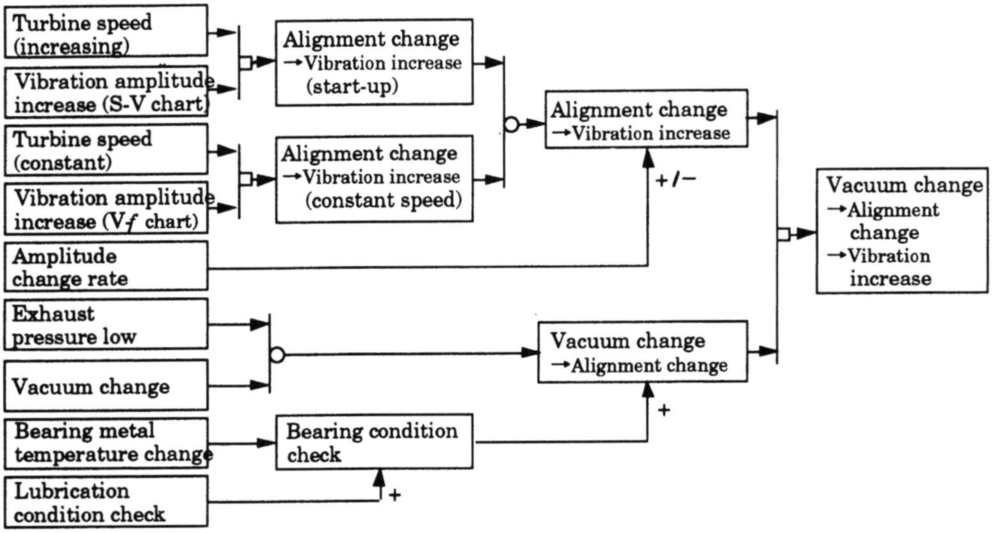

Fig. 8 Decision Tree for Detailed Diagnosis

As soon as the real-time diagnosis is finished, the system shows the result and corrective measures to take or the status item to be inspected on a CRT display.

The detailed diagnosis of the second-stage establishes the causes of the vibration identified by the real-time diagnosis. This detailed one normally starts working at the request of operators after they settle down the problem. Figure 8 shows a decision tree structure for a detailed cause given at the second stage in Fig. 3, and similar tree structures are established for the other causes too. For the detailed diagnosis, the certainty factor for a particular cause can be calculated by applying backward inference to the tree structure. This tree-like expression is better for operators' understanding, maintenance of the expert system and faster inference, compared with a simple pack of production rules. The detailed diagnosis normally takes one or two minutes to finish calculation for one cause.

Recent new systems use the combination of the decision tree structure and forward inference even for the real-time diagnosis instead of matrix-form decision tables, and also incorporate a model base such as computer programme to simulate the vibrational or rubbing response of the rotor. The latter enhances the inference accuracy of the diagnostics.

7. Display of Results

Expert systems are expected to explain to operators about not only the result of the diagnosis but its process. For easy and fast understanding, large-size colour CRT displays are usually used. The display tactics is very important, and the arrangements of windows, the size and colours of characters and graphs are carefully designed.

Figure 9 shows an example of such a coloured display. A schematic of a rotor-bearing system is shown at the top left corner. The window below shows by means of

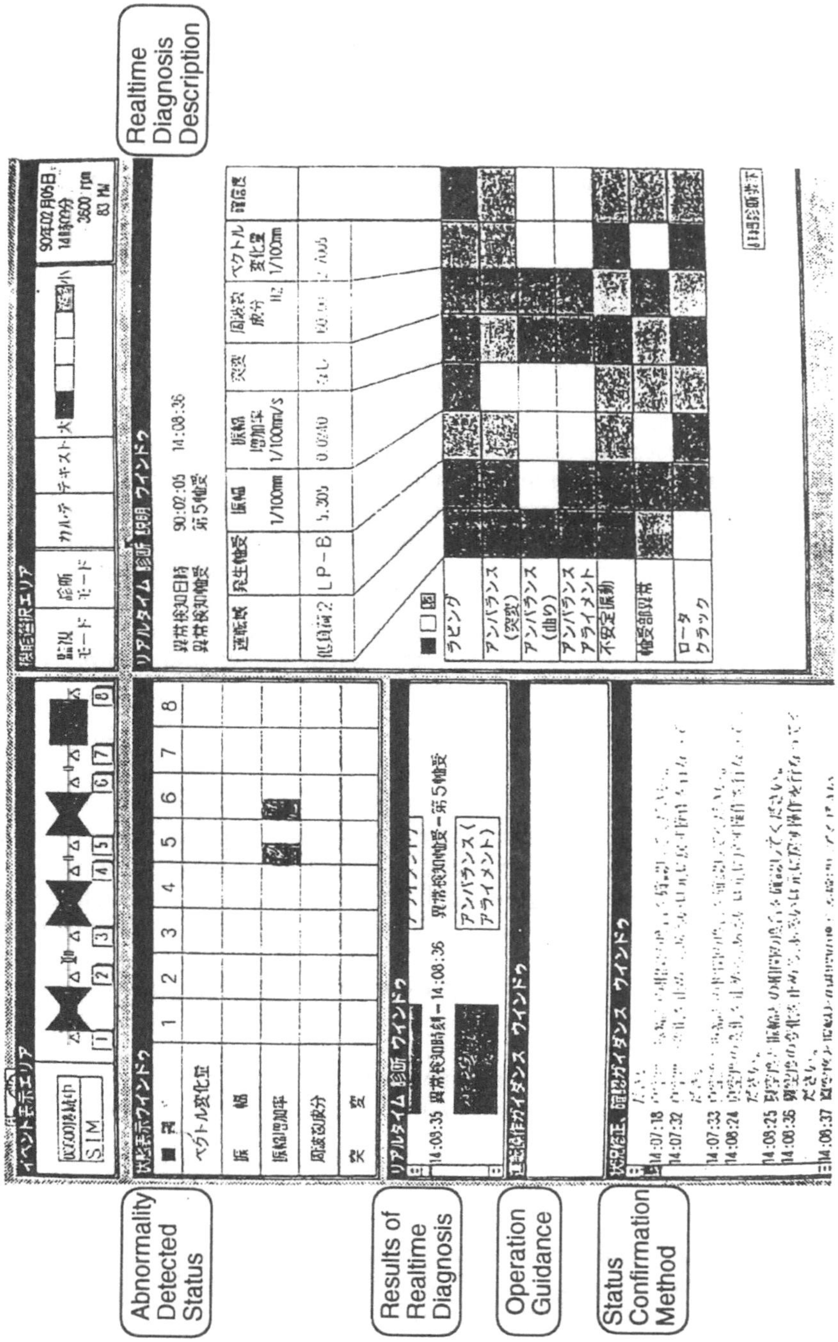

Fig. 9 An Example of CRT Display of Diagnosis and Guidance

bar graph that vibration level change with time exceeded the limit at Bearings No.5 & 6. The second lower window shows the result of real-time diagnosis which tells that rubbing is most probable, and unbalance vibration due to bearing alignment change is second most probable. The bottom left window tells operators to recover the vacuum of the condenser and also to check the previously recorded relationship between vibration level and condenser vacuum. The right half window explains the process of the diagnosis by means of the decision table. In the right end column, the slots of the high certainty factor are coloured red, and those of medium values, coloured amber, and those of low values, green.

8. Other Functions

(1) Operators can get the following information from the system on the CRT display any time to check the current status of the machine and the plant: time-domain vibration level, trend of vibration level, vibration vector and vibration spectrum.
(2) Unprocessed and processed data can be stored for quick reference. Operators want to check the old data of the machinery status when the same alarm signals triggered the diagnostic system before. Stored data can be also used to perform maintenance work efficiently and at a lower cost. Furthermore long-term trend of vibration data can be analysed and used to change the algorithm of diagnosis or the trigger limits for better accuracy.
(3) Operators can get access to the stored knowledge base and learn the cause and effect of vibrations of machinery, which corrective measures to take against which abnormal vibration and so on.

9. Concluding Remarks

The computer-based vibration disgnostic & operation support systems for power plants are in practical use in Japan, and tend to increase in number gradually.

The systems have achieved original goal which was initially established with the current science and technologies available. However the following problems are left to further investigation.

(1) Diagnosis-trigger level and alarm level for operation of better quality
(2) Higher reliability of transducers
(3) Speedup of diagnosis process
(4) Establishing how to use the expert system in the site
(5) Enhancement and extension of expert system

Acknowledgement

The author would like to thank the experts working in this field, particularly former members of Committee on Steam Turbine Diagnostic Systems of Japan Turbomachinery Association for their kind cooperation.

DIAGNOSTIC TECHNOLOGY FOR COMBUSTION TURBINES IN POWER PLANTS

A.J. Smalley and R.M. Baldwin
Southwest Research Institute, San Antonio, TX, USA

G.H. Quentin
Electric Power Research Institute, Palo Alto, CA, USA

H.R. Simmons
Southwest Research Institute, San Antonio, TX, USA

ABSTRACT

This paper addresses diagnostic technology for power generating combustion turbines. It covers measurement techniques and software to interpret measured data, including details of methods to analyze transient vibration data. The paper reviews diagnostic techniques for on-line condition assessment and automated diagnosis, together with methods for defining and solving specific operational problems. Measured quantities include vibration, low speed shaft run-out, blade tip clearance, blade temperature, case distortion, strain, and combustion flame condition. Analysis techniques include modal methods, detection of significant change, visualization, and hot section life management.

1. INTRODUCTION

Combustion turbines are a growing source of clean, efficient, electrical power. Over half the planned capacity increases in the United States for the next five years involve combustion turbines [1]. Generating companies in the

United Kingdom plan to install five or six combined cycle plants including the recently commissioned 1875 megawatt Teeside plant [2]. In Italy, ENEL will install a large number of combustion turbines for power generation. In the past, low thermal efficiency has limited combustion turbine applications to peaking, where they provide rapid start-up; however, with system thermal efficiencies well over 50 percent in combined cycle, base load application will grow.

2. NEED FOR MONITORING AND DIAGNOSTIC TECHNOLOGIES

Peaking service requires starting reliability more than thermal efficiency. However, combined cycle applications require high thermal efficiency combined with high availability and operating reliability.

Increasing turbine inlet temperature increases thermal efficiency. Turbine designers use high temperature alloys, protective coatings, and internal cooling to optimize blade and nozzle tolerance to high temperatures; however, operating near the temperature limit reduces tolerance to off design conditions (e.g., from cooling or combustion irregularities). A 25°C increase can reduce blade life 50 percent by some estimates [3]. To help refine operational and design tradeoffs, the need exists to detect blade temperature accurately.

Environmental regulations constrain design and operation of combustion turbines. Most manufacturers have chosen to develop the dry low NO_x (DLN) combustor [4] - a promising technology, based on lean premix and combinations of parallel or series combustion stages. Schorr [5] describes one manufacturer's DLN approach. DLN technology can achieve NO_x levels below 25 PPM, but sensitivity to load, and to changes or deterioration in condition make it complicated to control, and create a need to monitor directly the combustion process as a guide to design, control and operation.

Directly measuring emissions from the stack appears an obvious monitoring approach, but continuous emission monitoring systems (CEMS) are reportedly expensive to install and maintain. Many gas turbine operators prefer to infer or predict emissions, creating a need for documented empirical models of turbine emissions, with fuel air ratio, pressure ratio, air mass flow, and other measurement as input.

The drive to increase power can lead to blade vibration problems (periodic resonant excitation or aeroelastic flutter). The resultant need to characterize blade dynamic motion in a hot gas stream presents a particular measurement challenge.

The new generation of industrial turbines have only two fluid film bearing supports. In accelerating to speed, the turbine can lift 400 microns in its bearings, and negotiates two critical speeds with mid span whirl amplitudes of 250 microns (well balanced) to 1000 microns (significant unbalance). Localized high temperatures and operating transients can bend the case by 2000 microns [6]. The need exists to monitor vibrations, alignment, and clearance loss during these transients and to guide tradeoffs between tolerance to excursions (high clearance) and thermal efficiency (low clearance).

3. SUMMARY OF MONITORING AND DIAGNOSTICS TECHNOLOGIES

This section discusses hardware and software to meet these monitoring and diagnostic needs.

3.1 Vibration Monitoring

Well developed products (accelerometers, velocity sensors, and non-contact probes) can measure bearing and shaft motion. Commercial systems can acquire, display, and trend vibration data [7 and 8]. High exhaust end temperatures challenge the reliability of probe and cabling system, but locating cables in the oil drain limits their temperature, and redundant probes and cables at hot end bearings improve system reliability. Temperature sensitivity of zero and span calibration demand care in interpreting non-contact probes for shaft position. Systems exist to acquire and display vibration data during startup and shutdown, and Sections 4 through 8 of this paper address the need to characterize and interpret this transient vibration behavior.

3.2 Turning Gear Run-out Monitoring

A 25 micron shaft bow can cause excessive start-up vibration, but may cause only 5 microns low speed run-out at the bearings. Accurate monitoring of bearing run-out on turning gear can help predict high start-up vibration resulting from shaft bow. However, data acquisition at turning gear frequencies (e.g., 0.05 Hz) requires special attention to signal conditioning and data acquisition logic; for example, static voltages exceed the range of most analog to digital converters. Simmons, et al [6] describe their monitoring approach and document how shaft bow can grow by 5 microns per hour off turning gear, and how run-out falls by half every two hours once back on gear (see Figure 3.1).

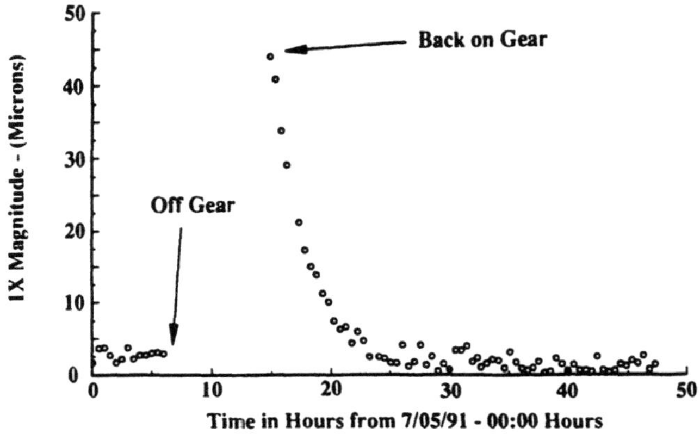

Figure 3.1. Typical gravity sag at ninth stage compressor from 8 hours off turning gear [6].

3.3 Blade Tip Clearance Monitoring

The challenges to measuring blade tip clearance include a high temperature environment, a fast moving narrow target (over 350 meters per second), a wide measurement range, and a long stand off. Available approaches include:

3.3.1 Touch Probe

Rotadata builds this device (Figure 3.2) and describe it in [9]. A high voltage terminal advances 2 to 5 microns each revolution until proximity to the tallest blade makes a detectable discharge. The system then retracts the terminal to a reference point and records the travel. It detects minimum clearance, with cycle time proportional to clearance, and inversely to speed.

3.3.2 Capacitance Probe

This probe from Rotadata, described in [9], measures capacitance between the fixed probe tip and individual blades. It offers high temperature capability and high frequency response. Signal processing must create a linear output, recognizing that capacitance varies inversely with clearance.

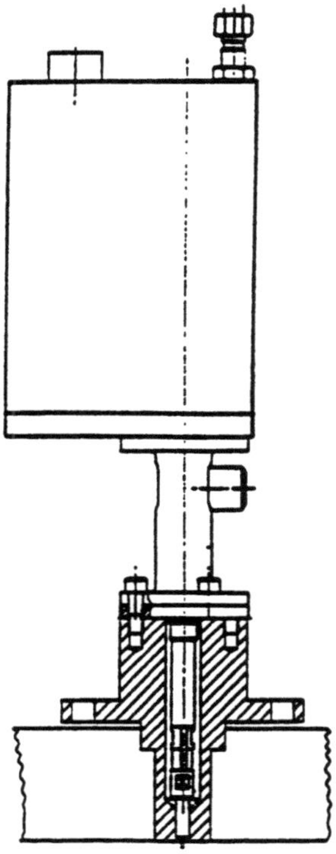

Figure 3.2. Touch probe installation [10].

3.3.3 Eddy Current Probe

Simmons, et al [10] describe the special eddy current probe from Bently Nevada Corporation shown in Figure 3.3. Its high frequency response, coupled with peak capturing electronics, detected blade tip clearance from 3 RPM to 3600 RPM. Figure 3.4 shows typical results. While resistant to temperatures up to 210°C, the present probe design is applicable only to the compressor section.

3.3.4 Optical Probe

Drinkuth, Alwang, and House [11] first developed this concept for minimum clearance detection. Simmons, Michalsky, Brewer, and Smalley, [12] describe high frequency circuitry with which it can measure individual blade clearances. The probe uses triangularization of a laser beam, and a position

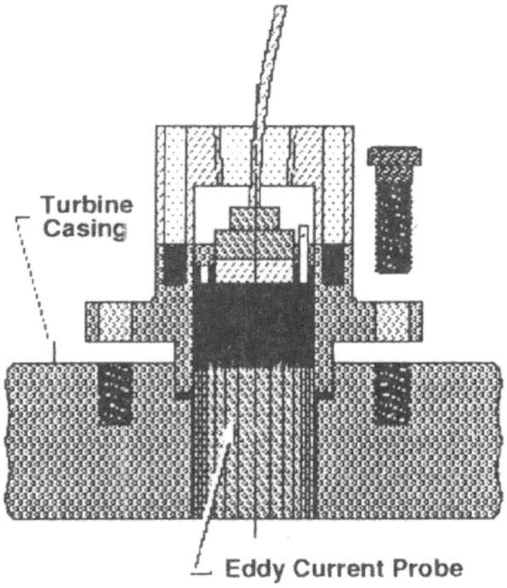

Figure 3.3. Eddy current probe for compressor blade tip clearance monitoring [10].

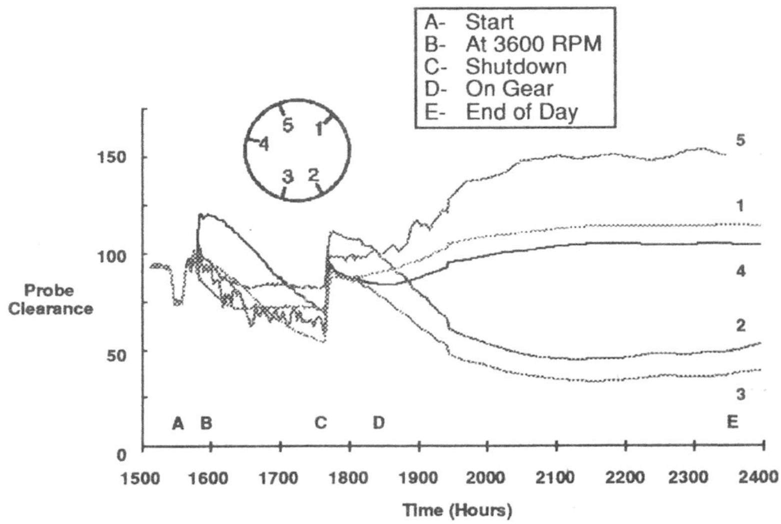

Figure 3.4. Ninth stage tip clearance: operation and shutdown [10].

sensitive detector (PSD) to measure the distance from probe tip to blade tip (Figure 3.5). When hardened by a sapphire prism, and cooling air, the optical probe can measure hot section blade tip clearances.

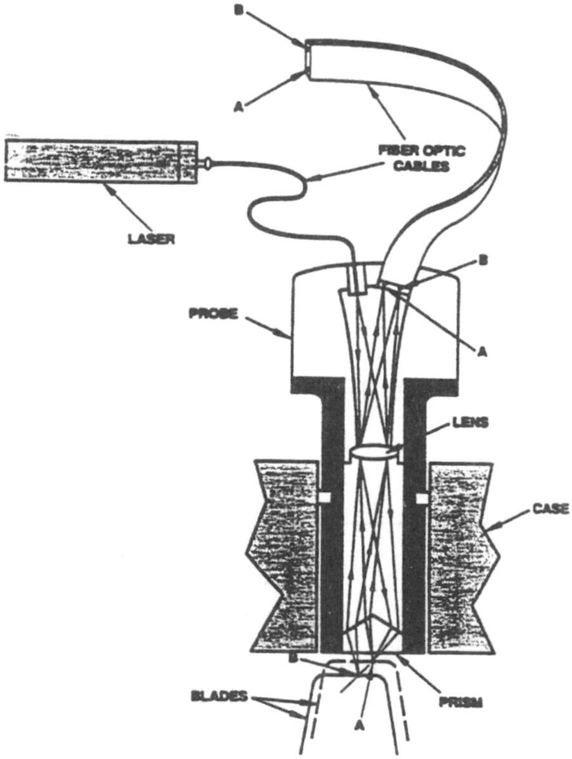

Figure 3.5. Schematic of optical blade tip proximity sensor [10].

3.4 Blade Vibration Monitoring by Time of Arrival

The literature contains examples of methods to detect blade vibration from deviations between a blade's actual and expected time of arrival. Roth [13] uses one or two fiber optic transducers to detect bending vibration in this way, pointing out that a single transducer detects vibrations only at non-integral orders of rotation. A second transducer enables detection of order related vibrations. Mase, Rozelle, and Zagar [14] detect multiple orders with a number of transducers (optical, variable reluctance, or eddy current), evenly spaced around the circumference. To detect small deviations in arrival time of the high speed blade requires a high frequency clock. Mase, et al show a 24 megahertz clock can resolve 25 microns vibration amplitude.

Nava, Paone, Rossi, and Tomasini [15] use just three unevenly spaced transducers, acquiring and analyzing the signals as a composite to detect higher order vibrations. Nava, et al have so far tested the method in the laboratory. Time of arrival technology requires significant refinement and applications experience to become an effective diagnostic tool for blade vibration.

3.5 Strain Gage Telemetry

Strain gages located appropriately in the vibration mode shape can sensitively detect blade vibration; however, temperature limitations and signal transmission difficulties limit their application to blading. Simmons and Smalley [16] show how strain gage telemetry can effectively detect and assess torsional vibration (see Figure 3.6).

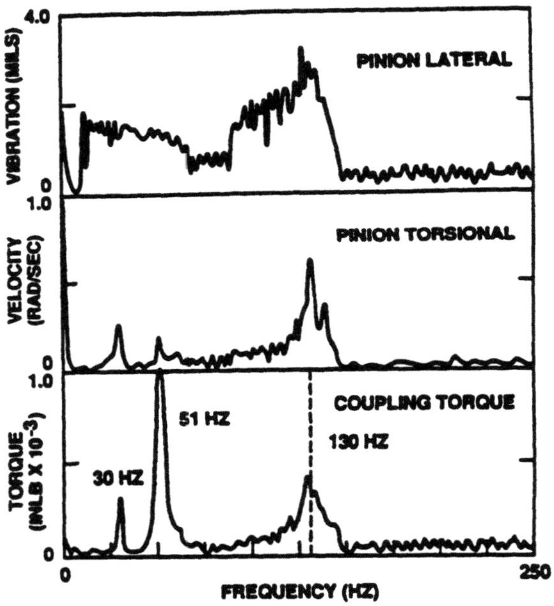

Figure 3.6. Comparison of torsional and lateral vibration during turbine run-up. Bottom trace is from telemetry of rotating strain gages [16].

3.6 Casing Alignment and Distortion Monitoring

Evans and Pantermuehl [17] and Bloch [18] describe methods for monitoring case-to-case alignment with a laser and PSD. The method requires a line of sight which limits laser monitoring ability within a single casing. Alignment monitoring with an externally mounted laser or optical transit encounters limitations because of the close fitting enclosure normally installed to control the thermal environment surrounding large combustion turbines.

The ADE fluid level transducers developed by CISE [19] can monitor vertical casing distortions at non-collinear points. Figures 3.7 and 3.8 from Simmons, et al [20] demonstrate the ADE probes monitoring case distortion on a large turbine.

3.7 Integration of Vibration Monitoring and Control Systems

Integrating vibration and other monitoring techniques into digital control systems offers benefits of cost reduction, simplification, and uniformity. The control system will need boards to acquire and process phased vibration information at data rates higher than normal for control functions, and a processor fast and large enough to manage vibration analysis. DeMoss [21] describes integrated monitoring and control functions offered by one supplier of gas turbines.

3.8 Monitoring the Condition of Hot Gas Path Components

Inspecting and replacing hot gas path components contribute substantially to operating costs. Ideally, component operating history would guide inspection intervals and damage rate experience would guide replacement decisions. Under EPRI sponsorship, Bernstein [22] has developed software to help manage life of critical components based on specific operating history. Algorithms calculate such quantities as thermal mechanical fatigue (TMF), damage accumulation, creep strain, and coating degradation, based on operating history, number of starts, fired hours, trips, and component temperatures. Figure 3.9 shows the encouraging correlation of predicted and observed cracks for first stage blades.

3.9 Component Temperature Measurement

Exhaust gas temperatures provide the normal basis for hot section component temperatures, as an alternative to directly measuring first stage blades and nozzles. Thermal infrared radiation measurement represents a promising direct

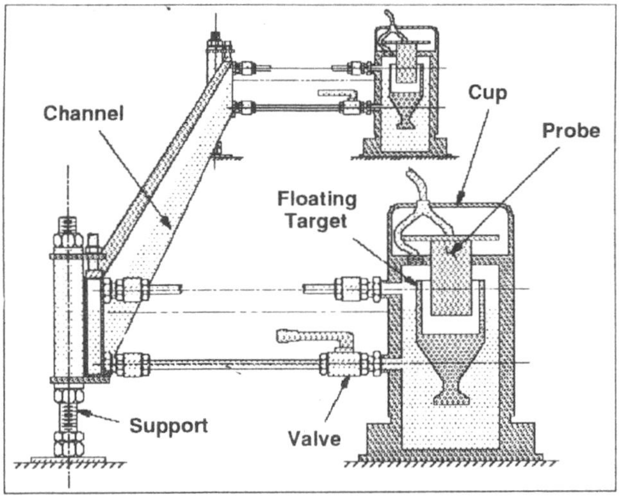

Figure 3.7. Schematic of ADE sensors for vertical deflection measurements [20].

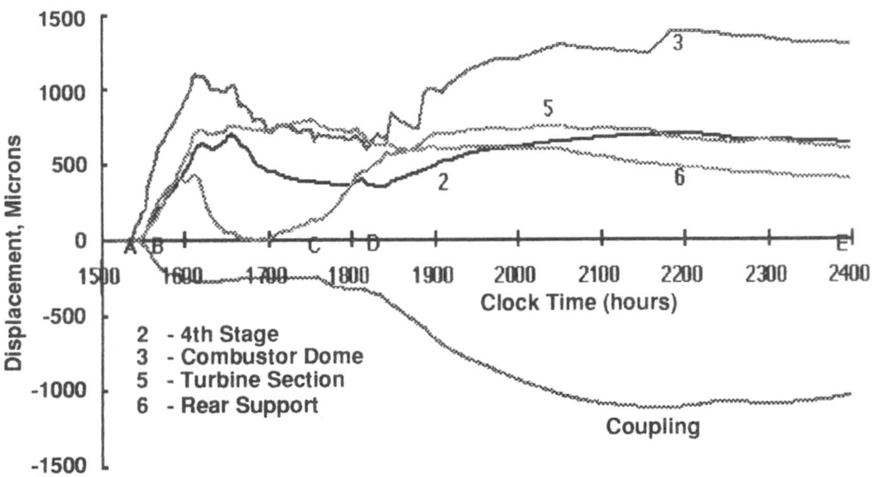

Figure 3.8. Turbine casing vertical growth (ADE measurements) [20].

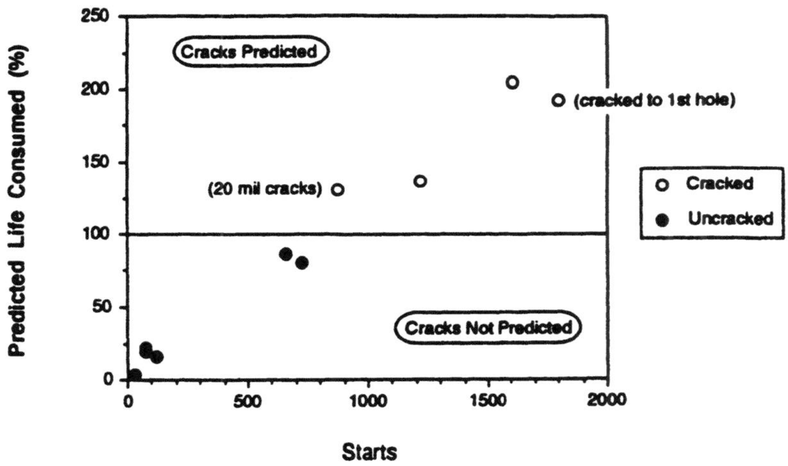

Figure 3.9. Predictions of thermal-mechanical fatigue life of first stages blades [22].

method, but imposes challenging requirements including a line of sight, access, a reliable sensor, and software to interpret results in light of interfering reflections. The Electric Power Research Institute (EPRI) [23] combines a commercially available optical pyrometer from Land Turbine Sensors Inc. with a cooled purged adapter and casing port to monitor first stage blade temperatures on a large industrial turbine. Figure 3.10 shows the measured 30°C variation between coolest and hottest blade. Meher-Homji, et al [24] describe similar variability on a 150 MW advanced gas turbine (from another EPRI program). De Lucia, et al [25] report ±4°C accuracy in laboratory tests of a system with an air cooled sight tube which reflects side entering radiation to the pyrometer.

3.10 Parametric Emissions Monitoring

Figure 3.11, reproduced from Røkke, Hustad, and Berg [26], shows how closely a parametric equation based on pressure ratio, air mass flow, and fuel air ratio reproduces measured emissions on several General Electric, Coberra, and Allison turbines. These results offer a promising alternative to continuous emissions monitoring from the stack.

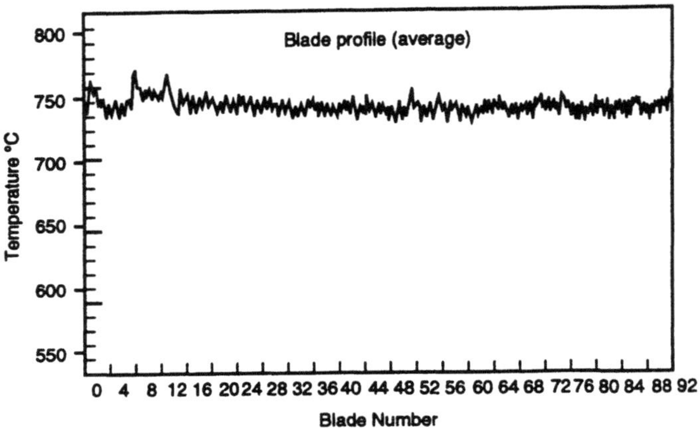

Figure 3.10. Blade temperature profiles for 92 blades [23].

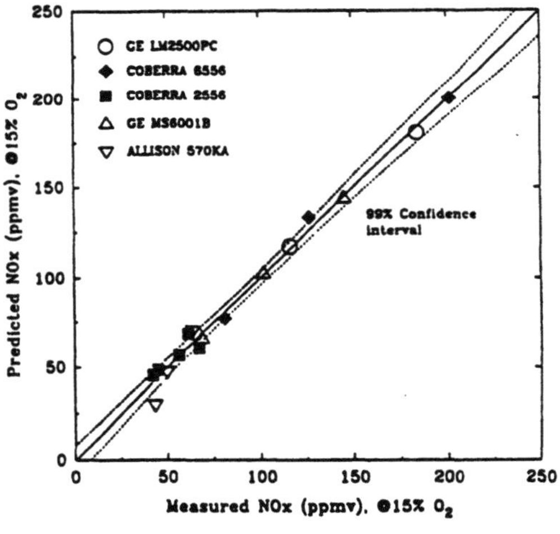

Figure 3.11. Predicted NO_x versus measured No_x [26].

3.11 Combustion Viewing System

Dolbec [27] describes EPRI work on a viewing system to monitor conditions in multiple combustors on a General Electric (GE) MS7001B turbine (Figure 3.12). Based on aeroengine technology, a 10 combustor system with 2000 optical fibers per probe showed igniter sparks, crossfire tubes, changes in flame behavior from ignition to full load, flame unbalances, and delayed ignition. A 5-hour test on a DLN combustor at GE's Combustion Laboratories, showed that a more rugged, cost-effective, prototype [28] could observe dilute transparent premix flames of the combustor, distinguish primary and secondary combustion zones, and monitor the staged transition from start-up to the full premix operating mode.

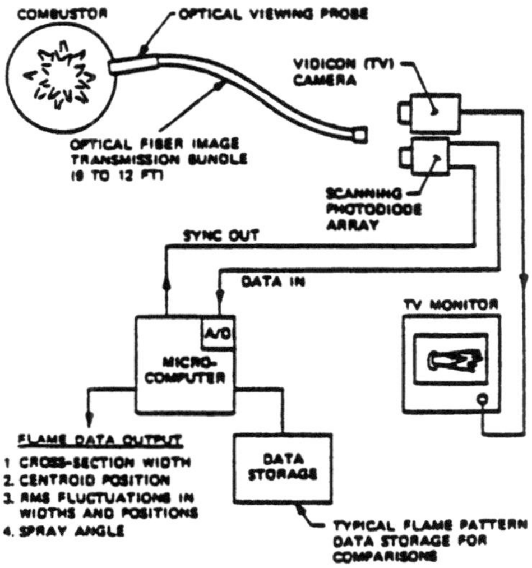

Figure 3.12. Combustor viewing probe for flame pattern analysis and monitoring [27].

4. DEVELOPING AND APPLYING TURBINE TRANSIENT VIBRATION CHARACTERISTICS

Peaking turbine operation tends to follow a series of transient conditions and requires condition monitoring approaches which differ from the more widely used steady-state vibration monitoring. Harris, et al [29] develop characteristics using data from acceleration, loading, and deceleration transients. Their preliminary results suggest that inferred modal characteristics and shaft position changes, normally repeat from run-to-run. Baldwin, Smalley, Harris, and Quentin [30] present concepts for diagnostic application of these characteristics. The remainder of this paper discusses characteristics obtained on four different turbines (Units 1A, 1B, 2, and 3) of the same basic model over an 18-month period, as a basis for assessing variability and feasibility of detecting significant change.

5. CHARACTERISTICS CONSIDERED

The assessment considers values for first and second critical speeds of turbine and generator; resonant vibration vectors at these critical speeds; bearing gap change during acceleration; thermal vectors; vibrations at speed and at load; and discontinuity in the exciter acceleration characteristics. Harris, et al describe the basic method of extracting resonant characteristics (guided also by Ewins [31]). Figure 5.1 summarizes the modal analysis steps. The analysis treats each mode as a single degree of freedom system and fits a circle to vibration vectors near the critical speed of interest. The frequency where phase angle, referred to the center, changes most rapidly around the circle (determined by algebraically differentiating a third order polynomial) defines the critical speed, and the circle diameter passing through this point defines the resonant vector. The maximum rate of angle change defines damping:

$$\zeta = 2/\left(N_c \left(\frac{d\theta}{dN}\right)_{max}\right) \qquad (5.1)$$

where:

- ζ = the damping ratio.
- N = the speed.
- N_c = the speed of resonance.
- θ = the angle from circle center to vibration points.

The turbine shaft tends to lift vertically as the tilting pad bearing film becomes thicker with speed. The displacement sensor looking down at each bearing measures shaft position,

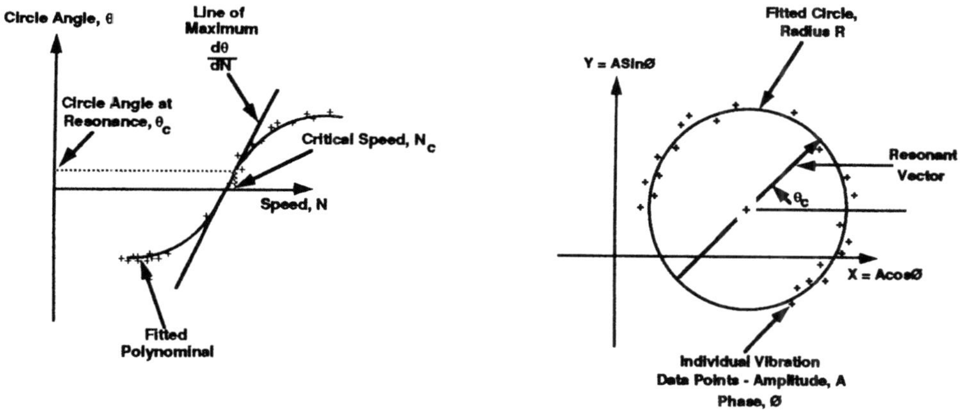

Figure 5.1. Schematic definition of parameters for modal analysis.

and its change in output from turning gear to full speed defines the change in bearing gap. The maximum change in vibration between unloaded and loaded conditions at speed defines the thermal vector. The maximum amplitude of vibration, under load, defines the at speed, at load vibration vector. Subsequent examples will define the exciter discontinuity mentioned above.

6. DATA ACQUISITION

Figure 6.1 presents a schematic of the turbine generator and data acquisition system. At 10 RPM intervals, the system transforms digitized vibration signals to give synchronous (and higher order if of interest) components from relative and absolute probes. The system selects data acquisition points by time once the unit reaches constant speed. Though not addressed in this paper, the system also acquires data for spectral analysis over multiple revolutions at more widely spaced intervals.

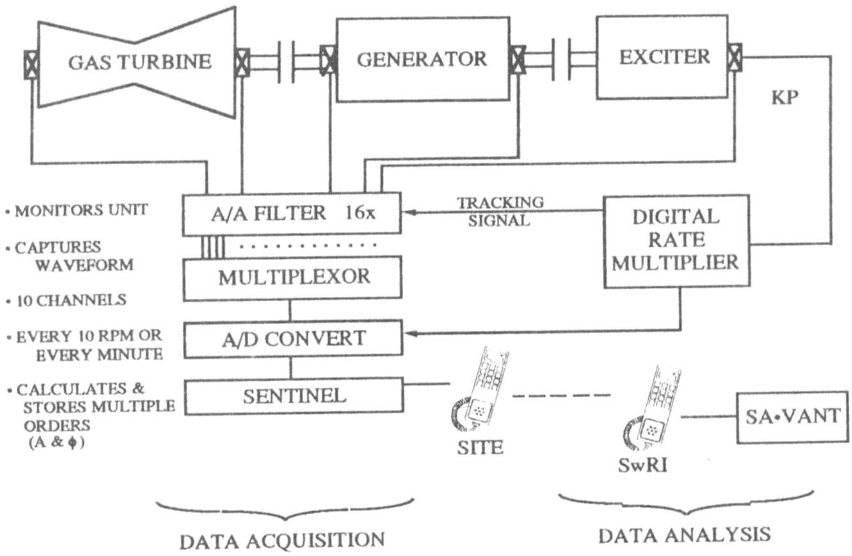

Figure 6.1. Schematic of turbine generator and data system.

7. VARIABILITY IN CHARACTERISTICS

7.1 General Vibration Characteristics

Figure 7.1 shows the major quantitative differences between acceleration and deceleration vibrations in a single run. While some features repeat qualitatively, subsequent data will confirm the necessity to treat acceleration and deceleration characteristics independently.

7.2 Turbine First Critical Speed

Figure 7.2 presents values for first critical speed on Unit 1A inferred from runs occurring over a 6-month period and shows all acceleration values in a narrow band centered about 993 RPM, and all deceleration values in a narrow band (with one exception) centered around 951 RPM. While apparently repeatable acceleration and deceleration mean values differ by over 40 RPM.

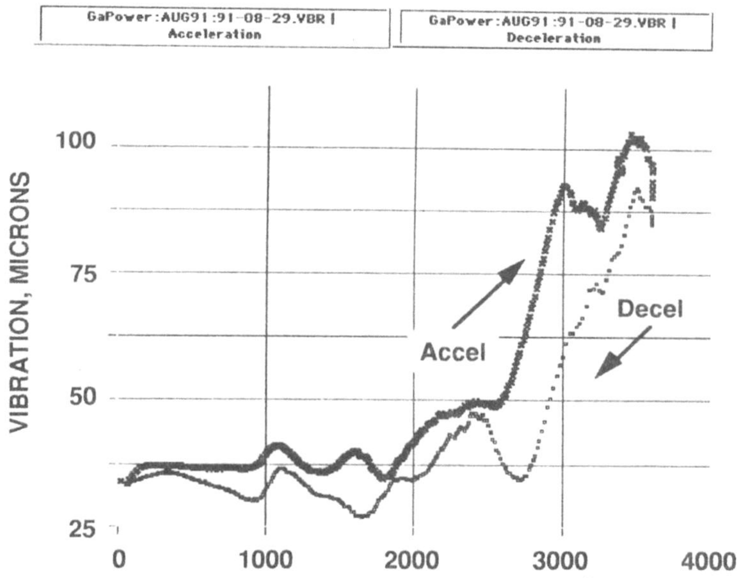

Figure 7.1. Comparison of run-up and run-down vibration versus speed - same unit, same run, compressor shaft probe.

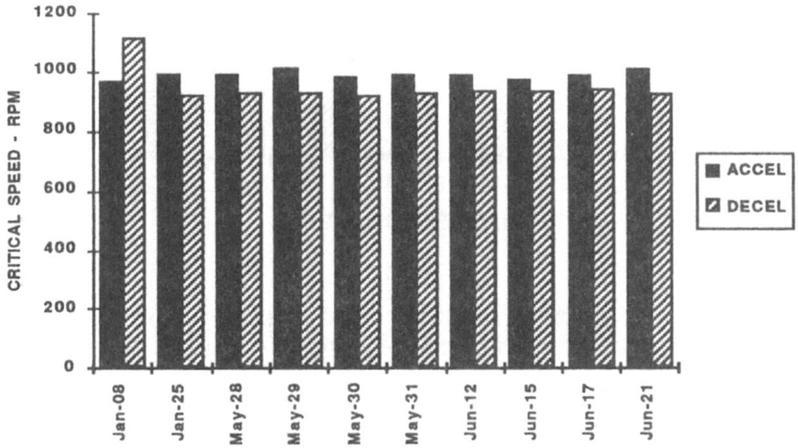

Figure 7.2. Variation of turbine first critical speed - Unit 1A.

7.3 Turbine and Generator Critical Speeds

Figure 7.3 summarizes mean values and standard deviations for turbine and generator critical speeds (two each) from Unit 1A. The standard deviations (13 to 44 RPM on acceleration and 12 to 115 RPM on deceleration), are generally smaller than the differences between acceleration and deceleration mean values. With mean values and deviations characterized for a unit, we can set tolerance bands on each critical speed. For example, a 2σ band for turbine acceleration first critical would run from 967 RPM to 1019 RPM. An inferred turbine first critical outside this band would trigger a first level warning and a value outside the 3σ band (954 to 1032 RPM) would trigger a second level warning.

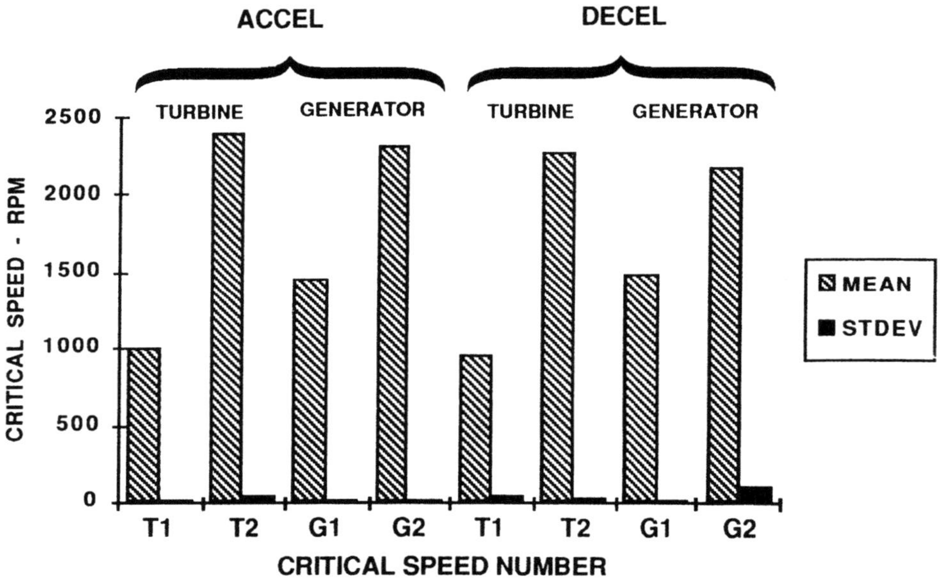

Figure 7.3. Mean and standard deviation for turbine and generator critical speeds - Unit 1A.

Figure 7.4 compares mean values for three critical speeds inferred from accelerating turbines at different sites (Units 1A and 2); in spite of the notable unit-to-unit repeatability, establishing statistical thresholds for each individual unit will ensure the most precise basis for monitoring critical speed abnormalities.

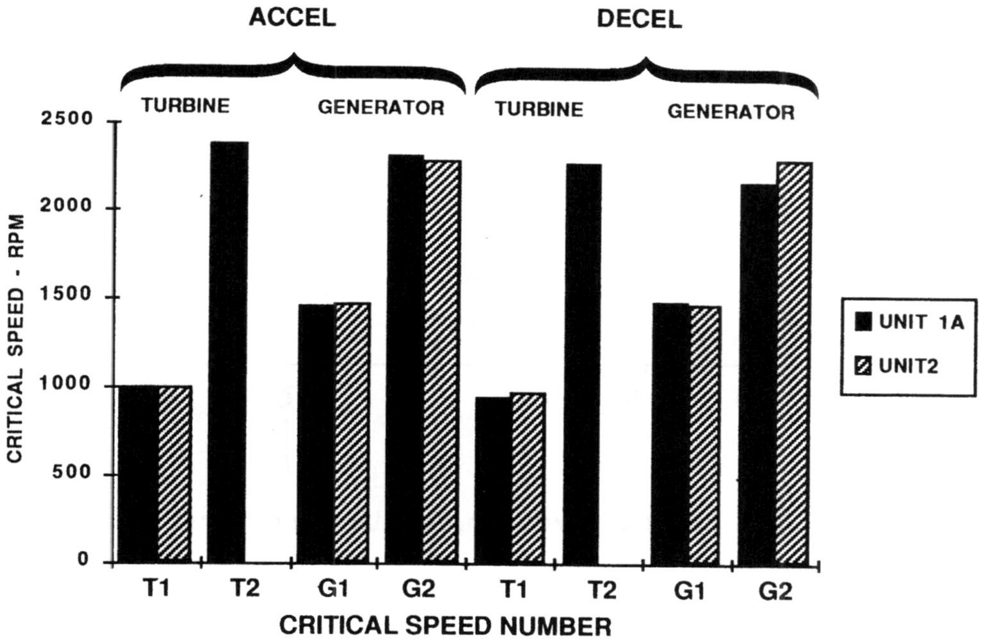

Figure 7.4. Comparison of critical speeds for Units 1A and 2.

7.4 Resonant Vectors

Figure 7.5 shows acceleration resonant vectors for first turbine critical speed lie in a 50 micron diameter circle, while deceleration values, in a 200 micron circle, are more widely scattered. Fifty (50) microns separate the centroids of accel and decel vectors. The 50 micron circle diameter holds promise for using the acceleration resonant vector characteristic for run-to-run monitoring, and for detecting significant change.

Figure 7.6 presents the magnitude of mean resonant vector (centroid) and standard deviation for four critical speeds extracted from eight runs of Unit 1A. The standard deviations should help size tolerance circles. As an example, for the turbine first critical, a "2σ" circle would be centered 76 microns from the origin with 80 microns diameter, and a 3σ circle similarly centered, would have 120 microns diameter. A resonant vector outside the 2σ circle might trigger a first level warning, and one outside the 3σ circle might trigger a second level warning. A 2σ circle which penetrates the turbine trip level would suggest a need to balance.

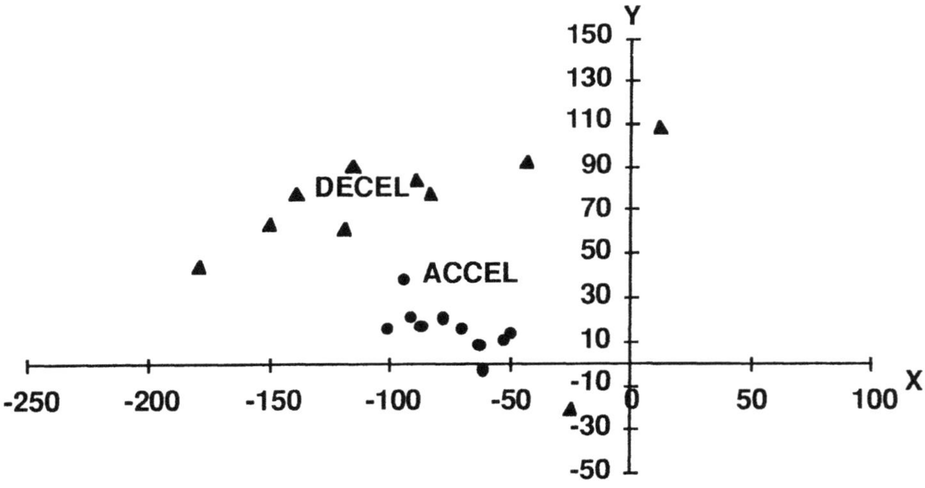

Figure 7.5. Resonant vector for turbine first critical.

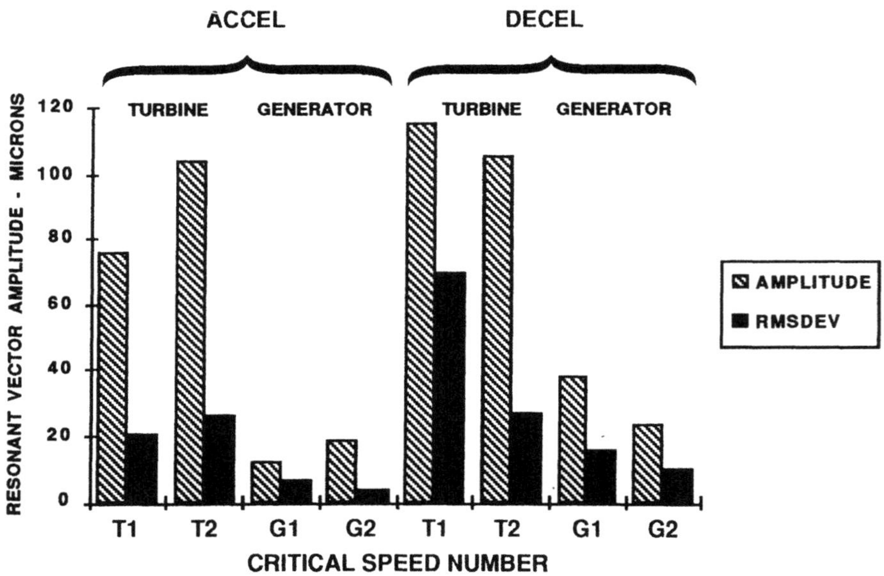

Figure 7.6. Resonant vector amplitudes and standard deviations for turbine and generator critical speeds - Unit 1A.

Monitoring deceleration vectors, with their larger standard deviations, will be less precise, but may offer the first evidence of a blade loss at full speed.

7.5 Modal Damping

Figure 7.7 presents mean and standard deviation for modal damping at the turbine and generator critical speeds. A significant loss in modal damping would indicate reduced energy dissipation, because either bearing conditions or the bearing support flexibility have changed. While modal damping clearly exhibits more variance than the critical speed values, a tolerance band based on standard deviation should provide an effective approach for detecting damping loss. A sustained (several runs) loss of damping, outside the 2σ tolerance band, should trigger concern.

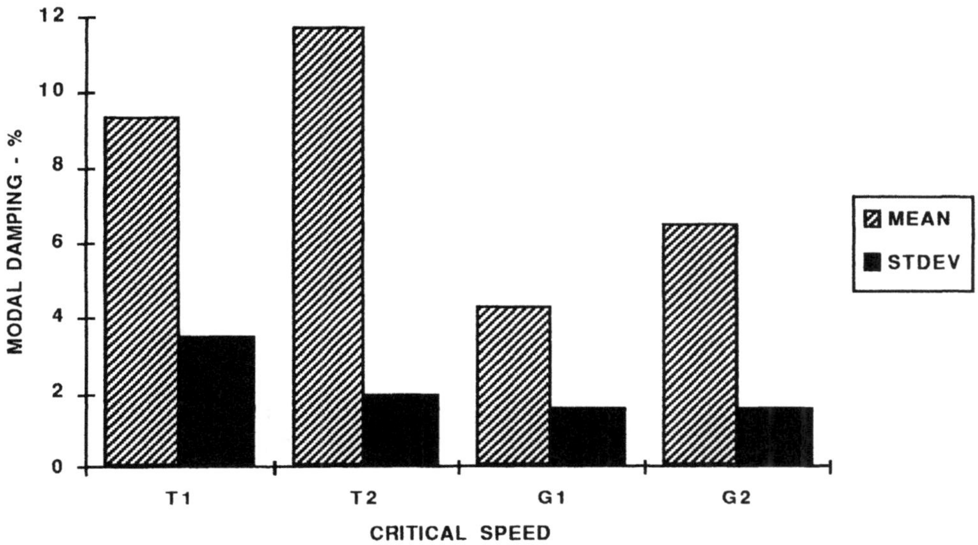

Figure 7.7. **Mean and standard deviation for turbine and generator modal damping.**

7.6 Bearing Lift

Figure 7.8 compares bearing gap changes for Units 1A and 1B. These units (at the same site) show broad consistency, but differ in detail. At bearing 3 (generator inboard), Unit 1A lift exceeds Unit 1B lift by over 50 microns. At bearing 4 (compressor), Unit 1B lift exceeds Unit 1A lift by over 40 microns. Misalignment of generator and turbine casings may explain these opposing trends in adjacent bearings, and detecting possible misalignment illustrates one use of the bearing gap change characteristic. However, bearing clearance differences or transient thermal distortions offer alternative explanations, and need further consideration [32].

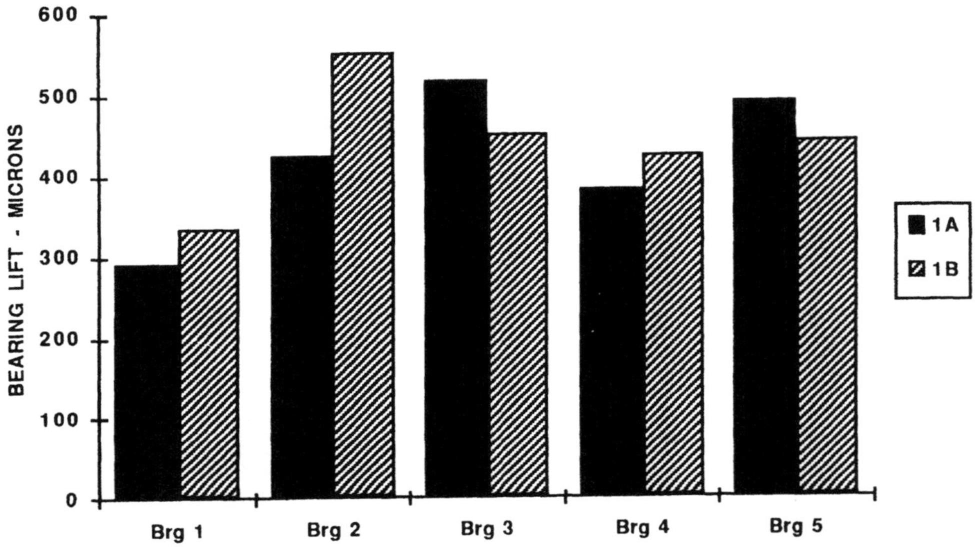

Figure 7.8. Comparison of mean bearing lift for two units (1A and 1B) at five bearings.

7.7 Thermal Vector

Figure 7.9 shows, for Unit 1A, that thermal vector (vibration change from unloaded to loaded conditions) on the two generator bearings varies substantially run-to-run. A large thermal vector may indicate thermal asymmetry caused, for example, by shorted turns or an asymmetrical shift in copper bars under load. The high run-to-run variability (variance probably exceeds the mean) indicates a need to develop more experience with this characteristic as a basis for abnormality monitoring. Fifty (50) microns suggests itself as an interim tolerance threshold for thermal vector, since none of the recorded values exceed this level. Most critically, neither loaded nor unloaded vibrations should exceed the trip level, and data discussed in Section 7.9 will show that Unit 1A's well balanced, low vibration levels at speed provide high tolerance to thermal vector change.

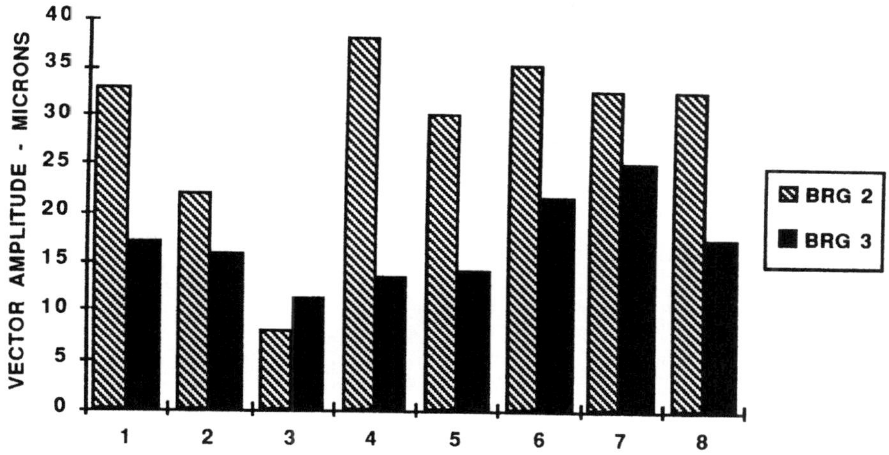

Figure 7.9. Thermal vector - generator bearings - Unit 1A.

7.8 Exciter Discontinuity

Figure 7.10 shows a step jump for both Units 1A and 3 in the variation of exciter bearing gap with speed. The jump occurs at 2300 RPM, with significant modulation below the jump, and smoother variation above it and is most pronounced for Unit 1A. Misalignment across the clutch between starter motor and exciter probably causes this discontinuity; the turbine overruns and disengages the clutch at about 2300 RPM. The vibration traces in Figure 7.11 show discontinuity at the same speed; Unit 1A vibrations modulate significantly below 2300 RPM and seem to correlate with shaft DC position. Above 2300 RPM, vibrations drop and show similar characteristics for both units.

7.9 Vibrations at Speed and Load

Figure 7.12 compares maximum turbine vibration vectors at load for multiple runs of four units. Low variability for each unit again suggests monitoring this characteristic relative to 2σ and 3σ tolerance circles. Comparing maximum vibration to trip settings (Figure 7.13) visually indicates severity, but for computerized monitoring, comparing vectors relative to tolerance circles offers more sensitivity and consistency. The standard deviation values in Table 7.1 suggest using 100 microns as a single 2σ tolerance circle diameter for all at load vibration monitoring.

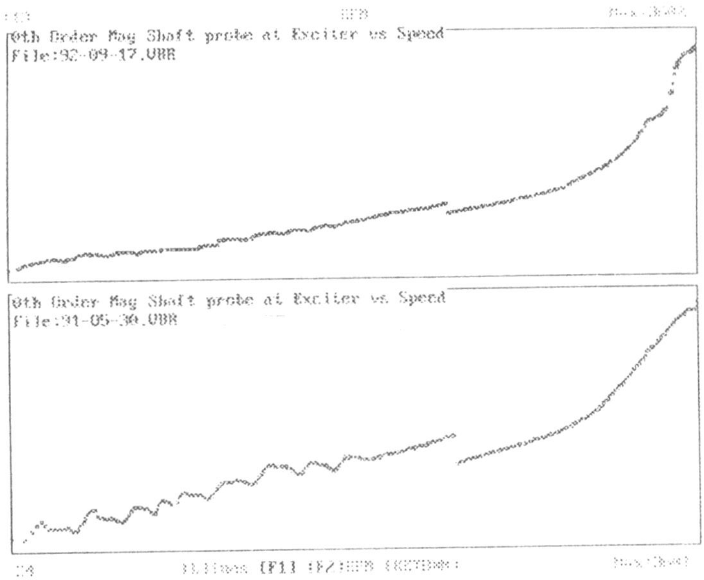

Figure 7.10. Exciter bearing gap change as a function of speed - Units 1A and 3.

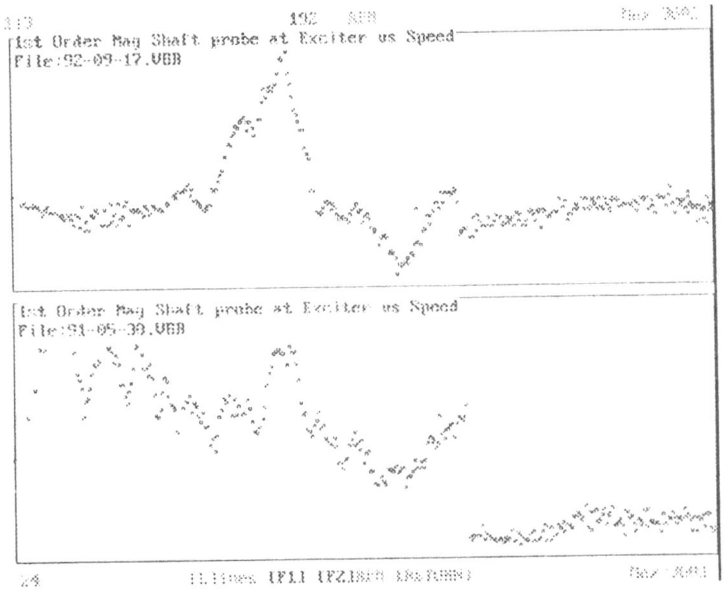

Figure 7.11. Exciter bearing shaft vibration as a function of speed - Units 1A and 3.

TABLE 7.1. Standard deviation of at load vector vibration for four units at turbine, compressor, and exciter bearings microns.

UNIT	TURBINE	BEARING LOCATION COMPRESSOR	EXCITER
1A	19.0	16.5	6.1
1B	10.9	19.6	12.6
2	1.7	4.0	25.7
3	23.4	12.2	5.0

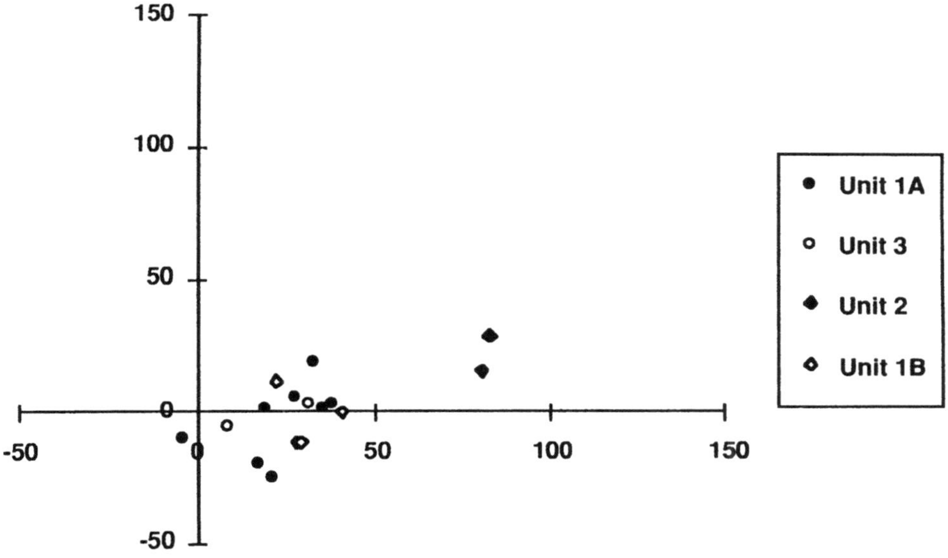

Figure 7.12. Turbine absolute vector vibration (microns) at speed -4 units.

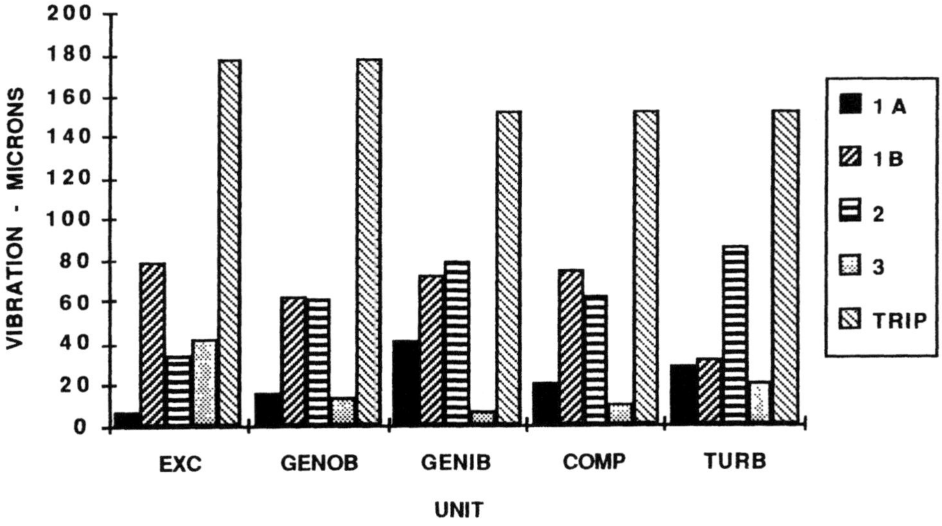

Figure 7.13. Mean absolute shaft vibration amplitudes for 4 units with trip level indicated.

8. OBSERVATIONS FROM VIBRATION CHARACTERISTICS DATA

The data presented so far suggest the following observations:

a. A number of characteristics hold promise, for detecting significant change, based on normal repeatability.
b. Parameter values from deceleration data tend to differ significantly from the same parameter value inferred from acceleration, and show more scatter.
c. Tolerance bands or circles based on mean and standard deviations offer promise for monitoring critical speeds, resonant vectors, damping, and at speed vibration vectors.
d. Tolerance levels of 2σ and 3σ offer two degrees of severity for evaluating deviations of characteristics from a particular run.
e. A 2σ tolerance circle for resonant or at speed vibration which penetrates the trip circle, suggests a need to balance.

f. Bearing gap change anomalies may help detect misalignment, but require further investigation; transient thermal conditions undoubtedly influence observed gap change.

g. The run-to-run variation in generator thermal vector indicates a need for more experience to establish appropriate criteria for this characteristic. Present limited data suggest 50 microns as an interim threshold level for thermal vector to trigger closer attention.

h. Inferred damping varies more widely (compared to its mean) than critical speeds; however, 2σ tolerance bands should still provide a basis for monitoring; evidence of sustained damping loss for two or more runs should trigger concern and attention for the unit in question.

i. In spite of high variance for deceleration based resonant vectors, detecting a significant change in this characteristic may offer the first evidence of a blade lost at speed.

j. Discontinuity in the exciter shaft position and vibration, when starting gear disengages, may be a basis for detecting excessive misalignment between starting gear and exciter.

In summary, transient vibration characteristics offer a promising basis for condition monitoring. Critical speeds and resonant vectors during acceleration and at speed vibration vectors, appear to offer the most repeatable reference for detecting significant change. Research will continue to assess the reliability of these criterion, and to actively establish more broadly based values for mean and deviation.

9. ACKNOWLEDGMENTS

Much of the work presented in this paper was conducted under research programs supported by the Electric Power Research Institute (EPRI). The authors wish to thank and acknowledge EPRI for their support, encouragement, and permission to publish the results.

10. REFERENCES

1. Riley, J.R., Gould, G.L., and Klover, R.A.: Can manufacturing capacity keep up with new orders for CTs?, Power Engineering (1990), 45-47.
2. Jeffs, E.: British energy situation after privatization, Turbomachinery International (1991), Vol. 32, No. 4, 60-72.
3. Gas Turbine Technology International Power Generation, 1992, 47-50.
4. Solt, J.C. and Tuzson, J.: Status of low NO_x combustor development, ASME 93-GT-270, International Gas Turbine and Aeroengine Congress and Exposition, Cincinnati, Ohio, 1993.
5. Schorr, M.M.: NO_x emission control for gas turbines: a 1991 update on regulations and technology, Turbomachinery International (1991), Vol. 32, No. 6, 24-30.
6. Simmons, H.R., Troxler, P.J., Smalley, A.J., Robinson, G.J. and Frischmuth, R.W.: Turning gear operation: its influence on combustion turbine rotor eccentricity and starting dynamics, ASME 93-GT-273, International Gas Turbine and Aeroengine Congress and Exposition, Cincinnati, Ohio, 1993.
7. Bently Nevada catalog, 1993.
8. Scientific Atlanta catalog, 1993.
9. Davidson, D.P. and Taylor, D.J.: Turbomachinery Instrumentation Seminar, Rotadata Limited, Derby, England.
10. Simmons, H.R., Smalley, A.J. Edlund, C.E., and Frischmuth, R.W.: Monitoring compressor blade tip clearance in combustion turbines, EPRI Steam and Combustion Blading Conference, Orlando, Florida, 1992.
11. Drinkuth, W., Alwang, W.G., and House, R.: Laser proximity probes for the measurement of turbine blade tip running clearances, Proceedings, 29th International Instrumentation Symposium, Albuquerque, New Mexico, 133-139.
12. Simmons, H.R., Michalsky, D.L., Brewer, K.E., and Smalley, A.J.: Measuring rotor and blade dynamics using an optical blade tip sensor, ASME 90-GT-91, Gas Turbine and Aeroengine Congress and Exposition, Brussels, Belgium, 1990.
13. Roth, H.: Vibration measurements on turbomachine rotor blades with optical probes, Joint Fluids Engineering Gas Turbine Conference and Products Show, New Orleans, Louisiana, 1980.

14. Mase, M., Rozelle, P.F., and Zagar, T.W.: Blade vibration monitoring technology, EPRI Steam and Combustion Blading Conference, Orlando, Florida, 1992.
15. Nava, P., Paone, N., Rossi, G.L., and Tomasini, E.P.: Design and experimental characterization of a non-intrusive measurement system of rotating blade vibration, ASME 93-GT-16.
16. Simmons, H.R. and Smalley, A.J.: Effective tools for diagnosing elusive turbomachinery dynamics problems in the field, ASME 89-GT-71, 34th ASME International Gas Turbine and Aeroengine Congress and Exposition, Toronto, Ontario, Canada, 1989.
17. Evans, B.F. and Pantermuehl, P.J.: Power and petrochemical vibration problems: quick solutions aided by microcomputer, 1984.
18. Bloch, H.P.: Use laser-optics for machinery alignment, Hydrocarbon Processing (1987), 33-36.
19. Lapini, G.L., Vadali, P., Zippo, M., Benante, R.: An automatic system for the control of rotor alignment in turbo generators, Conference on Monitoring, Surveillance and Predictive Maintenance of International Plants and Structures, (1989), 799-808.
20. Simmons, H.R., Smalley, A.J., Troxler, P.J., Adams, M.L., and Frischmuth, R.W.: Tools for troubleshooting misalignment of power plant turbomachinery, EPRI Troubleshooting Power Plant Rotating Machinery Vibrations Short Course and Symposium, San Diego, California, 1993.
21. DeMoss, S.H., A combined gas turbine control and condition monitor, ASME 91-GT-299, International Gas Turbine and Aeroengine Congress and Exposition, Orlando, Florida, 1991.
22. Bernstein, H.L.: Life management system for general electric frame 7E gas turbine, Reprinted from Life Assessment and Repair Technology for Combustion Turbine Hot Section Components, Proceedings of an International Conference, Phoenix, Arizona (1990).
23. Optical Pyrometer Measurement of Utility Gas Turbine Blade Temperatures, EPRI Technical Brief TB APS.33.7.88.
24. Meher-Homji, C.B., Lakshminarasimha, A.N., Dohner, C.V., Oncryas, I., Boehler, P.D., Lukas, H., and Cincotta, G.A.: Durability surveillance of advanced gas turbines: performance and mechanical baseline development for the GE frame 7F, ASME 93-GT-276, International Gas Turbine and Aeroengine Congress and Exposition, Cincinnati, Ohio, 1993.

25. De Lucia, M. and Lanfranchi, C.: An infrared pyrometry system for monitoring gas turbine blades: development of a computer model and experimental results, ASME 92-GT-80, International Gas Turbine and Aeroengine Congress and Exposition, Cologne, Germany, 1992.
26. Røkke, N.A., Hustad, J.E., and Berg, S.: Pollutant emissions from gas fired turbine engines in offshore practice-- measurements and scaling, ASME 93-GT-170, International Gas Turbine and Aeroengine Congress and Exposition, Cincinnati, Ohio, 1993.
27. Dolbec, A.C., Dohner, C.V., and Angello, L.C.: Advanced diagnostic instrumentation for combustion turbines, in Proc. of the American Power Conference.
28. Schreiber, H.: Gas Turbine Combustor Viewing System, EPRI Journal, 1988.
29. Harris, R.E., Simmons, H.R., Smalley, A.J., Baldwin, R.M., and Quentin, G.: Remote condition monitoring of a peaking gas turbine, ASME 91-GT-185, International Gas Turbine and Aeroengine Congress and Exposition, Orlando, Florida, 1991.
30. Baldwin, R.M., Smalley, A.J., Harris, R.E., and Quentin, G.H.: Development of a knowledge based advisory system based on vibration of a peaking gas turbine, ASME 92-GT-53, International Gas Turbine and Aeroengine Congress and Exposition, Cologne, Germany, 1992.
31. Ewins, D.J.: Modal Testing: Theory and Practice, Research Studies Press Ltd., Letchworth, Hertfordshire, England, 1986.
32. Simmons, H.R., Smalley, A.J., Frischmuth, R.W., Lapini, G., and Robinson, G.: Tools for diagnosing case deflections and alignment on a power utility combustion turbine, ASME 92-GT-179, International Gas Turbine and Aeroengine Congress and Exposition, Cologne, Germany, 1992.

VIBRATION DIAGNOSTICS OF ROTATING MACHINES IN NUCLEAR POWER STATIONS IN SWEDEN

Y. Vidhög

Forsmarks Kraftgrupp AB, Östhammar, Sweden

ABSTRACT

Vibration surveys have been important already during first commissioning and start up of the Nuclear Power Stations in Vattenfall (Swedish State Power Board).
We believe that discovered vibration problems should be corrected as soon as possible, before they result in increased maintenance costs and operation losses.
Preventive maintenance as well as predictive and proactive maintenance are important tools. Permanent vibration monitoring with e.g. vibration computers on important machines together with periodic measurements with computer based data collectors form a base for the vibration predictive maintenance work.
This paper describes:

- Diagnostic strategies on our large turbines as well as other machines, including measurement philosophy.
- Off-line measurements and diagnostic techniques for the analysis of vibration problems.

Case studies with good use of both permanent on-line monitoring as well as off line diagnostics are included.

1. INTRODUCTION

My experience is based on almost 18 years of work with vibration in Nuclear Power Stations in Vattenfall (Swedish State Power Board) both from commissioning and normal operation. The first 8 years, I was with the Maintenance Department at Ringhals, responsible for the vibration measurements and analysis at Ringhals 1-4, one BWR and three PWR, with totally 8 large steam-turbines.
Since almost 10 years in Forsmark, I work in the Technical Department with vibration diagnostics and troubleshooting acting as consultant to maintenance and operation departments. Forsmark has three units, all BWR:s, where unit 1 and 2 have two turbines each (2x500 MW, 3000 RPM) and unit 3 have one steam-turbine (1200 MW, 1500 RPM).
I have also been active in the ISO standardization work with standards for balancing and machinery vibration since many years .
According to my experience, many maintenance problems are already bought with the machine. Many vibration problems come from resonances in the structure and/or in the shafts. Other problems come from inadequate balancing quality or misalignment during the commissioning. Most of these problems are detectable already during commissioning before the maintenance problem occures and causes any costs and loss of operation time. The right time to prevent future problems is before the purchase, when making the specifications for the machine and include adequate vibration criteria in the purchasing contract. Vibration levels on installed machine, balancing quality, resonance influence, unbalance sensitivity are some examples of factors to be considered.
To run a machine close to a resonance means an increased sensitivity for normal disturbances. Higher accuracy in the alignment and balancing than normal is needed. Therefore on Ringhals 3-4, and Forsmark 1-3 a lot of efforts have been spent on early detection of incipient vibration problems.
Various types of modifications have been applied to move resonances, balancing on site and in our own balancing machine are some of the steps to decrease the vibration level and the sensitivity to unbalance.
Vibration measurements both using permanent vibration monitoring with eg vibration computers and periodic measurements now normally with computer-based data-collectors are important tools to follow up the status of all machines and to detect vibration changes.
This paper will mainly describe the procedures at Forsmark Nuclear Power Station, but will include in general the other Nuclear Power Stations in Sweden.

2. PERMANENT VIBRATION MONITORING.

2.1 GENERAL PHILOSOPHY.

The normal vibration monitoring on rotating machines have since long been based on velocity measurement on the bearings, normally with a velocity transducer measuring the bearing absolute vibration in mm/s RMS.

The velocity transducer have been chosen because it gives the vibration velocity direct, it is rugged and not so sensitive to disturbances. It will not need any electronics out at the machine, just the velocity transducer and then a shielded low impedance twisted lead cable. Amplifiers for low-frequency linearization, with balanced inputs can be mounted in a good environment such as the control-room and be attended during operation.

The transducer type is in most cases of Swedish make, type VT 11.17, but other makes and types exist.

The frequency range has normally been 10-1000 Hz (ISO-Severity) for permanent monitoring. On gearboxes, the frequency range is extended upwards to cover the gear teeth frequency better.

On machines with lower speed a special low frequency linearization is utilized to achieve a linear (phase and amplitude - linear) signal. Normally a frequency range from 1 or 2 Hz up to 2000 Hz is used.

Especially on pipes (like steam-pipes) with main excitation frequencies in the range 1-20Hz, velocity transducers with low-frequency linearization have been used with very good result for permanent monitoring. Accelerometer measurements in radioactive, humid and warm places, where low frequencies have to be measured are always difficult and need special arrangements like double-shielded cables, careful mounting of cables and electronics, and requires expensive maintenance. Very often problems with low-frequency disturbances or 50 Hz appears.

For measurements like loose-part monitoring on PWR:s where accelerometers are needed because the high frequency, we have had good result with accelerometers with impedance converters mounted near the accelerometers in the station and then the charge amplifier on long distance in the control room. Care should however be taken with the cabling, shielding and grounding to avoid 50 Hz electric noise.

We normally regard shaft vibration measurements (with proximity probes) as an extra addition (complementary) in special cases or for tests. We have sofar had several bad experiencies when only shaft proximity probes in the bearings have been used on for instance multi-stage centrifugal pumps. The transducers, when mounted inside the bearing, were close to a deflection mode shape node and did not give the expected monitoring or warning for increased vibrations.

The transducer position should be decided according to the application, near or far away from bearings.

On the large steam turbines both velocity transducers and shaft proximity probes are used for permanent monitoring. The shaft probes are used for balancing the 1st critical, to

check that the rotor is not bent, for diagnostics of vibration changes etc.
Accelerometers are used when their special abilities are needed for example:
- on the stator windings where we have high electric and magnetic field
- on small light structures, like thin pipes
- where very high frequency have to be measured

On several machines such as Turbines, Feed-water Pumps, Condensate Pumps, Clean-Water Pumps, Emergency Diesels and important Fans we have also mounted magnetic phase reference pickups. They are used to measure the vibration vector to follow up vibration changes and for balancing.

2.2. PERMANENT VIBRATION MONITORING EQUIPMENT ON LARGE TURBINES

Normally, the large nucler steam turbines in Sweden have a vibration computer for on-line monitoring of the vibrations.
Below is a short summary of the different turbines also showing what computer types that are permanentally installed for the moment (1993). The vibration computers will be described later.

Station	Unit	Turbines/ reactor	Speed RPM	MW/ each	Bearings per LP	Vibr Comp	Type
Barsebäck	B1	1	3000	610	1	Yes	TVM 10
-"-	B2	1	3000	600	1	Yes	TVM 10
Oskarshamn radialturb	O1	1	3000	460		Yes	TVM 10
Oskarshamn	O2	1	3000	620	1	Yes	TVM 10
-"-	O3	1	1500	1200	1	Yes	TVM 12
Ringhals	R1	2	3000	410	2	Yes	Prosig
-"-	R2	2	3000	455	1	Yes	TVM 300
-"-	R3	2	3000	460	1	No	
-"-	R4	2	3000	480	1	No	
Forsmark	F1	2	3000	500	1	Yes	TVM 10
-"-	F2	2	3000	500	1	Yes	TVM 10
-"-	F3	1	1500	1200	1	Yes	TVM 300

Table 1.

Most large nuclear steam-turbines in Sweden are of similar design, traditional BBC-turbines with one HP, three LP:s and then Generator and Exciter. The LP:s normally only have just one bearing each, while HP, Generator and Exciter have two bearings each, even though there are other types in Sweden. Ringhals 1 (R1), for example has two bearings per LP and are of a different design, Oskarshamn 1 (O1) is of another design with a twin-shaft counter-rotating radial turbine Ljungström-type).

2.3 TRANSDUCER ARRANGEMENT ON STEAM TURBINES

The turbines normally have two separate transducer- and measurement systems. One for alarm and trip that was delivered together with the turbine and connected to the turbine supervision system. The other system is for measurement, alarm and diagnostics and is connected to the vibration computer.

A) The basic protection-system (Amrein), included in the turbine delivery, with one velocity transducer per each bearing, normally in the vertical direction. The seismic velocity transducer measures absolute bearing velocity in mm/s rms. The vibrations from these transducers are displayed on meters, pen-recorders (normally line-recorders) and also on the operation computer together with alarm and trip signals. The alarm and trip is on total vibration (overall) level.

B) The other system has been installed for monitoring of startup/rundown and operation of the turbine and is used especially for analysis, diagnostics and alarming. Basically this system for the turbine set normally have two velocity transducers on each bearing, one in vertical and one in horizontal direction. Some turbines like for example Forsmark 3 (1500 RPM) also have a third velocity transducer per bearing, measuring in the axial direction.

All these seismic velocity transducers measures absolute bearing velocity in mm/s rms and are of Swedish make, type VT11.17. On the lower speed machines special low frequency linearization is performed.

The turbine-set also generally have 2 shaft relative (proximity) probes installed at each end of the rotors: HP, LP:s, Gen and Exc, with one probe in vertical direction and one in horizontal direction.

The proximity probes are not installed in the bearing. We try to position them far away from the bearings, so they will not be near the nodes (which often are close to the bearings). They are normally mounted on the steamseals of the LP:s in the end of each LP-rotor, as we are interested of the clearance in the seals and to detect rubs in the seals.

The shaft "proximity" probes measures shaft vibration relative to the housing, or the seals. The AC-level is the measured shaft vibration, with the unit microns peak.

The DC-level, which is the actual distance between transducertip and the shaft is measured separately.

We also have a phase reference transducer for measuring the speed and for vector measurements.

On the generator stator windings on the 3000 RPM generators we normally also have vibration measurements with accelerometers. These accelerometers measures the absolute vibration of the stator windings in the radial direction. The signal is converted to vibration velocity in mm/s rms. All these extra transducer channels are completely separate from the basic protection-system and connected to a separate measurement system.
Fig 1 and Fig 2 shows the normal transducer positions on a large steam turbine.

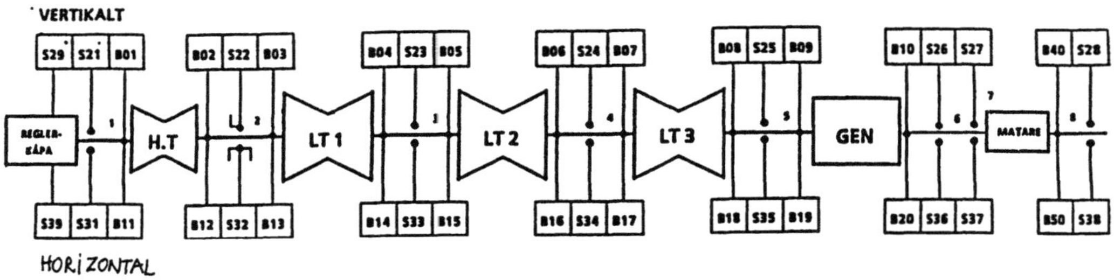

Fig 1: Vibration probe locations on a large steam turbine set.
B denotes proximity probes and S denotes seismic transducers.

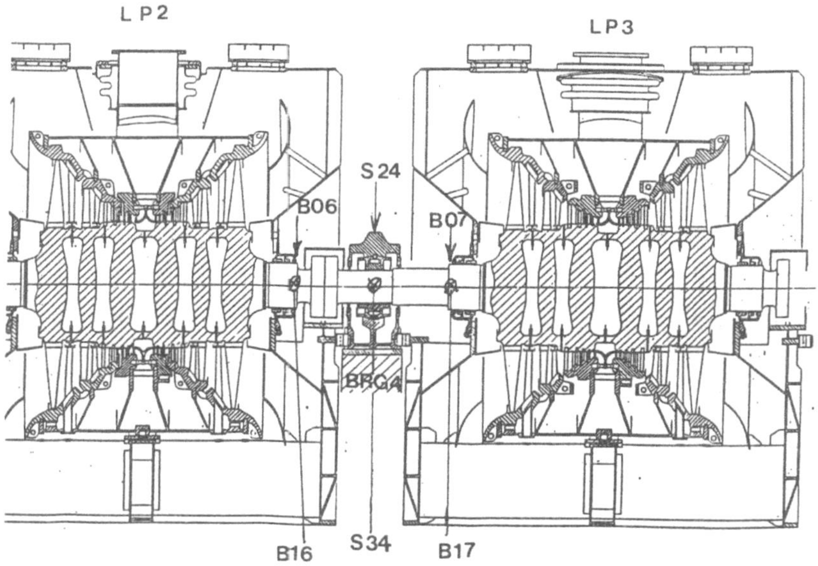

Fig 2: Example of vibration probe location on a LP-section of a steam-turbine
B denotes the proximity probes and S the seismic velocity transducers.

2.4. MEASURING SYSTEMS ON THE STEAM TURBINES.

The transducer system installed for monitoring, analysis and diagnostics are connected to two different measurement systems:

a) Manual backup system.
On all turbines these extra velocity transducers, and if present also the shaft proximity probes, are connected to a manual backup system. Each channel goes through a pushbutton switchboard to a Vector Filter, type Vibroflex (of Swedish make). Vectors 1xrpm, 2xrpm and up to 9xrpm can be measured.
The DC-level of the shaft proximity probes is measured on a panel Voltmeter.
In the pushbutton rack, there is also an oscilloscope where the vibration signals can be studied. Outputs for taperecorders and FFT are available.
Before the vibration computers were installed this system was used by the operators on a daily basis and also for vibration analysis. It now mainly serves as backup for the computer, but also for off-line FFT-analysis with the computerbased data collectors.

b) On-line monitoring with vibration computers.
The transducers from these extra channels are normally also connected to a vibration computer for on-line monitoring of the vibrations. (See also table 1.)
All turbines in Forsmark have had vibration computers since the commissioning more than10 years ago. Below the different type of vibration computers used in Forsmark are described.

2.5. VIBRATION COMPUTER TVM10.

Forsmark 1 and 2, with 2 turbines each, all together 4 turbines (3000 RPM, 500MW each) have the vibration transducers connected to one vibration computer (type TVM 10) on each station.
TVM 10 was installed in Forsmark 1981-82, and was designed by the turbine supplier (ABB-STAL). HP 1000 computers are used. Each TVM 10 can handle 2 rotors, with maximum 64 channels each. Both vibration channels and operation parameters are connected. The total overall vibration level, together with the filtered vibration vectors 1XRPM and 2XRPM are stored on a removable hard disc together with RPM, MW and MVAr and time.

This computer can measure quick enough to follow the run-up and run-down well and provides a good tool to detect and analyse changes of vibrations during normal operation. FFT measurements during normal operation on one transducer at a time can be done. Monitoring of overall level and the vibration vectors with alarms on level, vectors and trends are performed during normal operation.
The computer uses a circle around the normal vibration vector as alarm for vector change. We also have a multi-plane balancing program in this computer, which we use for the balancing of the turbines.

On TVM 10 the measurements have to be stopped when evaluations are done and diagrams are printed.
The TVM 12 as found in Oskarshamn 3 is a newer modified version of TVM 10. It has for example a little more sophisticated alarm sets, where alarms can be set for example for different speeds.

2.6. VIBRATION COMPUTER TVM300.

On Forsmark 3 we have a newer model of computer called TVM 300, also designed by the turbine supplier (ABB-STAL).
This computer was installed 1984 and the program was developed before the startup of F3 by the supplier in cooperation with Forsmark. Forsmark 3 has one turbine running on 1500 RPM.
Two separate HP1000 computers are used, one for measurement and one for analysis/diagnostics. Measurements, diagnostics and diagram plotting can be done at the same time.
TVM 300 can handle up to 32 different rotors with different speeds during normal operation, and to follow 2 rotors like turbine rotors with many transducers at the same time during a run-up and a run-down transient.
Vibration vectors 1XRPM, 2XRPM up to 5XRPM can be measured on each rotor. On the F3 turbine we normally store the first 4 multiples of the speed.
Maximum number of transducers for the computer is 256, with max 64 per rotor.
Both vibration signals as well as operation parameters can be measured. On F3 we store together with vibration signals several operation parameters like MW, MVAr, Speed, Cooling Water Temp, Condensor Vacuum, Bearing Oil Temp, Temp and Pressure from Seal and Leakage system, Temp before LP-turbine etc.
If we want to study the influence of some special parameter or vibrations on some new components we can easily connect extra transducers.
From commissioning, only the Turbine and the Feed Water Pumps were connected to the computer. Additional vibration transducers, both permanent and temporary installed on valves, pipes, condensor tubes, pumps etc have later been connected to TVM 300 and are measured together with operational parameters. This gives us a good possibility to follow up the vibrations on different components during different operation conditions.

On TVM300 the vibrations are stored in different buffers, and for the turbine we have chosen the following:
Short Buffer: - measurement every 5 minutes and special events, lasting 3 days. (gives a possibility to follow up carefully what have happened during e.g.a weekend)
Normal Buffer: - measurement every hour, lasting 2 months. (used for the normal monthly evaluations)
Long Buffer: - measurements every 8 hour, lasting more than 1 year (used for special evaluations of the whole operation season, showing any differences both between night

and day and between winter and summer with different Cooling Water Temp etc)
Run-up/Run-down Buffer: - measurements during run-up and run-down are stored
Interesting data are then stored on special tapes and saved.

FFT-measurements both during normal operation and during start-up can be performed on several transducers at a time and also one by one.
Different alarms for different speed ranges can be applied. Two different alarm levels for each type is possible, like H1=alert and H2=danger.
Level alarm on overall vibration, and for 1xRPM- and 2XRPM-components are normally used.
The vector alarm, as a radius of a circle centered on setable points, for both H1 and H2 alarm are used. Different trend alarms for both quick and slow changes can be applied.
Alarms for changes in the FFT spectrum are also available.

Ringhals 2, (a PWR) have also installed TVM 300. They measure both on their two turbines, 3000 RPM each, and on several pumps like Charging Pumps, Reactor Coolant Pumps etc.

2.7. TRIP LEVELS ON THE TURBINES.

In Forsmark the automatic trip for overall levels on the turbine on the basic protection system is set as follows:

a) 3000 RPM-turbines
Automatic trip = 9,0 mm/s rms (both run-up/run-down and normal operation)

b) 1500 RPM-turbines
Automatic trip = 5 mm/s rms (both run-up/run-down and normal operation)
Manual trip = 3 mm/s rms (full speed 1500 RPM and load)

When vibrations exceed 2 mm/s rms (Alarm) on our 1500 RPM turbines steps should be taken to reduce the vibrations. If they continue to increase the turbine should be manually tripped at 3 mm/s rms. The reasons for this are the low damping of 1st criticals during run-down which must be possible to pass through after a trip.

3. SOME SPECIAL MEASUREMENTS AND DIAGNOSTIC STRATEGIES WE USE ON LARGE TURBINES.

3.1 DURING REVISION.

- Measurements with strain-gauges to check that there is Zero-Moment in the couplings between the rotors. This measurement is more and more used.

- Run-out balancing.
  After overhaul we can compensate with balancing weights for the change of least 8 marked radial directions are taken and compared. The change of excentricity can then be compensated before starting the turbine after overhaul.

- Laser is more commonly used for alignment check of bearings and foundation.

- Balancing before startup after revision if no large jobs have taken place that can affect that special part of the rotor.
  Calculations using the vibration behavior before the shut-down. Balancing for both first critical and for full speed can be done.
  By compensating known unbalances before the startup, we can minimize the balancing time during startup and also achieve a better status of the machine.
  For example if we know the first critical is a bit high on LP1 and LP2, and we only are going to change the generator rotor during shutdown, we would compensate for these known unbalances already before starting up the turbine.

3.2. DURING STARTUP.

- Barring before starting the turbine.
  Barring is normally done for 24 hours after a revision when the rotors have been stopped. The barring time has been reduced since the commissioning as we now always hold at a certain speed (500 RPM or 250RPM depending on nominell turbine speed) to check if the rotor is bent. We then wait there until the rotor is straight. This procedure will save time, as the rotor straightens quicker at higher speed and will also give a better check and a possibility to correct if the rotor is bent.

- Check if the rotor is bent when running on a low speed.
  For our 3000 RPM turbines, we hold and wait on 500 RPM for the rotor to straighten, if it is bent. The 1xRPM-vector on the shaft probes are measured and checked for stability and compared with normal levels. For our 1500 RPM turbine we stop at 250 RPM. The speed should be as high as possible, but well below influence from the first critical.
  By doing this we can save time and prevent trip due to vibrations during running up. These normal "run-out"-levels are measured both on runup and rundown regularly. The measured 1XRPM-vector for the proximity probes at low speed can be used to verify if there has been any coupling change (change of excentricity) due to e.g. load transients. It is also a good help in the analysis of vibration changes during operation.

- Careful check of vibration vectors during running up and full speed.
  1XRPM-vectors and 2XRPM-vectors are evaluated regularely. We also measure 3XRPM- and 4XRPM-vectors on the newer turbines. We can analyse them if something out of normal is observed. Balancing is carried out if needed.

3.3. FOR BALANCING.

- Unbalance tests, with systematic test runs of an unbalance in each plane, have been performed and the result form the basis for the influence coefficient matrix used for the balancing. For Forsmark 1 and 2 we have performed the unbalance tests ourselves in Forsmark and calculated the unbalance sensitivities. We use the same matrix for all four turbines.
  For Forsmark 3 result from test on a similar machine is used for the balancing at full speed. However the actual measurement of the 1xRPM vectors for the proximity probes can be used directly for the balancing of the first critical which is the main issue of concern.

- The influence coefficient matrix is used for balancing but also for the analysis of vibration changes during operation.
  Balancing is often performed without any help from the manufacturer.
  We normally use trips to get a chance to improve the balance status of the turbine and we try to keep the levels very low as well. This is important, since experience shows that this enables us to detect changes early and to have margin for changes during the season without getting too high levels.

3.4. DURING OPERATION.

- Vibration measurements on the turbines with vibration computers enables us to have a good possibility to follow and detect changes of the turbines early.
  Measurement of the overall level and vector mesurements of the 1XRPM-vector are also stored (TVM 300) and can be analysed if needed.
  On our newer turbines (1500 RPM/TVM 300) we also monitor the DC-level from the proximity probes. The DC-level is useful for the analysis of rubbing in the seals and for the positions of the seals at certain occasions.

- The measurements with transducers in the different stiffness directions of the machine like horizontal and vertical direction are very useful when analysing the reason for the vibration. The positions of the proximity probes far away from the nodes also give useful and meaningful information away from the nodes. The readings are effective for balancing and for analysis of any rubbing in the seals.

- The coefficient matrix is used when analyzing vibration changes during operation.
  By using the matrix we can calculate for example which end of a LP a rub is taken place, where a coupling change has occured during operation or where we have got an unbalance change due to lost rotating parts. A suspected coupling change can then be verified with both vector measurements at run-out on eg. 500 RPM as described above and by indicating the coupling when the machine has been stopped.

4. SOME STRATEGIES TO REDUCE THE VIBRATION AND COSTS FOR MAINTENANCE AND OPERATION OF OUR MACHINES.

Based on practical experience and after cost-efficiency analysis we have found the following steps important:

A. Balancing in own balancing machine.
Vattenfall (Swedish State Power Board) have a balancing facility of their own. We purchased the balancing machine, a Schenk H5UB with a special longer foundation, to Ringhals 1983. The capacity is from very small rotors up to 5 ton, with variable speed to above 3000 RPM. Also Forsmark balance their rotors in this balancing machine.
One of the reasons for installing the machine was the problems we had when sending radioactive rotors away to normal industry for balancing. Normally only low-speed balancing could be performed at high cost. We wanted to be able to make high speed balancing and also balancing on variable speed to compensate for different modes.
Checking the unbalance state on new components and balancing quality if needed is normal procedure since many years. We found that many components in the warehouse (spare parts) were not so good as one would expect from a new or newly revised part. Especially impellers often had high dynamic unbalance and many had obviously not been balanced properly according to ISO 1940.
Balancing of hydraulic couplings with several shafts has been performed with good result in the balancing machine, building the set together step by step by prebalanced parts and then balance it together. The rotor set is not disassembled after final balancing.

B. Vibration check and balancing if needed at the facility of the manufacturers.
As an example we can mention that vibration check and balancing, if needed, is a normal procedure after motor service. The motor rotor is first balanced alone and then together with coupling before it is installed again on site.

C. Balancing on site when needed.
Balancing by site personal is a normal procedure since more than 15 years in Vattenfall (Ringhals and Forsmark). Examples of components that have been balanced on site are: Steam Turbines, Generators, RCP:s, Charging Pumps, Emergency Diesels, Feed Water Pumps and Hydraulic Couplings, Condensate Pumps/Motors, Ventilation Fans etc.

D. Alignement using laser.
This technique is more and more commonly used during the recent years.

E. On-site evaluation of lubricating oils.
With Oilview, a test equipment for oilanalysis from CSI, we can now on site monitor degradation of the lubricant, contamination of the lube system as well as mechanical wear. The oilanalysis can be performed easy and periodically on site by the maintenanace

people. This is proactive as we can find and fix problems before they cause damage, due to for example oil degradation and oil contamination. It is also predictive as we can monitor the concentration of ferrous wear particles.

F. Checking of resonances and making modifications if needed.

Different type of measurements can be performed to detect if there is resonances in the structure and/or shaft, which affect the running. Some examples are given below:
- shaker tests (an easy way to test for resonances and also to monitor the modeshape)
- excitation tests with rubber hammer (easy check for resonance frequencies)
- modal analysis with more sophisticated equipment if needed
- monitor the run-down and vibrations as a function of speed, with vector monitoring
- critical speed calculations with computer programs
- unbalance tests to check the unbalance sensitivity and the residual unbalance

G. Using frequency calculation programs as a tool in fault analysis.

By using a frequency calculation program (CSI) we can easy calculate the different frequencies that different type of damages create for bearings, motors, gears and belts. Data from over 10000 different type of bearings are included. Frequencies from electric faults on motors, like broken rotor bars or excentric rotor mounting, can easily be analysed. With these programs we can then analyse which of the bearings are damaged and need replacement and have a tool for the decision of what type of maintenance work is needed.

H. Testing of all new machines during commissioning for early detection and correction of faults.

Vibration measurements on all machines during the first startup. By detecting vibration problems early we can prevent from future increased maintenance costs and operation losses. If a resonance is the cause of the vibration problem then we try to modify and move the resonance away. According to our experience there is normally no problem to fullfill the vibration criteria from ISO 2372 class A on a newly installed machines, if it is balanced according to ISO 1940. If the vibrations are higher very often some resonance is close and increase the vibrations.
There are often problems with for instance parallell pump-mounting, where one pump is running and the other is standby. Standing still vibrations often occur if there is excitation of structural resonances.
Another machine mounting that often also causes problem is seismic mounting on rubber feet. Very often the design has not taken into account all degrees of freedom in the calculations and the result is often that some resonance is affecting the running giving high vibrations.

Modifications have been performed on many of our machines (Ringhals and Forsmark). Below are some examples of modifications to the structure and/or the shaft:
- stiffened bearing with a plate
- stiffened foundation
- increased mass on bearing
- mass on top of motor
- changed diameter of rotor
- increased mass of rotor
- changed type of rubber feet
- stiffened floor under machine
- stiffened piping between pumps
- seismic mounting on rubber

I. Periodic vibration monitoring of all machines since commissioning of the stations..
From the beginning, more than 15-20 years ago, the periodic measurements where done with handheld vibrometers or vector filters for measurement of overall level, manual filtering and vector measurements. Later with use of RTA-analysis (FFT).
Since 1987, Forsmark use computer based data collectors for the preventive/pedictive vibration measurements. We use CSI 2110-instruments together with the CSI MasterTrend programs.
Ringhals and Barsebäck have also the CSI system for vibration measurements and analysis. Oskarshamn have used another system based on a tape recorder and frequency analysis with computer (type Nåiden).
The vibration measurements can be used as a maintenance tool for the preventive and predictive maintenance.
- to decide type of action on the machine
- to choose the machine to be revised
- to correct faults early before major damages occur, and correct early when only minor adjustments or repairs are needed. This saves spare parts and maintenance costs.

By keeping the vibration levels low we can early detect faults and vibration changes.

5. CASE STUDIES WITH USE OF PERMANENT ON-LINE MONITORING OR OFF-LINE DIAGNOSICS.

CASE 1: SUDDEN VIBRATION CHANGE DURING OPERATION OF A LARGE STEAM TURBINE.

Due to permanent on-line monitoring a sudden change of unbalance was discovered. The turbine (500 MW, 3000 RPM) was running on full load and had been running for some weeks since the overhaul. Due to low power demand, at that time, turbine no 2 was stopped for the moment. The turbines on the other units did not show any vibration changes at that time.

A vibration change on the 1XRPM-vector of 0.5-2.0 mm/s rms was seen on the bearing velocity transducers and change of 5-25 microns peak on the shaft proximity probes. The total level on the basic alarm and trip system was well below alarm limit but we had vector alarm on some transducers on the TVM-system. On the pen-recorders from the original vibration system one could just barely notice a small step.There had not been any change of operation parameters at that time and no known disturbance on the power net.

Figure 3 shows diagrams from some transducers plotted from TVM and also the overall level from the pen-recorder. For exact positions of the transducers see fig 1 and 2.
The largest change was noted on the proximity probes on the LP1-rotor. This rotor had been out during the overhaul, so the couplings on both sides had been opened recently. We were however told by the turbine manufacturer at that time, that there could be nothing wrong with the couplings.
Calculations using the influence coefficient matrix on the vibration change, pointed on a vibration change in the generator side of LP1. The size of the change was possible to come from a coupling change but also possible from a lost blade.
To be able to analyse further we decided to start the other turbine and to be able to then stop this turbine. The run-down would give more information of what had happened, and if needed possibility to make some measurements or inspections.
During run-down, the first critical was passed with much higher levels than normal. The out-of through, excentricity measured on 500 RPM, pointed on a coupling change between LP1 and LP2.
By checking the couplings with dial-indicators it was verified that we actually have had a coupling change at that coupling. This was however not possible to prove until we had run down with the turbine.
Permanent monitoring of vector measurements and analysis of the vector change was very important for the detection and analysis of the problem.
Afterwards it was found, that there actually had been some electrical coupling work going on in the switch gear at the time of the vibration change, and because an electrical reactor for the moment was out of order, the voltage change was a little higher than normal. This resulted in the coupling change. It seems that the affected coupling not yet was in its normal stable position and therefore we got the coupling change.

CASE 2: CHANGED VIBRATION CONDITION AND SUSPECTED BEARING DAMAGE ON LARGE COOLING WATER PUMP.

The Cooling Water Pump, is a vertical pump running on 360 RPM with power 515kW. There are 4 pumps running in parallell on each unit. The vibration levels on the pump and the drive side of the motors are normally very low, typically 0,2-0,5 mm/s rms. Levels above 1 mm/s indicates that something is out of normal. One day one of the operators noted a strange sound, when he was passing this pump. We were called out to make an analysis.
We used our CSI-system. Highest vibration was noted seen on the drive end of the motor

in axial direction where the total level was found to be above 4 mm/s rms. The radial vibrations were much lower 1-3 mm/s rms. Other bearings on the motor and pump were increased but below 1.5 mm/s rms.
The sound from the area between the pump and the motor reminded of a reciprocating machine (indicating many frequency components). The frequency spectrum showed a large number of peaks at much higher frequencies than normal. The spectra did not have the pattern of a normal worn bearing. Several spectras with different frequency ranges were taken, and the waveform was also stored. Analysis of the waveform showed spikes with a delta frequency of 33.2 Hz.

Figure 4 shows the spectra and the waveform from the bearing.
The frequency spectra showed a number of frequencies also with delta frequency 33.2 Hz. The frequency calculation program was used for the different bearings in the machine and we found that the frequencies probably came from the innerring in the SKF 6330 bearing in the drive end of the motor.
Figure 5 shows the result of the frequency calculations for that bearing.
We therefore suspected one damage in the inner ring of that bearing or maybe a crack in the inner ring. The bearing was exchanged and the vibrations went back to normal low levels. The old bearing was cut in two pieces for inspection and we found, as expected, one damage in the inner ring surface, a spot with about 20 mm diameter.
The reason for the damage was probably a small initial damage due to handling from the beginning. This initial damage would then make this spot more sensitive for spalling, thus giving the damage we found on the inner ring.

8. SUMMARY

Vibration measurements and diagnostics are the base for predictive maintenance and condition monitoring of machines. By improving the vibration status of the machines we believe it is possible to reduce the costs for the maintenance considerably and also minimize the cost for operation losses. Both permanent on-line vibration monitoring and off-line diagnostic techniques are important tools in the maintenance work.
Improved procedures and more sophisticated equipment in the maintenance work will considerably decrease the vibrations and reduce the wearing and damages.
Predictive and proactive maintenance are important tools.
Diagnostic strategies and measurement philosophy are described together with some case studies, where we had good use of both permanent on-line monitoring and off-line diagnostics

Fig. 3: Case 1. B03,B04,B05 (shaft vibration) ,recorder plot (Amrein) bearing vibr.

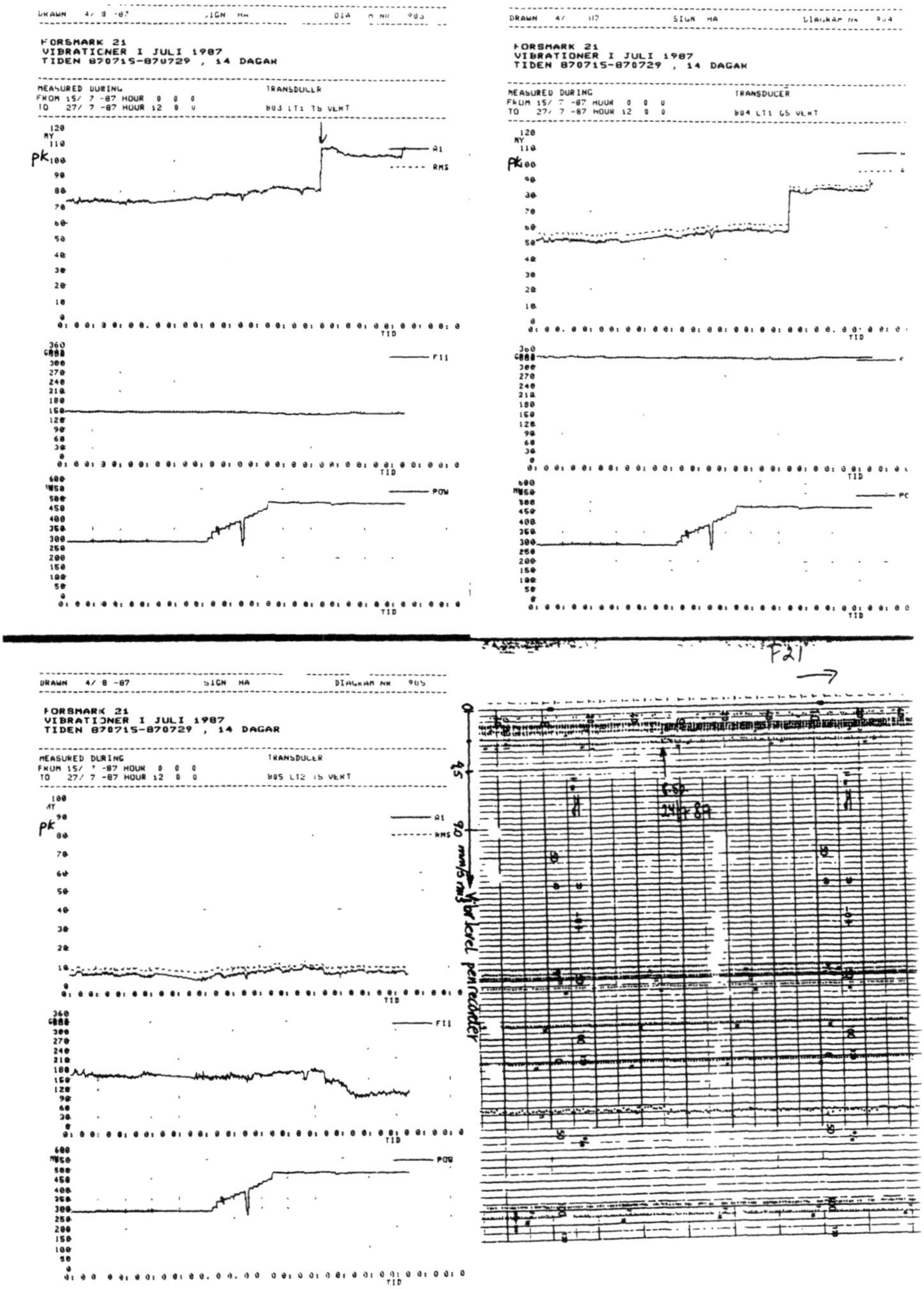

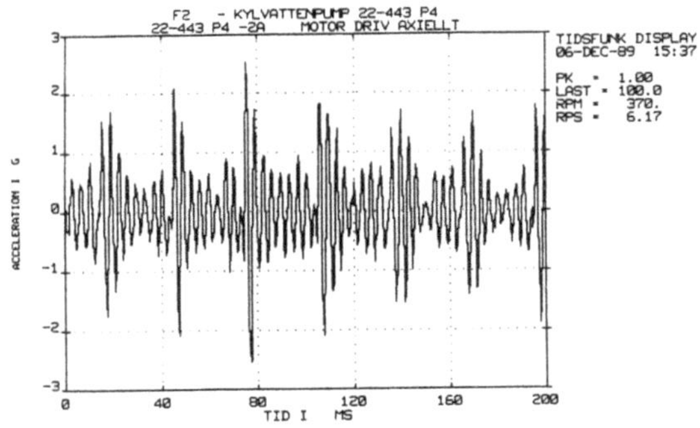

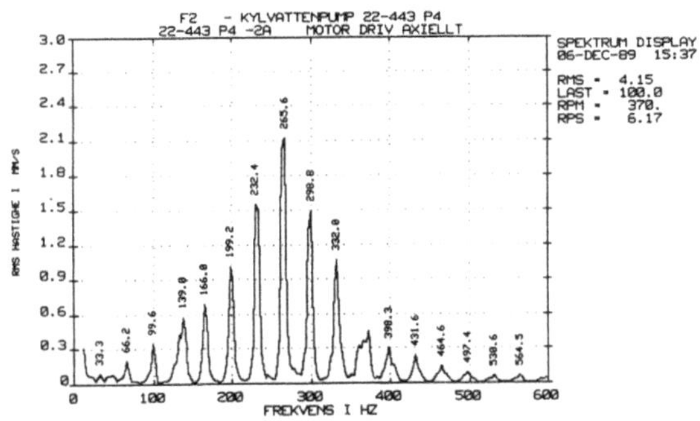

Frekvensanalys
****

Maskin: (F2) KYLVATTENPUMP 22-443 P4
Mätpunkt: 22-443 P4 -2A --> MOTOR DRIV AXIELLT
Dat./Tid : 06-DEC-89 15:37:33 Sort: mm/s RMS

TOPP NR	FREKVENS (Hz)	TOPP NIVÅ	MULT. NIVÅ	TOPP NR	FREKVENS (Hz)	TOPP NIVÅ	ORDER MULT.
1	33.34	.0876	5.41	11	277.54	.1460	45.01
2	49.25	.0937	7.99	12	298.84	1.6655	48.46
3	66.21	.2070	10.74	13	332.04	1.1476	53.84
4	99.55	.3572	16.14	14	361.30	.3635	58.59
5	133.36	.4839	21.63	15	367.41	.4244	59.58
6	139.00	.6163	22.54	16	371.08	.4815	60.17
7	166.04	.7390	26.93	17	398.29	.3174	64.59
8	199.25	1.1240	32.31	18	431.58	.2465	69.99
9	232.43	1.7958	37.69	19	464.56	.1504	75.33
10	265.64	2.4439	43.08	20	497.44	.0953	80.67

TOTALNIVÅ
4.1503

Figure 4: Case 2. Frequency spectra and waveform from motor drive end brg axial.

Bearing frequency calculation.

LAGERFREKVENSER FOR SKF 6330
****

Tabellvärden . (TABELLVÄRDEN)

K (FTF):	.40
CA (BPFI):	5.41
KA (BPFO):	3.59
R (BSF):	2.37

MULTIPLAR	RPM-MULT.	HÅLL. (FTF)	ROT. (BSF)	YTTRE (BPFO)	INRE (BPFI)
1	6.17	2.46	14.59	22.13	33.37 Hz
2	12.33	4.92	29.18	44.25	66.75
3	18.50	7.38	43.76	66.38	100.12
4	24.67	9.83	58.35	88.50	133.50
5	30.83	12.29	72.94	110.63	166.87
6	37.00	14.75	87.53	132.75	200.25
7	43.17	17.21	102.11	154.88	233.62
8	49.33	19.67	116.70	177.00	267.00
9	55.50	22.13	131.29	199.13	300.37
10	61.67	24.58	145.88	221.26	333.74

MULTIPLAR	YTTERRINGSSKADA (BPFO) MED SIDBAND			INNERRINGSSKADA (BPFI) MED SIDBAND		
1		22.13			33.37	
	15.96		28.29	27.21		39.54
	9.79		34.46	21.04		45.71
2		44.25			66.75	
	38.08		50.42	60.58		72.92
	31.92		56.58	54.42		79.08
3		66.38			100.12	
	60.21		72.54	93.96		106.29
	54.04		78.71	87.79		112.46
4		88.50			133.50	
	82.34		94.67	127.33		139.66
	76.17		100.84	121.16		145.83
5		110.63			166.87	
	104.46		116.79	160.71		173.04
	98.29		122.96	154.54		179.21
6		132.75			200.25	
	126.59		138.92	194.08		206.41
	120.42		145.09	187.91		212.58
7		154.88			233.62	
	148.71		161.05	227.45		239.79
	142.55		167.21	221.29		245.95
8		177.00			267.00	
	170.84		183.17	260.83		273.16
	164.67		189.34	254.66		279.33
9		199.13			300.37	
	192.96		205.30	294.20		306.54
	186.80		211.46	288.04		312.70
10		221.26			333.74	
	215.09		227.42	327.58		339.91
	208.92		233.59	321.41		346.08

Frekvenser i Hz

Figure 5: Case 2. Bearing Frequency Calculation.

VIBRATION MONITORING AND MECHANICAL FAULT DIAGNOSTICS ON LARGE ROTATING MACHINERY IN ENEL POWER PLANTS

A. Benati
ENEL S.p.A. DCO - AITC, Rome, Italy

C. Frigeri
ENEL S.p.A. DSR/CRA, Cologno Monzese, Italy

G.C. Silvestri
ENEL S.p.A. DPT/STE, Pisa, Italy

ABSTRACT

The paper shortly describes the diagnostic methods and points out the relevant activities that in the most recent time have been carried out systematically and in particular those activities related to an extensive program for the vibration signature (base line) of turbine generator sets. For the purpose the machines design was modified in order to install new transducers and improvements was achieved in developing mathematical models, data processing software, acquisition systems, including mobile laboratories.

Advanced diagnostic oriented vibration monitoring systems (SMAV) have been specified by ENEL and two prototypes will be installed at the end of this year on turbo generator sets. The main characteristics of "SMAV" are described and some indication of the future development of such systems are also mentioned; these development are mainly related to special alarm logic and data reduction for an easy information recovery.

The strategies that were developed for steam turbine plants and that caused the above mentioned actions are being to be adapted and applied to turbogas plants.

1. Introduction

It is generally recognized that vibration analysis in rotating machinery is an extremely effective means to diagnose incipient malfunctions that in some cases might even have disastrous consequences if not discovered in time.

At the end of 1992 ENEL had 119 fossil fuel units in commercial operation for a total of 30,300 MW installed with over 13 million hours of operation and 28 gas turbine units for a total of 2760 MW; these units can be subdivided in 6 classes according to size, as follow:

GENERATING UNITS IN ENEL POWER PLANTS		
Class	MW	N° Of Units
2	< 100 Without RH	7
3	< 100	17
4	100 ÷ 200	28
5	200 ÷ 400	57
6	660	10
7	Gas Turbine (> 25 MW)	28

Furthermore there are 2 x 320 MW and 4 x 660 MW units under commissioning.

Given the amount of machinery, beginning in the 1970s ENEL began to set up experimental, analytical and computer facilities to increase plant reliability and availability.

In more recent years a more systematic effort has been made at defining systems, procedures and strategies to obtain and optimize information for correct operation and maintenance of rotating machinery, particularly turbine generator units.

The use of modern data acquisition and monitoring system and vibration analysis methodologies has effectively enabled to improve turbine and generator reliability, probably reducing the gravity of breakdowns and certainly reducing the frequency and duration of outages due to accidental causes through early detection of anomalies.

Improved operation and lower maintenance costs were also achieved by being able to keep plants with malfunctions or incipient faults in service. This allowed planning outages and maintenance operations while still guaranteeing safety through appropriate monitoring actions.

2. Evolution in Time of the Vibration Monitoring Systems on the Turbine Generator Units

2.1 Turbine Generator Units

The turbine generator units are equipped originally with shaft rider vibration transducers (velocimeters) that measure the absolute shaft vibration, generally visualized and recorded as global peak-to-peak value. The smaller machines are equipped with accelerometers that measures the supports vibration.

Two threshold levels are provided: an alert level and a shut down alarm for machine protection.

It is evident that the above mentioned instrumentation cannot be utilized for diagnostic purposes. For this reason it was decided to install two vibration probes, 90 deg. apart from each other. Such probes, called "VIPROX", each including an eddy-current proximity transducer and a velocimeter (or an accelerometer) measuring the support vibration, enables measurement of both relative and absolute shaft vibration and the relevant orbits, as well as the center loci of the journal in the bearing.

Such instrumentation is becoming standard for most machines in service and for all the newly built turbine generators, also in view of an extensive use of mobile laboratories for diagnostic activities and, for the most important machines, the installation of an Advanced Vibration Monitoring System (SMAV).

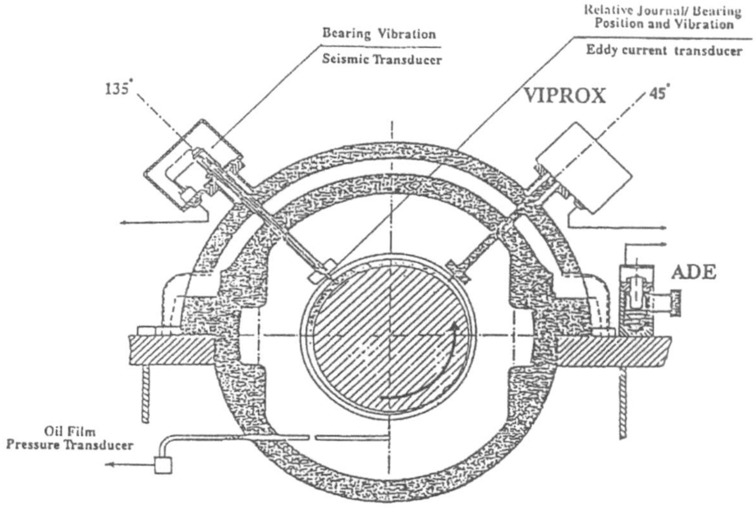

Fig.1 **VIPROX Installation on a Typical Turbogenerator Bearing, with ADE for Alignment Measurement**

A standard panel of connectors will be installed on each of these TG sets. The panel collects all the signals of interest for the dynamic behavior of the shaft line and will make the connection of mobile labs extremely quick. A few prototypes have already been installed.

The installation of these new transducer systems requires installing modified bearings to allow mounting and replacement of the sensors even with the machine in operation.

At present the situation of the instrumentation with regard to the 320 and 660 MW thermal units can be summarized as illustrated below.

INSTRUMENTATION ON GENERATING UNITS		
Size	320 MW	660 MW
Total N° of TG units	53	14
Units with not yet modified bearings	11	6
Units with modified bearings	42	8
Units with installed VIPROX	18	2

It is planned to complete the bearing installation activies during the outages for general overhaul, revamoing, repowering or modification to improve environmental influence.

2.2 Gas Turbine Units

ENEL has eighteen 90 MW gas turbine generator units used for peak service, ten 120 MW gas turbine units, recently went into service, both for peak and basic service, and another 16 units of the same size now being built for basic service.

These 120 MW GT units are installed for repowering in combined and the simple cycles

In monitoring of the dynamic behavior of the gas turbine units, proximity probes are used for relative shaft-support vibration measurements; sometimes velocimeters for the measurement of absolute vibrations of supports.

This instrumentation is adequate for monitoring the machines during operation due to the relative simplicity of the shaft lines and by the limited number of hours in service of the peak units.

Lastly, increasing the size and complexity of gas turbine units and keeping in mind that them will be used for basic service and will operate for several thousand hours per year,ENEL has been induced to apply to the gas turbine units similar criteria to those used for the steam units, also with regard to problems related to dynamic behavior.

Although the standard instrumentation is considered adequate for monitoring the machines of the last generation during commercial operation, it has been planned to istall instrumentation analogous to that available on the steam turbine units. This will add capability in diagnostic activities. For characterization activities (dynamic base lone) only, an altimetrica system (ADE) will be installed for the verification of the shaft line alignment.

For the future large gas turbine units, ENE will evaluate the advisability of installing advanced monitoring systems that supply more complete information and allow more detailed analyses for diagnostic purpose.

3. Diagnostic Facilities and Tools at ENEL

A digital data acquisition and processing system was developed for diagnostic purposes.

The system performs acquisition on a maximum of 64 dynamic channels, plus an additional 64 channels of slowly variable signals, both in steady operating conditions (up to one acquisition per 1 s) and at transient speed (one acquisition every 10 rpm).

Fig. 2A - Mobile Laboratory

Amplitude and phase versus time or frequency, Nyquist and center loci plots are provided by the data processing programs, together with spectra and a number of vectorial or arithmetical operations on the data. These diagnostic resources have been installed on a mobile lab, where it is possible to accommodate other software and instrumentation for special purpose investigations (multiplane balancing program, experimental modal analysis, etc.).

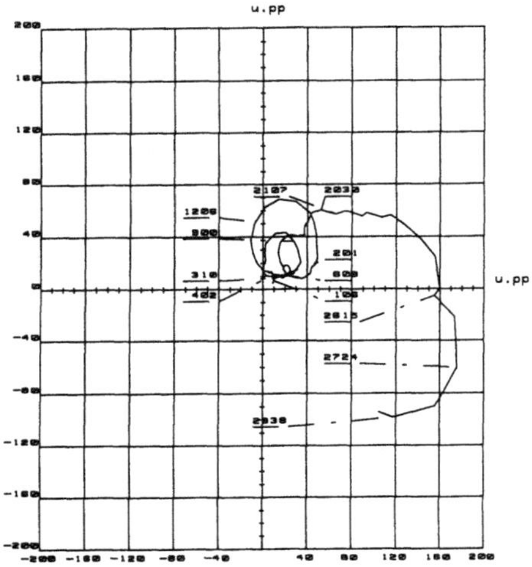

Fig. 2B - Nyquist plot

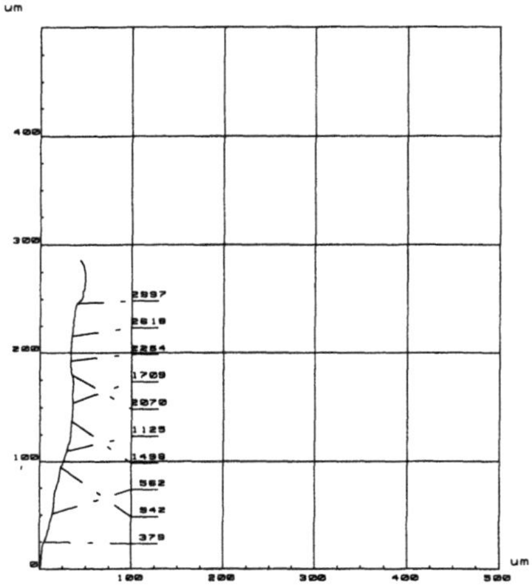

Fig. 2C - Center loci plot

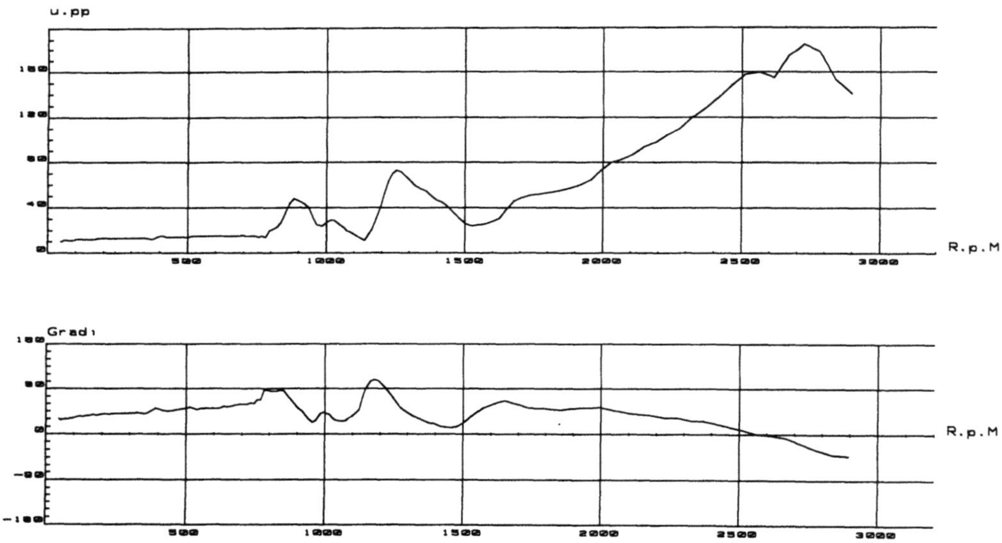

Fig. 2D - Run Down - 1 X rev Bode plot

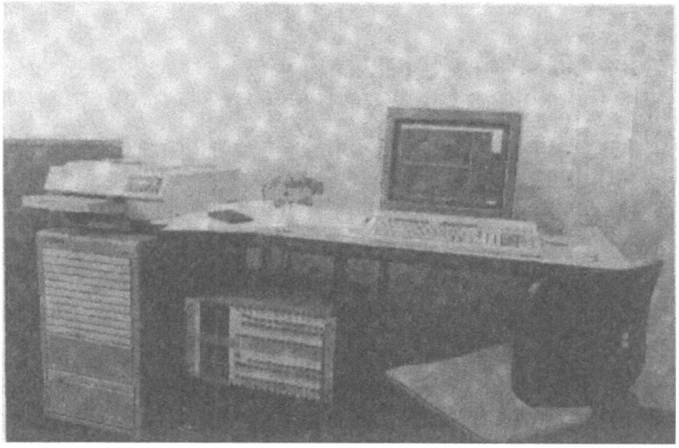

Fig.3 - Modal Analysis System

Other smaller portable acquisition systems have been developed for quick diagnostic operation or short time acquisition/monitoring purposes.

Mathematical models of the shaft line dynamics have complemented in some cases the experimental investigations, in order to verify and quantify the hypothesis of malfunction. Moreover, they are also used to perform sensitivity analysis or to provide the influence coefficients for the field balancing of rotors.

Two codes are available at present: the code DYTSO5 [1], (NUCLEAR ELECTRIC); the code PALLA [2] (Milan Polytechnic University).

DYTSO5 and PALLA are finite-element codes that can calculate, with different options, the linear steady-state vibration response of a shaft under excitations which may be synchronous or multiples of the rotating speed.

4. ENEL's Activity in the Field of Machinery Dynamics

ENEL's technical departments tackle problems relating to machinery dynamics as a function of their specific task: -new plant construction, production and research.

The activities carried out in collaboration among the departments also regard the development and introduction into the company of innovative techniques and means. Significant examples that can be cited are the definition of systems and measurement procedures (characterizations) specifications, diagnostic actions on machinery in operation or newly designed machinery, etc.

4.1 Activities on Plants Under Construction

It is planned to carry out measurement campaigns in order to obtain the dynamic base line of the new units while they are being to put into service; This for any subsequent diagnostic activities.

The purpose of these campaigns is also to check for correct behavior of the shaft line at the first startup, verify the design hypotheses and, if made necessary by the presence of anomalies, intervene with suitable corrective actions.

Carrying out these activities is the job of specialized units present within ENEL which, with the collaboration of the manufacturers, operate with experimental and analytical methodologies. The characterization of new machines are carried out with the same specification used for the machines in operation. In the characterization process, as is also foreseen by the base line specification to point out meaningful diagnostic information, extreme operating conditions are explored and some of the possible causes of anomalous dynamic behavior (such as rotor bending due to heat sensitivity, for example) are simulated by acting on the values of some process quantities.

In some cases masses, calculated from models, are placed on the balancing planes. This in order to evaluate the influence coefficients that can be later utilized for balancing operations and to better understand the dynamic behavior also, through the exact knowledge of the critical speeds obtained by an adequate excitation.

The attention dedicated by ENEL to the new machines has enabled early identification of problems already in the first startup phase, allowing quick implementation of the necessary modifications [3].

4.2 Activities on the Plants in Operation

In the area of thermal power production, ENEL has equipped itself with an organization and procedures to deal with the problems connected with vibrations of rotating machinery both in the case of malfunction and for predictive purposes.

4.2.1 Auxiliary Rotating Machinery

For the auxiliary machinery (feed pumps, fans, etc.) ENEL is carrying out, by means of its own means, a complete reference characterization for the various types of machines of the different manufacturers. This basic characterization initially serves as a reference for all the machines of the same type and is aimed at evidencing and evaluating any anomalies utilizing the auxiliary instrumentation of the power plant. If necessary, mobile labs are used, as well as the most suitable methodologies and the aid of specialized technicians.

The results of this activity will also be used to decide the most effective and least costly system for monitoring the auxiliary machinery.

4.2.2 Turbine Generator

For the turbine generator unit the following strategy was adopted.

According to an ENEL specification a complete dynamic characterization of each shaft line is made right after the first startup and on the occasion of the restart after a general overhaul.

This collection of data on the machine in different operating conditions defines its dynamic behavior in a complete way and serves as a basis of comparison in the course of the history of the machine to verify the evolution of its vibration behavior for diagnostic purposes.

Generally dynamic characterization activities are planned before a maintenance shutdown for the purpose of obtaining data for defining corrective actions to be carried out in the course of the overhaul (rotor balancing, parts replacement, rewinding, etc.). The adopted technical specification allows having a homogeneous collection of data on the different machines, independently of who performs the measurements. It also permits the creation (now in progress) of a computerized, centralized file on the behavior of the different machines, facilitating comparisons between machines of the same type and reducing the time needed to identify and diagnose a malfunction.

The programming, examination and interpretation of the results of the characterizations is done at a central level by specialized personnel who also coordinate any diagnostic actions relating to the dynamic problems and operate in collaboration with the units present within ENEL (research and experimental centers) or utilizing resources available at outside structures (university institutes, engineering societies).

To date, 27 of the 320 MW and 660 MW turbines have been characterized (40% of the total).

4.3 Research Activity

Some research activities, many of them related to the introduction in the power plants of "SMAV" (Advanced Vibration Monitoring System), are currently in progress. The activities are carried out in cooperation with the other ENEL technical departments. An outside research center (CISE) and a university (Milan Polytechnic) are also involved in important ENEL research activities, which can be summarized in the following items of activity:

- New instrumentation development and qualification.
- Data validation methods and systems.
- Data storage strategies.
- Statistical data processing.
- Development of an integrated environment for storage and processing of large quantities of data.
- Integration of identification techniques and numerical models of the shaft line dynamics.

The quality of the instrumentation is essential to preserve from frequent false alarms and also to guarantee the quality of the stored data. Therefore the transducers and the specially developed or assembled hardware are carefully qualified.

In order to obtain a further improvement of the data quality, a research activity has been started this year to develop a prototype of an expert system for the validation of the data produced by the different acquisition systems and SMAV.

SMAV would produce approximately 1.5 Mbytes a day, with the machine in normal operation, and 0.6 Mbyte during a run-up or a run-down. Clearly data compression is unavoidable.

The statistical data processing is one of the main concerns of the present research, because it is aimed at many essential goals: to provide the reference data for alarm generation and selective storage, to reduce the information spread in a very large amount of data to a few significant parameters, and to obtain a diagnostic tool through correlation analysis.

The relevant software is an advanced state of development, thanks to cooperation with Milan Polytechnic, and a first set of test data has been gathered by a mobile system installed in two power stations [4].

Another research topic is integrating parameter identification techniques and mathematical models of the shaft line dynamics, by measuring both support and absolute shaft vibrations. The aim of the research is to overcome the uncertainties connected to the schematization of the supporting structure of the turbine generator rotors.

5. General Approach for the Diagnosis of Mechanical Malfunctions

The comparison of the vibration data obtained in the presence of a malfunction with the corresponding vibration data obtained in normal conditions during a dynamic charcterizaion (base line) is the basis of the diagnostic activity.

5.1 Dynamic Characterization

The Technical Specification of the dynamic characterizations foresees the collection, according to standardized criteria, of the following data, which will be kept on magnetic storage media and with a unified format.

The characterization report will include some general information such as the major events that occurred during operation, charts relative to the machine alignment and balancing, etc.

The characterization is subdivided into a certain number of tests significant of the behavior of the machine:

- Run-out with and without lift pumps with machine on turning gear.
- Cold startup. Hot run down and startup including in overspeed.
- Operation at 3000 rpm without excitation, load transient and operation at different power levels.

Also recorded, if significant, is the machine response to the variations of some plant parameters (SH steam temperature, hydrogen temperature, steam temperature at the seals, journal oil temperature, steam condenser vacuum, etc.).

Each test is complete with different graphs of the static and dynamic quantities as a function of the time and speed of rotation of the machine. In particular the first and second harmonics of the relative and absolute vibration of the shaft and the absolute vibration of the support are reported.

The static and dynamic components of the journal oil pressure, the average value of the position of the shaft in the bearing during the run-up and the orbit of the center of the rotor.

5.2 The diagnostic method

The diagnostic method is generally based on the combined use of experimental data and mathematical models and consists of the following basic activities:

A) Comparison of the vibrations in the presence of a malfunction with the corresponding vibrations in normal conditions (base line).
B)Testing in various operating conditions.
C) Identification of the most likely forcing systems acting on the rotor as an effect of the malfunction using the corresponding influence coefficients, previously obtained utilizing calibrated models.
D)Analysis of the physical congruence of the results.

For the instability problems, the investigation starts looking at vibration spectra, shaft center orbits and at all those plant parameters that are related with static load on bearings.

The above-mentioned codes are used in this kind of investigation in conjunction with a specialized code for bearing calculation.

5.3 Some Significant Examples o diagnostic activities

We report some recent examples of the way dynamic problems are tackled.

5.3.1 Cracked Generator Rotor

A gradual increase in vibrations and an anomalous sensitivity to load variations was detected by the plant instrumentation on the generator of a 320 MW unit [5].

The additional vibration measurement performed with a mobile acquisition system, compared with the dynamic characterization performed two years previously after overhaul, showed a considerable increase in the vibrations at the passage of the first critical speed and an increase in the second harmonic component in correspondence to half the first critical speed (Fig. 4).

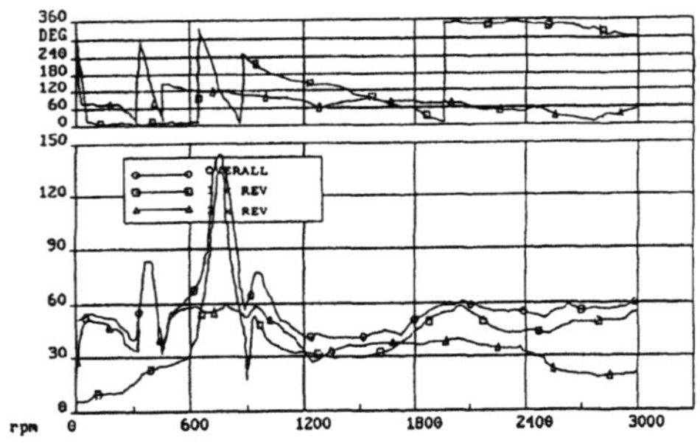

Fig.4a- Relative shaft vibration measured at the second generator bearing (brg.no 10) during a previously run-down.

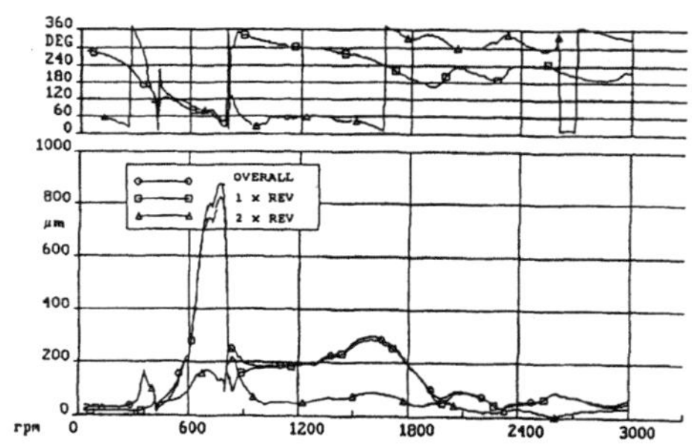

Fig.4b- Relative shaft vibration measured on the second generator bearing (brg.no 10) , during the run-down before generator out of service

These symptoms led to the hypothesis that a crack was present on the shaft. The rotor was removed from service and replaced with the spare.

Tests were conducted in the balancing facility and a dynamic simulation of the rotor was done by means of a mathematical model.

The results of the tests and mathematical simulations supported the hypothesis of a crack, and made it possible to define the size and position of the fault with good approximation.

Subsequent controls confirmed the presence of the crack, which had an angular extension of 180 and affected almost half the section of the rotor.

Later, as a precautionary measure, continuous vibration monitoring was set up for four identical units while awaiting completion of the investigations and studies on the origin of the crack.

Fig. 4C - Crached Generator Rotor

5.3.2Other Examples

On various occasions rubbing was diagnosed on the seals of HP-IP rotors (Fig.5), in other cases misalignments of thermal origin of bearings leading to instability from oil film.

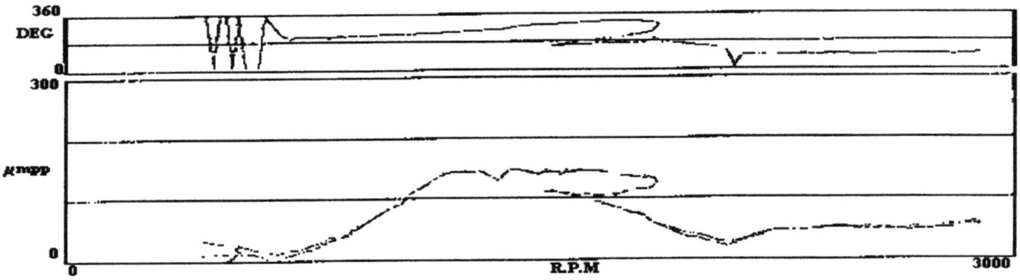

Fig.5 Rubbing One Hp Turbine Seals During a Start Up

Various vibration problems derived from rotor bending in some cases of thermal origin (rotor anisotropy, short circuits between loops) (Fig. 6); in other cases the curvature of the rotor shaft line was due to a poor coupling made in correspondence to the joints.

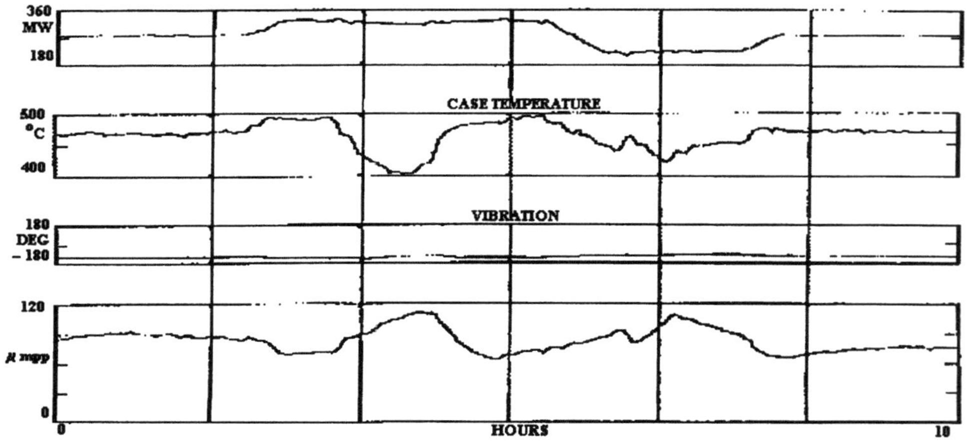

Fig.6 - Thermal Bending in an Hp Turbine Rotor

6. The Advanced Vibration Monitoring System - SMAV

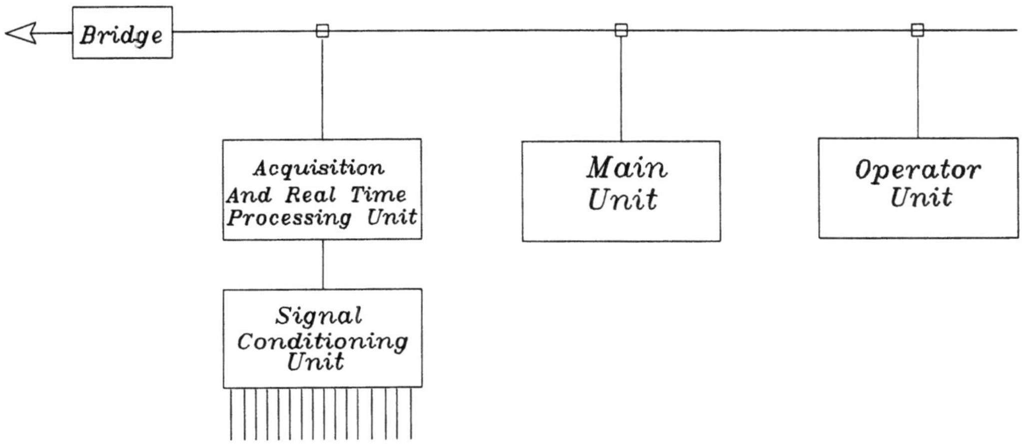

Fig.7 - Advanced Vibration Monitoring System - SMAV

SMAV, which is to be installed in the most important plants, utilizes the above-mentioned transducer setup [see 2.1].

The system can be defined through the following blocks:

- Signal conditioning unit
- Acquisition unit
- Main unit
- Operator workstation

The main unit manages the activity of the other parts of the system, the communication among the different parts of the system and between the system and the external environment (supervisory system).

The other functions of the main unit are:

- Initialization of the whole system.
- Data processing, compression and storage.
- Processing and output of real time and stored data by means of a powerful graphics program. This function is specialist-oriented for diagnostic purposes.
- Alarm generation.
- Scheduled automatic self-diagnosis and calibration of the system.

The signal conditioning unit is very simple. Only one standardized board every four channels is used for dynamic and slowly variable signals.

A high performance acquisition unit is adopted, owing to the diagnostic use of the data collected by the monitoring system.

The main functions of the unit are:

- Synchronous and asynchronous acquisition, up to 64 dynamic channels and 64 static channels.
- Harmonic analysis (three harmonics plus average value) or, when required, spectral analysis (360 lines, 1 Hz resolution).
- Compensation of amplitude and phase errors introduced by signal conditioning electronics and rotation of measuring axis to obtain the values in vertical and horizontal directions.
- Automatic switching in the different operating modes:
- Harmonic analysis
- Spectra
- Constant time interval (1 second) (at constant speed).
- Constant rpm interval (c.a 10 rpm) (during run-up and run- down).

The system switches from "harmonic analysis" to "spectra" mode if the true RMS value is greater than the RMS value obtained by the calculated harmonic components.

The operator workstation is a low-performance one and may be part of the supervisory system. Through the main unit, a specialist can modify some pages for the operator workstation or even generate new, non-standard ones.

Some pages may be aimed at pointing out a particular behavior of a machine suspected to be due to an incipient fault. This feature, in conjunction with a special alarm logic, makes early fault detection possible.

Data are temporarily or permanently store as show in fig. 8.

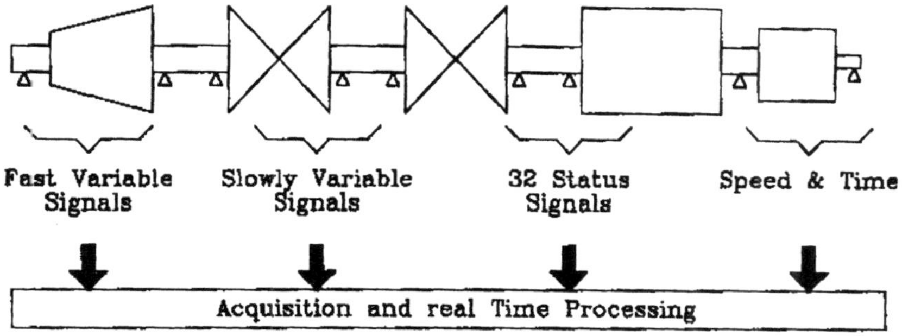

	Sheet Kind	Sheet #	Content	Information Time Interval	Max Time Interval per Sheet	Total Time	Operating Conditions		
							Turning Gear	Stand By Speed	Operating Speed
Temporary Storage	a	1	360 Acquisition	1s	6'	30'	No	No	Yes
		2 3 4							
		5	360 Acquisition	1s	6'				
	b	1	360 Means Value of the Last 5 Acquisition from a5	5s	30'	4h	No	Yes	Yes
		2 7							
		8	360 Points From b7	5s	30'				
	c	1	360 Means Value of the last 5 Acquisition from b8	20s	120'	24h	Yes	Yes	Yes
		2 ..	.....	.. ..					
		12	360 Points From c11	20s	120'				
Final Storage	d	1	360 Means Value of the last 5 Acquisition from c12	120s	12h	1 Month	On Request	On Request	Yes
		2 64	..	..	...				
		65	360 Points From d64	120s	12h				

On Request → to Final Storage

Fig. 8 Strategie for Temporary and Final Data Storage (Constant Speed)

7. Conclusions

The activities in the field of diagnostics of rotating machines in power plants at ENEL are carried out by the three technical departments that cooperate in the more complex diagnostic activities and mostly in the development of new systems and technical solutions.

Outside contributions (Manufacturers, CISE, Politecnico di Milano , etc.) were and still are determinant for achieving ENEL's goals in the field.

This paper has briefly described the developments in the field instrumentation and sensors, the diagnostic tools and methods and the activities of ENEL in this field and the new Advanced Vibration Monitoring System.

The activities of the technical department have been described, attempting to give a general view of the different tasks and matters for cooperation.

Even if it is impossible at the present moment to know the exact amount of money saved by ENEL through diagnostic activities on rotating machinery in the last years, a few examples (two are mentioned in this paper) are sufficient to demonstrate that without an appropriate diagnostic approach some machines would be shut down for a long time for not urgent repair instead of waiting for a scheduled overhaul.

REFERENCES

1. DYTS04 USER'S GUIDE-CEGB-RD/L/P-15/80-JOB No. VE316
2. P.A.L.L.A. - A Package to Analise the Dynamic Behaviour of a Rotor-Supporting Structure Sistem
3. C.Frigeri, M.Pastorino, G.A.Zanetta:
   Investigations on the vibrational behaviour of gas turbogenerator units at the first start-up: aims and methodology. A case history on a 120 MW unit
4. G. Diana, A. Vallini, A. Vania,G.A. Zanetta:
   On-line diagnostic techniques in condition monitoring
5. G.L. Lapini,M. Zippo,M. Bachschmid,A. Collina,A. Vallini:
   Experimental tests and model based calculation for the diagnosis of a crack in a 320MW generator
6. R. Benanti,C. Frigeri,A. Vallini,G.A. Zanetta:
   Diagnostic Oriented Flexural Vibration Monitoring Systems for Large Turbogenerator Units.
   AIPnD Int.Conf. on "Monitiring and Predictive Maintenance of Plants and Structures" - Florence (ITALY) 17÷20 May 1992

NEW INDUSTRY STANDARD TECHNOLOGY APPLIED TO CONDITION MONITORING IN THE POWER INDUSTRY

J. Courrech
Bruel and Kjaer, Naerum, Denmark

ABSTRACT

Despite being referred to as an entity, the Power Industry is a broad concept when speaking of machine condition monitoring. It can be small isolated hydraulic turbines one refers to, or huge steam turbines, or many gas turbines, or even diesel generators. Moreover, besides the power generation, a multitude of classical machines such as pumps, fans, blowers contribute to the operation of the plant.

This paper will present how new technologies allow to fulfil the requirements of all types of power generation plants for on-line monitoring of both vibration and process parameters; small geographically isolated plants with few measurements as well as big turbo generator sets with many parameters to monitor and correlate. It will show how different types of faults can be individually monitored using specific analysis techniques, allowing an immediate diagnosis as soon as an alert is detected. Examples from existing installations will illustrate the paper.

1. INTRODUCTION

The power industry has been a driving force behind the evolution of condition monitoring techniques over the past years.This paper will describe how new technologies allow to solve all the major challenges in condition monitoring for all types of machines used in power generation whether in conventional, nuclear or hydroelectric plants.

From a vibration monitoring point of view , there is little difference between a conventional power plant or a nuclear one: a steam turbine is a steam turbine, whether the steam comes from a fuel fired boiler or other sources. Similarly, a reactor cooling pump is just another pump. What is significant is the type of machines involved : steam turbines, gas turbines, or hydraulic turbines. The faults expected and their symptoms differ. Significant as well is whether the machines are manned or not, whether network communication is necessary and whether diagnosis expertise is available .

Traditionally **safety monitoring** has been the main concern in the power industry, meaning monitoring the condition related parameters of a machine in order to shut it down automatically as soon as a predefined limit is exceeded. However, periodic measurements have proved that more sophisticated techniques did allow a much earlier warning of developing faults. This has resulted in implementing in on-line systems **predictive monitoring** based on these techniques as a complement to the traditional safety monitoring.

2. STEAM TURBINES MONITORING

One of the main concerns in steam turbines is the detection of developing cracks in shafts. This applies more generally to all the newer types of turbomachineries with long shafts with distributed masses . Two techniques are available:

-The monitoring of both phase and magnitude of the firsts harmonics of the rotation speed during continuous operation

-the monitoring of the magnitude and phase of the first harmonics of the rotation speed during run up and coast down.

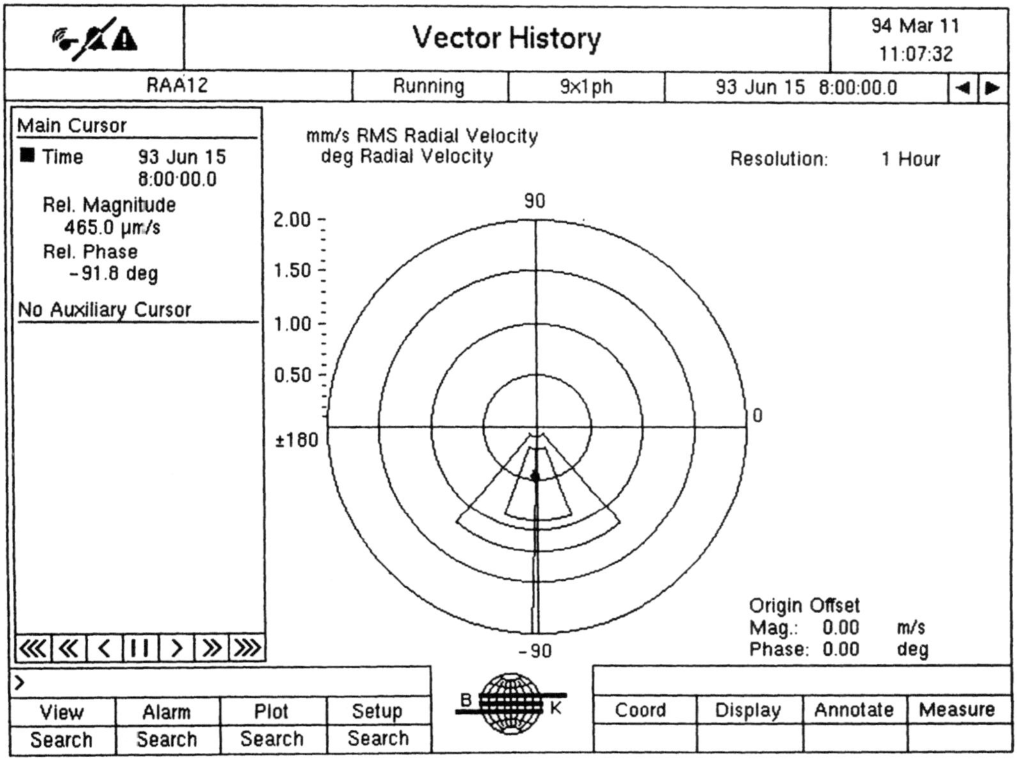

Fig. 1, History of the evolution of the first harmonic vector, with alert and danger limits. Measurements performed on a steam turbine in continuous operation mode

The first technique consists into following continuously during operation magnitude and phase of the vibration up to the 4 first harmonics of the rotation speed in order to detect any changes, as the one which would occur if a transverse crack did generate a change in stiffness twice per rotation. This technique is called **"Vector Monitoring"** as the vibration is here followed as a vector rather than a single scalar. This technique is extremely powerful and is compulsory for any efficient on line monitoring of turbomachineries. However other sources may create these phase and magnitude shifts , such as load condition changes or other faults such as seals wear.A good monitoring system will however be able to differentiate between the different load and operating conditions and no false alarm will occur in this case.Figure 1 . shows measurements performed on a steam turbine in Finland . It can be seen on this vector history plot how stable both magnitude and phase of the fundamental frequency are during fault free operation. Such measurements can be performed permanently on-line on all channels in parallel, giving assurance that no crack can occur without being detected.

The second technique concerns monitoring the vibration during non stationary operation, namely during run up and coast down. A fault such as a crack , being a change in the mechanical impedance will manifest itself by changes in the critical speeds, changes of amplitude and decreases in the frequency. Any structural modification of the stiffness or the damping of the rotor /bearing/foundations system will manifest itself by changes in the frequency response function between exciting forces inside the machine and vibration measured on the bearings. These changes can adequately be observed using the fundamental frequency created by the residual unbalance as exciting force during run up and coast down. The different critical speeds and resonance are excited in turn and any change from one run up to the next will be detected. This technique is called **"Profile monitoring"** and its originality lies in the fact that with newer technology each point can have a reference profile and alert and danger alarms for increases as well as decreases for both magnitude and phase of up to 10 harmonics of the rotation speed. In addition .,if required , the complete FFT spectrum can be stored for each measurement points during a complete run up or coast down allowing later to do a slice along any frequency component or range of interest.

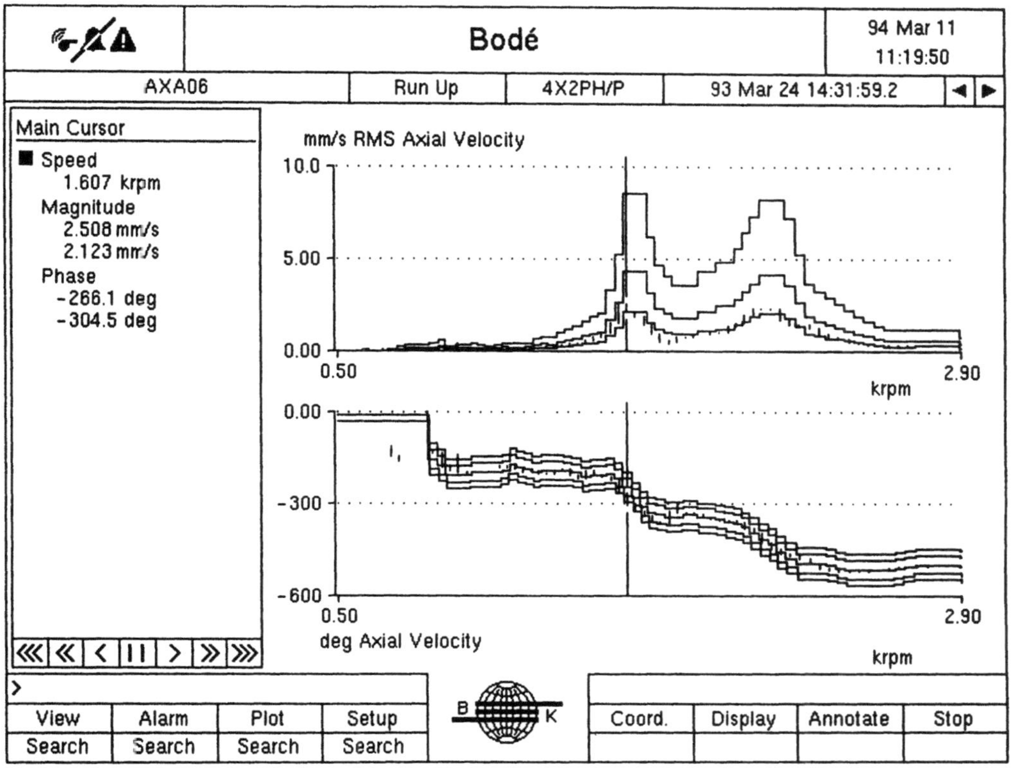

Fig. 2 , Profile of magnitude and phase during a steam turbine run up. The continuous line indicate alert and danger limits generated from a reference profile. The dotted line indicates the actual profile. Each vertical line visualises the minimum and maximum amplitude and phase respectively achieved during the speed change interval

Figure 2 illustrates a bode plot (magnitude and phase of the first harmonic of the rotation speed function of the machine speed) measured on a steam turbine where both magnitude and phase are compared to references and limits. Many such Profiles are stored and can be visualised as a profile history plot. It is worth noting that such systems **switch automatically** from vector monitoring during continuous operation to profile monitoring function of the rotating speed as soon as the change of machine state is detected. These functions which traditionally were available for investigation and diagnostic are now **REAL** Monitoring functions where any limit exceedence ends in a trip relay if desired.

Each measurement point can be monitored with up to a dozen different analysis techniques. This means that, rather than as usual just monitor the unbalance condition, it becomes now possible to monitor other frequency bands such as around the blade passing frequency, which is a very early indicator of journal bearing wear.In addition , by choosing cleverly different analysis bands or technique concentrating on specific faults, the diagnosis will be directly apparent by just looking at which symptom (measurement type) has given an alert. Figure 3 shows an example of different measurements selected to do predictive monitoring on a gas turbine.

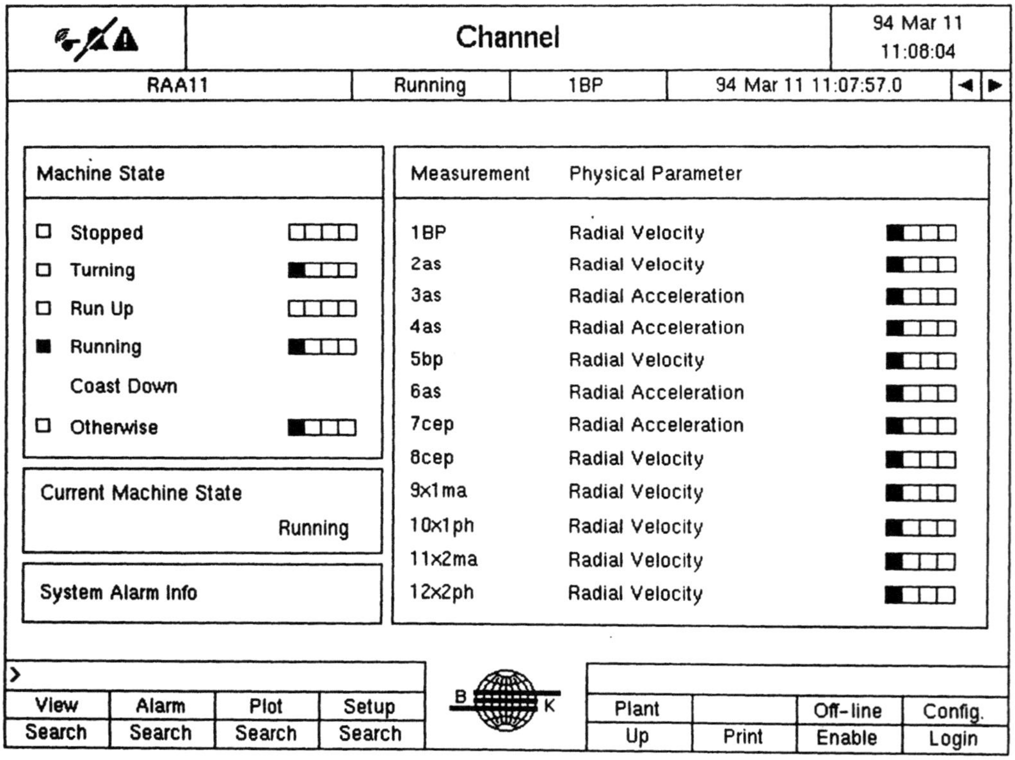

Fig. 3, Example of different measurements performed on a single measurement point on a gas turbine. The first measurement in capital is a continuous measurement (safety), whereas all other ones are for predictive monitoring, including frequency spectra, cepstra and vector monitoring.

3. GAS TURBINES MONITORING

Gas turbines differ often in so far that they have an intermediate gear box between themselves and the generator. Often with journal bearing smaller ones may however have rolling element bearings. Gear boxes monitoring is basically different from basic rotor dynamic forces monitoring.The frequencies concerned by the usual gear faults such as gear wear, are much higher (tooth meshing frequency and higher harmonics) and local faults such as tooth crack create modulation showing up as side band families rather than deterministic single components. All these information are in an FFT spectrum but difficult to extract and interpret. **Cepstrum analysis** allows to represent a complete family of sidebands with a single component and will separate and quantify the different families. Cepstrum has now changed from a diagnostic gadget to a monitoring function. Each point has reference and limits and any change is detected , visualised with the corresponding alerts and trended.

Cepstrum analysis not only applies to detection of cracked teeth but will similarly detect cracked turbine blades.The cepstrum analysis shown on figure 4 is an on-line monitoring measurement on a gas turbine.

Envelope analysis is an other technique basically consisting in demodulating high frequencies such as tooth meshing frequency or resonance excited by lower periodic frequencies such as bearing faults impacts or gear wheel rotation speed. Well known as a rolling element bearing fault detection technique, it has a more general application for detection of local gear or blade faults. Both envelope and cepstrum analysis are typically used for predictive monitoring rather than safety monitoring and, being extremely sensitive to the mentioned faults , give a very early warning of their development.

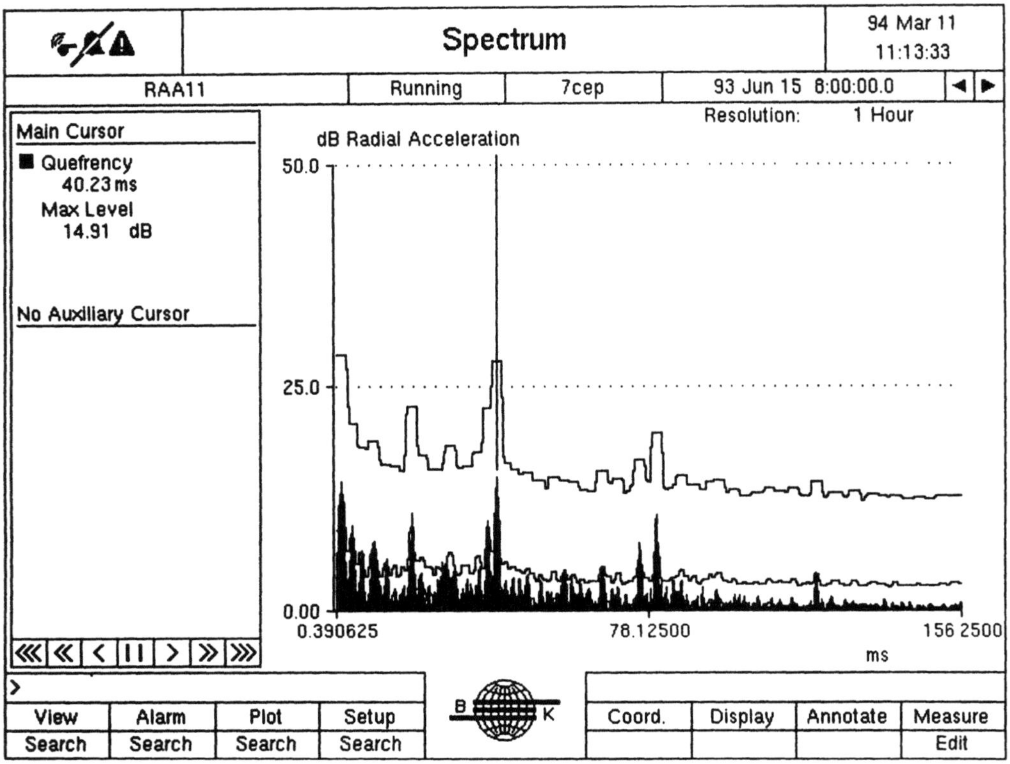

Fig. 4, Example of cepstrum monitoring. These measurements are performed on a gas turbine and exceedence of the alert limits are detected.

4. LEAK DETECTION

Leaks in pipes are a common problem in power plant . A leak. as minor as it may be , will excite radial resonance of the pipes which will be detectable with an accelerometer. The same system, which monitors bearing vibrations, will be able to monitor leaks either in a predifined frequency band or by high frequency spectrum comparison, preferably constant percentage bandwidth . In the CPB spectrum (constant percentage bandwidth) , the resolution in the high frequencies of interest is more coarse than in an FFT spectrum and thereby simplifies the detection and the trending.

5. AUXILIARY EQUIPMENT MONITORING

Auxiliary equipment are often too numerous and wide spread in a plant to be allocated an on line monitoring system, but they are nevertheless critical for the correct operation of the plant. The solution is to perform intermittent monitoring using a data collector. The new systems available combine now in a common database machines monitored on-line and the ones monitored off-line in a single software and user interface. New technology allows to implement in a data collector more and more advanced analysis methods such as envelope or cepstrum or vector monitoring together with ever simpler operating procedures .Typically no more than pressing two pushbuttons is necessary for collecting measurements in a route, which eliminates the need for trained operators and garanties that measurements get done independently of shifts or dedicated operator's availabily.

6. EXPERT DIAGNOSTIC SYSTEMS

Often diagnostic expertise is not available on site and despite a very efficient predictive monitoring no decision can be made as whether and what to repair. New **expert systems based on neural network technology** can now combine the use of rules applied by international experts and the ability to learn new diagnosis and rules as a human brain would. Such systems when working on existing rules , make use of all rules applicable to a given machine and weights them according to their relevance with the detected symptom. Standard rules can be copied and modified by the user to fit best its specific machines, and new symptoms and rules can integrally be generated by the user. Symptoms are based on all available analysis techniques and rules can combine as many of the symptoms as desired. This way, process parameters are combined with vibration data to optimise the diagnostics.When a new case occur, a new diagnostic with its coresponding symptom can be directly learned by the system without creating artificial rules. Figure 5 shows an example of a diagnostic report performed on a motor driven pump .

Such expert systems can operate either off-line where routes can automatically be diagnosed when unloading, or with on-line systems where the complete batch of machine is cyclically diagnosed automatically .

Machine Diagnosis Software 7106 Report

Machine : E5 S9

Diagnostic Results

Component	Fault	Certainty
S9 E5 A4	Bearing Flaw	75
S9 E5 A4	Outer Ring Flaw	4
S9 E5 A4	Inner Ring Flaw	0
S9 E5 A4	Cage Defect	0

Fig. 5, Example of a diagnostic performed by a diagnostic expert system on a motor driven pump. The only element judged faulty is the rolling element bearing 3.

7. HYDROELECTRIC PLANTS

Hydraulic turbines are slightly different from other above mentioned .They are often slower rotating, vertical and bearings both radial and thrust are of major interest as well as shaft condition and possible cavitation. Figure 6 shows an example of monitoring system on such an hydraulic turbine.

What happens often in mountain areas is that hydraulic turbines are very numerous but geographically wide spread. No man power is available on site and the installation of central vibration would be of no use. In these configurations it becomes of the utmost importance that the vibration monitoring device posses sufficient memory to store all the events occurring and can communicate directly, for example by modem, with a central computer. This allows a single operator to collect in a single location, at the rate required the event which have happened on all sites . These events have been stored in the internal memory of each vibration monitors until they got unloaded, but at the same time safety monitoring continue to function and any danger limit exceedence could have, if so desired, tripped the machine down.

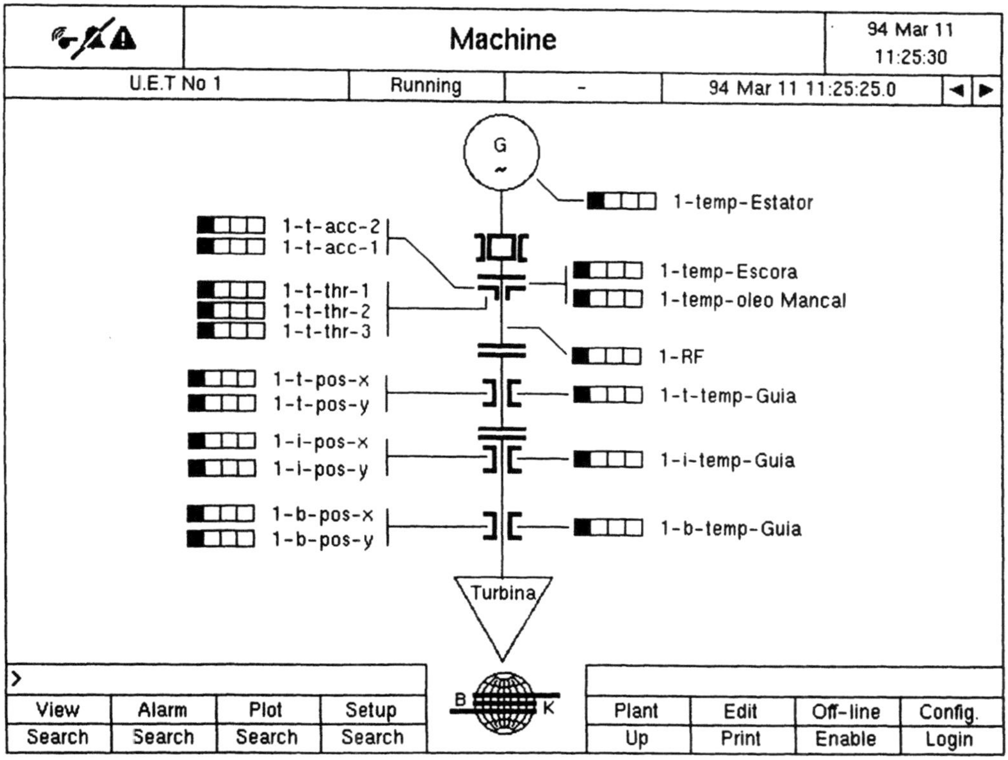

Fig. 6, Example of monitoring on an hydraulic turbine.

8. CONCLUSION

In short, with the newer technology including digital signal processing, multi users operating systems such as UNIX, and network and modem communication, there are no limitations left to the condition monitoring performance of any type of machines. Expert systems can be customised to fit the diagnoses and symptoms required in the power industry. Communication with other process control systems, which is a consequence of the use of relational data bases , allows to monitor machine efficiency simply or control the trip of a turbine based on a combination of vibration and process parameters, rather than just trip and thereby creating more damages than would have occurred letting the machine run a little while longer.

INFLUENCE OF MODELING OF A ROTOR SYSTEM ON ACCURACY OF IDENTIFICATION OF ITS UNBALANCE CHANGES

J.M. Krodkiewski, N. Zhang and J. Ding
The University of Melbourne, Parkville, Australia

ABSTRACT

The paper presents a procedure for identification of unbalance changes of rotor-oil bearing system during its operation. The procedure bases on the monitored trajectories of journals with respect to their bearing cases and the non-linear mathematical model of the rotor system. To assess applicability of this procedure, the influence of measuring errors and accuracy of modeling of the rotor is investigated. Results lead to conclusion that this procedure could be applied to on line identification of plane at which change in unbalance distribution has taken place.

1. INTRODUCTION

The unbalance forces of a rotor system are always the main cause of vibration of the rotating machine. The work for the determination and correction of the unbalance forces has been continued for the past decades. The approaches such as Least-Squares Method[1], Two-or-Multi-plane Balancing[2,3], Spray Automated Balancing[4] and so on have been developed. For the balancing of large turbo-generator sets, A. Craggs presented a method[5] to determine the unbalance force vector from the measurements of displacements, bearing forces, phase angles and the coefficient matrix, which is composed of mass, stiffness and damping matrices of the rotor system.
After a rotating machine starts its service, the unbalance forces of its rotor may change due to the loss of a blade, wear or even corrosion. The presented method permits to identify this kind of unbalance change without stopping the machine's service. The unbalance changes are assumed to take place at only one plane of a rotor system(one or a few of its blades at one plane are lost during operation). The unbalance changes are then determined from the measured relative motion of the rotor journals with respect to the bearings before and after the unbalance changes have taken place.
In this paper, it is investigated how the errors in the modeling of a rotor system influence the accuracy of the identification of its unbalance changes.

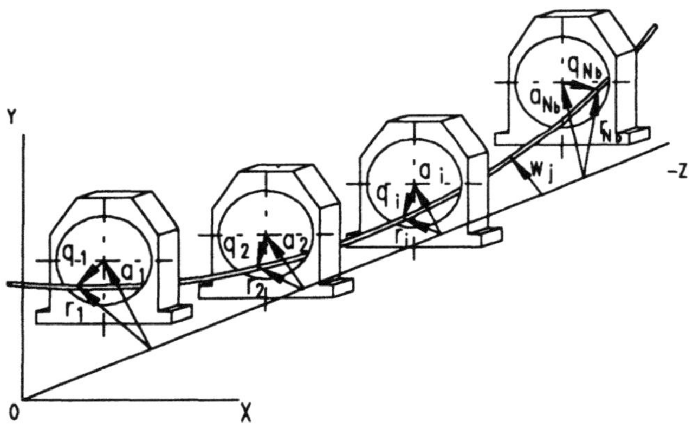

Fig. 1 Coordinate system for a multi-bearing rotor system

2. Description of Identification Method

The vibration response of a multi-bearing rotor system, shown in Figure 1, is governed by the following equation:

$$[M]\{\ddot{r}\} + [K]\{r\} = \{H\} + \{P\} + \{U\}e^{i\omega t}. \tag{1}$$

The mass matrix and the stiffness matrix of the rotor system are denoted by $[M]$ and $[K]$ respectively. The displacement vector $\{r\}$, which is partitioned into elements referring to the bearing stations and the unsupported stations, represents the instantaneous

position of the N stations of the rotor with respect to the absolute system of coordinates XYZ. The elements of vector $\{r\}$ referring to the bearing stations contain the known configuration parameters which are independent of the operation speed. On the right hand side of the equation, $\{H\}$ is vector of the bearing reaction, $\{P\}$ is vector of the static load and $\{U\}e^{i\omega t}$ represents the unbalance force of the rotor system for a constant speed ω.
For the steady state motion of the rotor system, both the solution of eq.(1) and the periodic bearing forces, can be approximated by the Fourier's series.

$$\{r\} = \sum_{k=-\infty}^{\infty} \{R_k\}e^{ik\omega t}, \tag{2}$$

$$\{H\} = \sum_{k=-\infty}^{\infty} \{H_k\}e^{ik\omega t}. \tag{3}$$

Therefore, upon substituting the above results into eq.(1), the following equation is obtained:

$$\sum_{k=-\infty}^{\infty} [(-k^2\omega^2[M] + [K])\{R_k\} - \{H_k\}]e^{ik\omega t} = \{P\} + \{U\}e^{i\omega t}. \tag{4}$$

Since the above relationship must be fulfilled for any instant of time, it follows that,

$$[K]\{R_0\} - \{H_0\} = \{P\}, k = 0, \tag{5}$$

$$(-\omega^2[M] + [K])\{R_1\} - \{H_1\} = \{U\}, k = 1, \tag{6}$$

$$(-k^2\omega^2[M] + [K])\{R_k\} - \{H_k\} = \{0\}, k = 2, 3, \cdots, \pm\infty. \tag{7}$$

The equation (5) allows to develop the identification procedure of the current configuration vector $[a_1 a_2, ..., a_{N_b}]$ (Fig.1) which coincides with the one derived in [6]. If one assumes that the exciting force has only one harmonic term $\{U\}e^{i\omega t}$, the eq.(6) can be used for the identification of the unbalance change. For simplicity, the equation is re-written as follows:

$$[A]\{R_1\} - \{H_1\} = \{U\}, \tag{8}$$

where

$$[A] = -\omega^2[M] + [K], \tag{9}$$

$$\{R_1\} = \left\{ \begin{array}{c} Q_1 \\ W_1 \end{array} \right\}. \tag{10}$$

In eq.(10), the elements of vector $\{Q_1\}$ referring to the bearing stations as well as the non-zero elements of vector $\{H_1\}$ can be obtained from the measured motion of the rotor with respect to the bearings.

$$Q_{1n} = \frac{1}{T}\int_0^T q_n(t)e^{i\omega t}dt, H_{1n} = \frac{1}{T}\int_0^T H(q_n(t), \dot{q}_n(t))e^{i\omega t}dt, n = 1, 2, \cdots, N_b. \tag{11}$$

Hence, if the measurements of the relative motion between the rotor and the bearings are taken both before $\{R_1^b\}$ and after $\{R_1^a\}$ the unbalance change takes place, the relationship between the unbalance change and the measurements can be found as:

$$[A]\{\Delta R\} - \{\Delta H\} = \{\Delta U_r\}, \tag{12}$$

where $\{\Delta R\} = \{R_1^b\} - \{R_1^a\}$, $\{\Delta H_1\} = \{H_1^b\} - \{H_1^a\}$.
The unbalance changes at those planes which refer to bearing stations can always be considered as zero and the above equation can then be partitioned as:

$$[A_{11}]\{\Delta Q\} + [A_{12}]\{\Delta W\} - \{\Delta H\} = \{0\}, \tag{13}$$

$$[A_{21}]\{\Delta Q\} + [A_{22}]\{\Delta W\} = \{\Delta U\}. \tag{14}$$

where $[A_{ij}]$ are submatrices of matrix $[A]$, the vector $\{\Delta Q\}$ represents the changes of rotor's response corresponding to the bearing stations, $\{\Delta U\}$ represents the unknown changes of unbalance at the unsupported stations and $\{\Delta W\}$ represents the unknown changes of rotor's response at the unsupported stations.
Upon determining the unknown vector $\{\Delta W\}$ from eq.(14) and then introducing it into eq.(13), one can obtain that

$$[C]\{\Delta U\} = \{\delta\}, \tag{15}$$

where

$$[C] = [A_{12}][A_{22}]^{-1}, \tag{16}$$

$$\{\delta\} = \{\Delta H\} - ([A_{11}] - [A_{12}])[A_{22}]^{-1}[A_{21}]\{\Delta Q\}. \tag{17}$$

The matrix $[C]$ and the vector $\{\delta\}$ can be determined from the system parameters and the measured displacement $q(t)$.
The dimension of matrix $[C]$ is $N_b \times N_u$. Since N_b, representing the number of bearing, is usually not equal to N_u, representing the number of unsupported stations, equation (15) has no unique solution with respect to the unknown changes in the unbalance. However, one may assume that the sudden change in unbalance is due to loss of blades in one plane only, say, at n_u plane. In this case, the vector $\{\Delta U\}$ of unbalance changes becomes:

$$\{\Delta U\} = \{0, 0, \cdots, \Delta U_{n_u}, \cdots, 0\}^T. \tag{18}$$

Substituting eq.(18) into eq.(15), one can obtain a set of N_b equations with one unknown ΔU_{n_u},

$$C_{kn_u}\Delta U_{n_u} = \delta_k, k = 1, 2, \cdots, N_b. \tag{19}$$

If the guess is correct, it means that n_u is really the plane at which the unbalance has been changed, any arbitrarily chosen equation of the above set(eg. $k = n_b$) can be used to compute the change in unbalance ΔU_{n_u}.

$$\Delta U_{n_u} = \frac{\delta_{n_b}}{C_{n_b n_u}} \tag{20}$$

and all the other equations must be unconditionally fulfilled.

$$C_{kn_u}\Delta U_{n_u} = \delta_k, k = 1,2,\cdots,n_{b-1},n_{b+1},\cdots,N_b. \tag{21}$$

If the above set of equations is not fulfilled, the unsupported station n_u must have been chosen incorrectly. The above "trial and error" method can be repeated for each unsupported station until the plane at which the change in unbalance distribution occurred is detected. Performing numerical computations, one has to take into account the fact that due to numerical errors the equation (19) will never be fulfilled absolutely. Hence, to detect the faulty station by means of the numerical computation it is convenient to introduce an error function.

By elimination of ΔU_{n_u} from eq.(19), one can obtain the following relationship

$$\frac{\delta_{n_b}}{\delta_k} = \frac{C_{n_b n_u}}{C_{kn_u}}, \tag{22}$$

where $n_b = 1,2,\cdots,N_b$; $k = 1,2,\cdots,n_b-1,n_b+1,\cdots,N_b$; $n_u = 1,2,\cdots,N_u$.

The error function which bases on the above relationship was developed as follows:

$$J_{n_u} = \left(\frac{2}{N_b(N_b-1)}\sum_{k=1}^{N_b}\sum_{n_b=1}^{N_b}\frac{\epsilon_{n_u}^2}{(C_{n_b n_u}/C_{kn_u})^2}\right)^{\frac{1}{2}}, \tag{23}$$

where

$$\epsilon_{n_u} = \frac{C_{n_b n_u}}{C_{kn_u}} - \frac{\delta_{n_b}}{\delta_k}, n_u = 1,2,\cdots,N_u. \tag{24}$$

If n_u is chosen correctly, the error function corresponding to the blade-loss station should be the smallest one among all the evaluated error functions.

For the detected plane, the magnitude of the unbalance change can be determined by averaging the quantities which are obtained from the eqs. (19),

$$\Delta U = \frac{1}{N_b}\sum_{n_b=1}^{N_b}\frac{\delta_{n_b}}{C_{n_b n_u}}. \tag{25}$$

3. DESCRIPTION OF ROTOR SYSTEM

The investigations of the influence of the modeling of a real rotor system on the accuracy of the identification of its unbalance changes were carried out by means of numerical simulation of the four-bearing rotor system shown in Figure 2. The installation consists of a flexible rotor, four identical three-sleeve journal bearings and a rigid foundation.

The rotor assembly was modeled by means of finite element method. By making use of the condensation method, 4 bearing stations and 5 unsupported stations (9 stations in total) of the rotor system were retained for the final assembly of the system equations

Table 1. Assumed system configuration (bearing transverse positions)

Bearing number	$x[mm]$	$y[mm]$
1	0.	0.
2	0.	-0.18
3	0.	-0.18
4	0.	0.

Table 2. Assumed changes in unbalance distribution (M–magnitude of unbalance [N], ϕ–phase [deg.])

Station	1		3		7		9	
Case	M	ϕ	M	ϕ	M	ϕ	M	ϕ
1	5.55	-129.9	9.24	73.77	10.0	-129.3	8.08	58.84
2		0.						
3				0.				
4						0.		
5								0.
6	15.							
7			20.					
8					0.			
9							0.	

of motion. The journal bearings were modeled by means of the Reynolds equation, which is solved by the finite differential method[7].

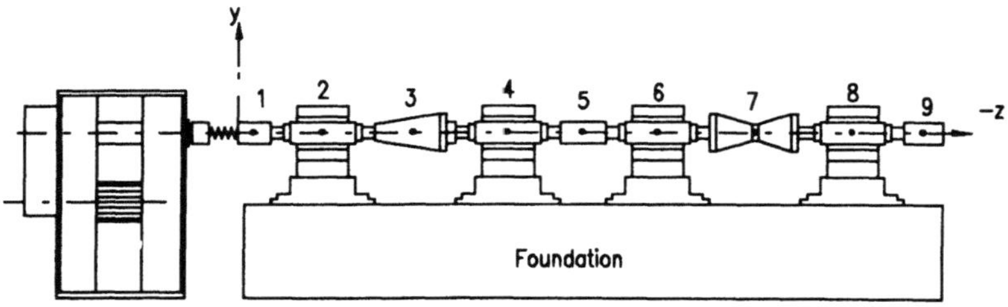

Fig. 2 Schematic diagram of the rotor-bearing installation

Based on analysis of the nonlinear vibration of the rotor system, the response of the four-bearing rotor installation was simulated for several different operating speeds. The assumed system configuration is listed in Table 1.

The mathematical model (1) of the system described was used to simulate its response before and after the change in unbalance has taken place.
This simulation was carried out for 9 cases of the residual unbalance distribution. The

unbalance distribution in case 1 is considered as the initial one, while case 2 to case 9 is considered as the changes of unbalance distribution with respect to the first case. The differences in the residual unbalance between case 1 and the other cases are shown in Table 2. For instance, in case 2, the phase of the unbalance force at station 1 was changed from $-121.9°$ (refer to case 1) to $0°$.

4. SIMULATION OF ERRORS IN MEASUREMENT AND MATHEMATICAL MODEL

In general, there are two main sources of errors.
(a) measuring errors,
(b) differences between the real rotor system and its mathematical model.
The measuring error was simulated by superposition of white noise on the system response. The white noise was generated from the following function based on Central Limit Theorem[8],

$$\Delta q = \sigma(\sum_{k=1}^{12} \beta_k - 6), \tag{26}$$

where β_k–random variable with a uniform distribution between 0 and 1,
$\sigma = \gamma_1 \sigma_r$–variance of random variable Δq,
γ_1–factor of noise level,
σ_r–root mean square of response q.
To investigate the influence of modeling accuracy, the simulation of response q was carried out for the mass matrix $[M]$ and stiffness matrix $[K]$ whereas the identification was based on the following mass and stiffness matrices.

$$[M_d] = [M] + \gamma_2[M], \tag{27}$$

$$[k_d] = [K] + \gamma_2[K], \tag{28}$$

where γ_2 is the factor reflecting accuracy of the system modeling.

5. INFLUENCE OF MEASURING ERRORS

To analyze the influence of the measuring errors, the response of the system was simulated in all nine cases specified in Table 2 for angular speed of the rotor equal to $500 rad/s$. Last 10 periods of steady stable solution were used in the identification procedure. Three different levels of noise which correspond to $\gamma_1 = 0.0$, $\gamma_1 = 0.01$ and $\gamma_1 = 0.03$ were added to these responses. Those disturbed responses were implemented to identify the plane and change in unbalance. Results of the identification are shown in Table 3.

For all cases, the stations at which the unbalance change has taken place, were identified correctly. The error function (23) which corresponds to these solution was at least 10 times less than all the other error functions. The error in measurements results in accuracy of assessment of magnitude of unbalance change.

Table 3. Unbalance changes for different level of the measuring errors γ_1(ST–station, ΔU–unbalance changes [N])

case	assumed		identified					
			$\gamma_1 = 0.0$		$\gamma_1 = 0.01$		$\gamma_1 = 0.03$	
NO	ST	Δ U	ST	Δ U	ST	Δ U	ST	Δ U
2	1	9.70	1	9.83	1	9.70	1	9.38
3	3	11.1	3	11.2	3	11.5	3	12.6
4	7	18.1	7	17.8	7	17.9	7	17.6
5	9	7.94	9	8.00	9	8.13	9	7.41
6	1	9.45	1	9.54	1	9.72	1	9.79
7	3	10.8	3	10.5	3	11.2	3	11.9
8	7	10.0	7	10.1	7	10.7	7	11.3
9	9	8.08	9	8.05	9	8.30	9	8.03

Table 4. Error functions for different number of periods

	error function		
station	2-periods	5-periods	10-periods
1	2.264	0.169	0.075
3	5.043	2.347	2.280
5	11.399	4.741	4.553
7	69.651	18.684	17.485
9	66.483	17.617	16.469
average	30.968	8.712	8.172

For $\gamma_1 = 0$, the average accuracy of assessment of unbalance change was 1.05% whereas for $\gamma_1 = 0.01$ and $\gamma_1 = 0.03$, the accuracy was 2.95% and 6.7% respectively.
Accuracy of the identification depends to large degree, on the number of periods of the steady state motion used in the identification procedure. This relationship is illustrated in Table 4. The data displayed in Table 4 corresponds to case 2 and $\gamma_1 = 0.01$. The identification was carried out for three different number of periods (2, 5, 10).
For number of periods equal to 2, the ratio of the smallest magnitude of error function (station 1) to any other is less than 0.45. For number of periods equal to 5 and 10, this ratio is less than 0.072 and 0.032 respectively.

6. INFLUENCE OF MODELING ACCURACY

The simulated responses of the system considered due to the unbalance listed in Table 2 were used to identify unbalance changes. The identification procedure was based on matrices $[M_d]$ and $[K_d]$(formulae(27) and (28)) for three different magnitudes of factor γ_2(γ_2=0, 0.01 and 0.05) and the factor $\gamma_1 = 0$. Results of this identification, for angular

Table 5. Assumed and identified unbalance changes for different accuracy of modeling($\omega = 500 rad/s$, ST–station, Δ U–unbalance changes[N])

case	assumed		identified					
			$\gamma_2 = 0.0$		$\gamma_2 = 0.01$		$\gamma_2 = 0.05$	
NO	ST	Δ U	ST	Δ U	ST	Δ U	ST	Δ U
2	1	9.70	1	9.70	1	10.6	1	11.1
3	3	11.1	3	11.1	3	10.7	3	11.3
4	7	18.1	7	18.1	7	16.8	7	17.5
5	9	7.94	9	7.90	9	7.99	9	8.35
6	1	9.45	1	9.50	1	9.51	1	9.91
7	3	10.8	3	10.8	3	9.46	3	9.85
8	7	10.0	7	10.0	7	10.1	7	10.6
9	9	8.08	9	8.10	9	8.00	9	8.34

Table 6. Assumed and identified unbalance changes for different accuracy of modeling($\omega = 209 rad/s$, ST–station, Δ B–unbalance changes[N])

case	assumed		identified					
			$\gamma_2 = 0.0$		$\gamma_2 = 0.01$		$\gamma_2 = 0.05$	
NO	ST	ΔU	ST	Δ U	ST	Δ U	ST	Δ U
2	1	9.70	1	12.7	1	12.9	1	14.0
3	3	11.1	3	11.1	3	11.2	3	–
4	7	18.1	7	21.2	7	21.3	7	22.1
5	9	7.94	9	8.06	9	8.41	9	9.81
6	1	9.45	1	9.58	1	9.63	1	11.4
7	3	10.8	3	11.0	3	11.1	3	11.7
8	7	10.0	7	10.0	7	10.1	7	10.7
9	9	8.08	9	8.06	9	8.31	9	9.86

Table 7. Combined effect of measuring errors and accuracy of modeling on identified unbalance changes($\gamma_1 = 0.03, \gamma_2 = 0.05$)

Case	assumed		identified	
NO	ST	Δ U	ST	Δ U
2	1	9.70	1	9.88
3	3	11.1	3	13.2
4	7	18.1	7	18.6
5	9	7.94	9	7.84
6	1	9.45	1	10.3
7	3	10.8	3	12.6
8	7	10.0	7	11.9
9	9	8.08	9	8.48

velocity $\omega = 500 rad/s$, are shown in Table 5.
In all three cases the plane where the change in unbalance has taken place was identified correctly. Relative error of the identified changes in unbalance depend on the accuracy of modeling and its average magnitudes is 0.15%, 3.% and 6.% for γ_2=0, 0.01 and 0.05 respectively.
If operating speed is close to critical speed, results of identification are less accurate. This case is illustrated in Table 6. The angular speed of the rotor($\omega = 209 rad/s$) was chosen close to one of critical speed ($\omega = 210 rad/s$).
In this case the average magnitude of relative errors in identified unbalance changes are 6.5%, 8.4% and 20.% for γ_2 equal to 0, 0.01 and 0.05 respectively.
Combined effect of measuring errors and errors in modeling of the rotor system is shown in Table 7. Identification procedure was carried out for angular speed equal to $500 rad/s$, $\gamma_1 = 0.03$ and $\gamma_2 = 0.05$.
The relative error in the identified unbalance changes was less that 19% for all cases considered in this simulation.

7. CONCLUSIONS

The developed procedure for the identification of unbalance changes of the rotor system is of practical importance since it is based on the experimental data which is usually monitored during operation. This procedure was used to identify changes in unbalance which are only twice as large as the standard residual unbalance. Simulation of more than 50 cases allows to conclude that the influence of noise can be significantly reduced if the identification is performed on signals monitored for more than 10 revolutions. If accuracy of modeling of the rotor system is less than 10%, application of this procedure results in acceptable identification of the unbalance change. Although reliability of this procedure depends on accuracy of the rotor system mathematical model, the analysis provided make a fair promise that it can be successfully used for identification of unbalance changes.

References

1. Goodman, T. P., "A Least-Squares Method for Computing Balance Corrections", *Journal of Engineering for Industry, Trans. of ASME,* Vol.86, August 1964, pp. 273-279.

2. Lund, J. W. and Tonnesen, J., "Analysis and Experiments on Multi-Plane Balancing of A Flexible Rotor", *Journal of Engineering for Industry, Trans. of ASME,* Vol.94, February 1972, pp. 233-242.

3. Ramamurti, V. and Anantaraman, K. W., "Two-Plane In-Situ Balancing", *Journal of Sound and Vibration,* 134(2), 1989, pp. 243-352.

4. Smalley, A. J., Baldwin, R. M. and Schrick, W. R., "Spray Automated Balancing of Rotors: Concept and Initial Feasibility Studies", *Journal of Engineering for Gas Turbines and Power, Trans of ASME,* Vol. 111, October 1989, pp. 659-665.

5. Craggs, A., "A Procedure for Balancing Large Turbo-Generator Set via A Finite Element Model",*Journal of Sound and Vibration,* 108(2), 1986, pp. 349-352.

6. Krodkiewski, J. M. and Ding, " Theory and Experiments on a Method for On-Site Identifying Bearing Alignment Changes of Multi-Bearing Rotor System", *Journal of Sound and Vibration,*, Vol.164, pp. 281-294, 1993

7. Parszewski, Z.A., Roszkowski, M. and Gluza, J., "Pressure Distribution and Hydrodynamic Forces of Journal Bearing Partial Sleeves", *Arch. Mech. Engng.,* 4, 1967, p.455.

8. Freund, J. E. and Walpole, R. E.,"Mathematical Statistics", *Prentice-Hall Inc.,* 1980, pp.256-257.

SOME THEORETICAL PROBLEMS AND CONCEPTS IN DEVELOPING THE TECHNIQUES OF MACHINERY

MONITORING AND DIAGNOSIS

Xu Min
Shanghai Jiao Tong University, Shanghai, China

ABSTRACTS

In recent years, the condition monitoring and fault diagnosis techniques of machinery and equipment have attracted a great attention and made a rapid progress. In this paper, from the aspects of diagnosis theory and expert system, signal processing and pattern recognition, emphasis on the study of fault mechanism and some concepts related, auther intends to express some theoratical problems and some suggestions in devoloping the techniques of machinery monitoring and diagnosis briefly.

---

* This work is supported by SPC 8 – 5research project

* * Chairman of Fault Diagnosis Society, China, CSVE.

1. Introduction

Recently, the condition monitoring and fault diagnosis techniques of machinery and equipment (CM&FD) have unprecendently made a great progress and attracted a great attention from industry at home and abroad. This is because of not only some catastrophes, such as radioactive leakage accidents of nuclear reactors (1979.3.29 USA, 1986.4.27 USSR), aeronautics and space accidents (1987 Japan, 1986.1.28 USA), ship sinking accidents (1987 Britain, Philippines), generator station explosion accidents (1986.10, 1988.2, China), ... ect., which caused heavy economic losses, but also frequent emergencies of severe assidents which caused lack of confidence in modernized industry and new technology, a great panic among the people, harmful consequences in politics, and a heavy pressure on industry-related officials and personnel.

As is known to all, with the contemperory mass industrial production as its application field and close relation to various engineering machines, today's condition monitoring and fault diagnosis techniques come to be developed into a highly feasible and hightech-associated dicipline which embraces not only theories (diagnosis theories) and experiments (fault mechanism study), but also methods (analysis methods) and plannings (modern monitoring instruments and diagnosis system).

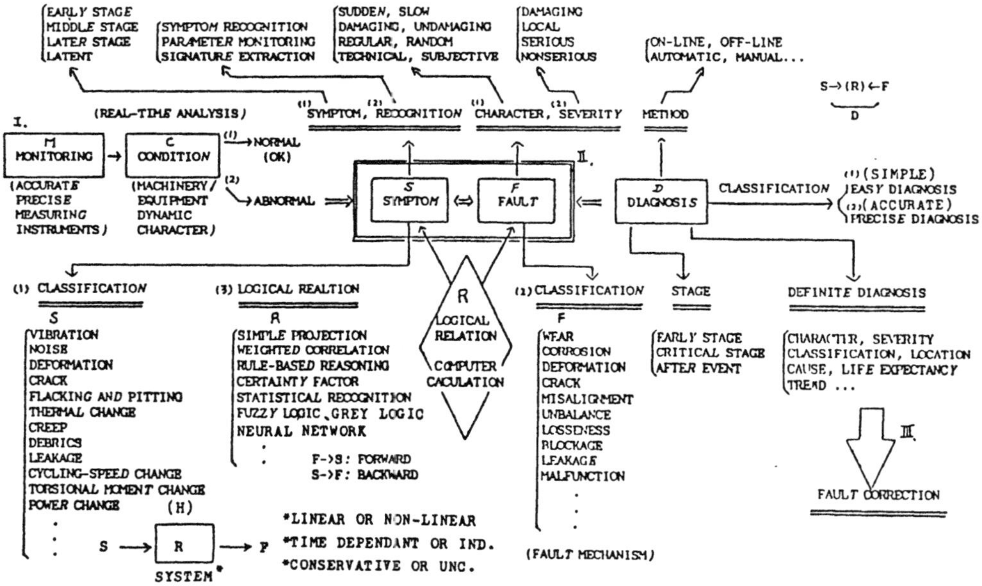

Fig.1 Diagram of Main Content and Related Knowledge of CM&FD Configuration

Fig.1 shows the main content and relevant knowledge (systematical configuration as an example) of the CM&FD. It encompasses: dynamic characteristics (including strength), condition parameter classifications, model & recognition, various symptom effects and fault mechanism study, various logical reasonings and statistical recognition relations, precise measurement and real time analysis techniques. It also contributes to the contemperory management science related to the contemperory economy management and maintenance organization system.

But, for an advanced system today, three functions must be carried out and can not be replaced from one to the other. They are monitoring function, diagnosis function and management function. The first function is to know the working condition of machinery itself, normal or abnormal. We need real time measurement and precise analysis. The second function is to make sure the fault classification, the location of fault, the fault level of severity and the fualt character as well. Monitoring is one thing, diagnosis is another. The third function is to take account of economic production, to optimize the operating parameters of machine in order to get more benefits from the process of production. It's really a high level of an advanced system for CM & FD developed today.

2. Expert System and Diagnosis Theory

While the advent of computer technology has brought the world to a new era, the development of expert system has pushed the computer application to a higher stage. Expert system can solve many complicated problems in a variety of prospective fields. It uses the domain specific knoweledge to simulate the reasoning procedure of human expert. It has a collection of ambiguous, uncertain, fuzzy knowledge, and learning & expansion abilities as well. The encyclopaedia-like system can help a novice solve very complicated problems. Therefore, it has greatly improved the quality and advancing speed of the technology, and become a very attractive developing direction.

The acquisition and expression of expert experiential knowledge, the knowledge hierarchy, the knowledge blocking and effective operation, the establishment of in-depth and formalized knowledge, multi-goal strategy, high-speed matching reasoning, ... etc. are all becoming the developing problems and emphases of study, not achieve completely yet.

With the advantages of non-linear neural element, parallel processing of layer matrix, self-learning and self-associate-memory functions, the neural network (NN) diagnosis method has been applied to CM & FD system in recent years, particularly the NN back propagation method (BPM). Some theoracal prolems of NN are waiting for further study, such as: the decision of hidden element number and hidden layer number, the matching of transfer function (Sigmoid or Hyperblic tangent or others), the value of connection weight, ... etc..

Fuzzy logic diagnosis method or grey logic diagnosis method has the superiority of simplicity of calculation, the easy application, the clear concept, ... etc..

But the decision of subordinate degree function for FLD, the choice of relational function for GLD, need more practical experiences.

For most of diagnosis methods, in order to get more accurate diagnosis results, in other words the high confidence, the pre-experience knowledge is fairly needed. Some times, it is a prerequisite for taking a diagnosis method before going to a path of uncertainty.

3. Signal Processing and Pattern Recognition

Signal processing and pattern recognition are very active and rapidly developed in monitoring and diagnosis techniques as an important branch. If you regard the diagnosis as a structure, you should consider signal processing and pattern recognition as the infrastructure. This is also true in expert system. Let us take symptom parameter extraction as example: with highly sensable and distinguishable signatures (including the effectively processed signatures), diagnosis accuracy can hardly be assured even though the methods used afterword are rigorious and reasionings are elaborate. Therefore, it should be greatly concerned.

Conventional time domain processing functions, such as tiome domain wave form, peak value, periodic energy, root mean square, kurtosis factor, peak number, axis orbit, probability distribution, ... etc., and frequency domain functions, such as signature analysls, harmonic analysis, sideband analysis, cepstrum analysis, spectrum energy analysis, ... etc., are still very useful. Various new analysis and recognition methods, such as phase analysis, frequency composition, holography spectrum, AR-ARMA model parameter identificition, FFT-AR method, AR-PE method, CE-AR method, SPM method, HT method, adaptive filter, Wigner distribution, statistical pattern recongnition, classifiers, ... etc., have successively emerged and made great progress, and very encouraging as well. Fig. 2 shows the relationship of most signal processing functions, with the common functions and recently-developed processing and recognition methods, signature set description of fault condition promises an another improvement of diagnosis techniques.

It is nearly as a law that a certain mode (or pattern) of sigal processing must definitely have a correct match for a certain fault recognition

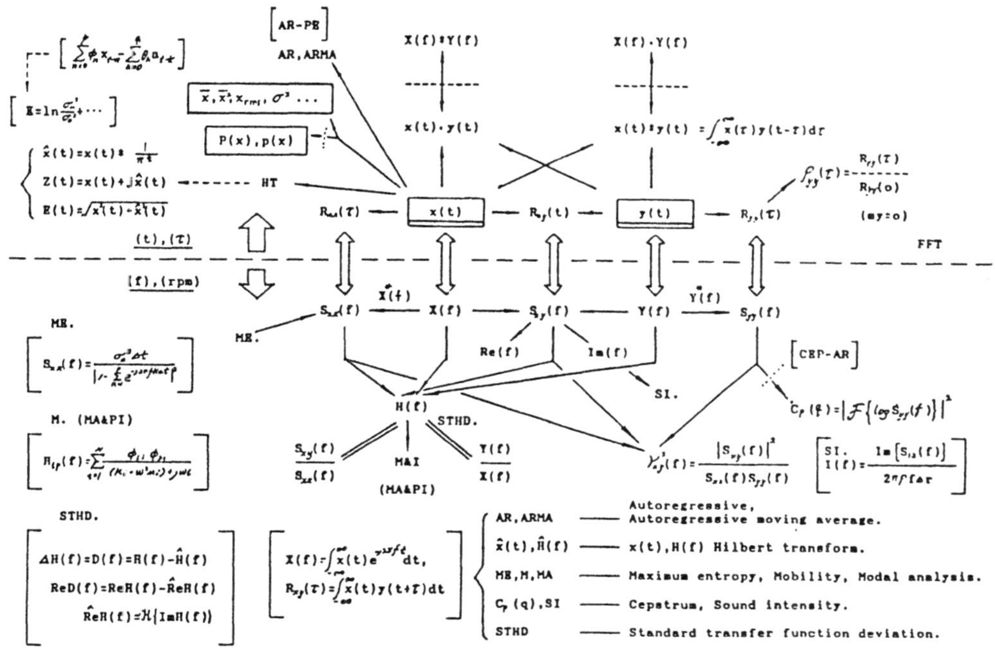

Fig.2 Diagram of Signal Function Relationship

correspondingly. For examples, the mode of shock envelope modulation is quite a good match for rolling bearing fault recognition, the mode of harmonic frequency of meshing frequency is quite a good match for gear fault recognition, the mode of shaft center orbit of oil film whirling is quite a good match for bearing oil film oscillation, the mode of holography spectrum of shaft rotating is quite a good match for distinguishing 50 Hz synchro cycles between electric one and shaft vibration; ... etc.. The more correct matching between the mode and fault recognition you have, the more precise diagnosis of machinery faults you will obtain. There is no universal mode or pattern which can match for all kinds of fault recognition.

4. Emphasis on the Study of Fault Mechanism

During the processing of machinery fault diagnosis techniques, not only the clear understanding of machine structure itself is required, but also the mastering of its running characteristics is important. Moreover, we should deeply know the origin or evidence which cause machine damages. Taking rotating machinery as example: if vibration fault is the main cause of the machine damage, we should find vibration cause of each part and study vibration-associated characteristics, effect of thermal deformation on vibration fault, wear and tear mechanism, fluid vibration characteristics, symptoms and possibilities of random failures, and the behaviour course of them exactly. We should also study synthetical and systematical effects based on individual fault mechanism. The study of individual fault mechanism and system fault mechanism is a very complicated but an important task. It should be considered as the basement of the technique foundation. There is much room for the work of experimental research and theoretical research in the field of CM & FD.

5. Conclusion

For an advanced system of machinery CM & FD, three funations must be carried out separately. They are monitering function, diagnosis function and management function. Monitoring is one thing, diagnosis is another. Management function is to optimize the operating parameters of machine in order to get more benefits from the process of production.

For most of diagnosis methods in order to get more accurate diagnosis results and higher confidence, the pre-experience knowledge is fairly needed, before going to a path of uncertainty.

The more correct matching between the mode (or pattern) and fault recognition you have, the more precise deagnosis of machinery faults you will obtain. There is no universal mode which can match for all kinds of fault recognition.

We muat pay more attention to the study of fault behaviour course for unknown effects as the basement of diagnosis technique foundation in ovder to get more accurate fitting data.

How to evaluate the quality of a CM & FD system before or after mounting on spot, it's still a unsolved problem.

Most of viewpoints maintioned above are suitable for the diagnostics of rotating machine in power plants. All the poupose of this paper is to get some impression as reference for readers from another reticle research angle.

REFERENCES

[1] Xu Min, 'Some Theoratical Problems and Suggestions in the Development of Equipment Diagnosis Techuiques' Sino-Japan Chemical Equipment Diagnisis Conf. Proceedings, Dec., 1992, China.

[2] Xu Min, H. Yan, J. Chen, 'Application of Neural Network to the Fault Diagnosis of Ball Bearing', Proceedings, MOVIC'92, Japan.

[3] Dong S. Kim and Others, 'Machinery Diagnostics for Rotating Machinery Using Back Propagation Neural Network' Proceedings, IMMDC'91, Las Vegas.

[4] Liu Yong-Hong, Xu Min and Yan Hui, 'The Fault Diagnosis Method by Grey Relational Analysis of ABo-Mode and the Comparison with Fuzzy Diagnosis Method', Proceedings, IMMDC'91, Las Vegas.

ON-LINE IDENTIFICATION AND DIAGNOSIS OF UNBALANCE FORCES OF THE ROTOR OF ROTATING MACHINES

Fu Zhi-Fang, Zhou Hai-Ting and Rao Zhu-Shi
Shanghai Jiao Tong University, Shanghai, China

ABSTRACT

This paper deals with a method detecting the unbalance forces of the rotor of rotating machines. Using this method, the amplitudes, phases and locations of unbalance forces of the rotor can be determinde simultaneously in the running case. In this paper, the load identification method [1,2] is used to identify the unbalance forces of the rotor on the basis of Modal Analysis of the rotor system. The method presented in the paper provides a theoretical basis for on – line identification of unbalance forces of the rotor of rotating machines. This method can also be used to balance the flexible rotor without trial weights.

Key words: Fault Diagnostics, Dynamic Balance of Rotor, Load Identification.

1. INTRODUCTION

The out of balance of the rotor of a large rotating machine will lead to severe vibration and serious fault. If the unbalance forces of the rotor can be detected at all times in the running case, the severe vibration and fault of machines can be prevented in advance and the downtimes of machine can be reduced, as well as the balance of machine becomes very easy and economized. The general techniques used to predict the unbalance of rotor under the operating condition rely on the experiential judgement on the measured vibration of bearings or on its spectrum analysis. These techniques can only be used to determine the unbalance state of the rotor characteristically and can not be used to determine the unbalance forces quantitatively under operating conditions.

In this paper, a method is presented whereby the amplituds, phases and locations of the unbalance forces of the rotor can directly on-line determined. The identification of unbalance forces of the rotor involves the inverse solution of the equations. The parameter identification techniques [1] is used in the paper. Because of the complexity of the rotor-bearing system of large rotating machines, the dynamic substructure analytical technique is used in the paper. The measuring accuracy of the vibration signals and anti-disturbance are the key problems for the unbalance forces identification. The preprocessing of the vibration signals is also discussed in the paper. Not only the theoretical analysis of the method is carried out, but also some practical problems and working conditions of rotating machines, such as the problems about the effect of oil film of bearing, inadequate measuring point, useful information extraction and anti-disturbance of vibration signals, etc. are discussed in the paper.

2. THEORETICAL BACKGROUND

According to the theory of modal analysis, the displacement responses of the rotor caused by the unbalance forces can be calculated by following equation

$$D = HT \tag{1}$$

Where D is the vector of displacement responses of the rotor system, H is the matrix of Frequency Response Function (FRF) between the applied points of unbalance forces and measuring points of responses of the rotor system, T is the vector of unbalance forces.

For a certain practical rotor system, the FRF can be obtained by test. the reletionship between FRF and modal parameters can be expressed as

$$H = \sum_{r=1}^{n} \frac{\phi_r \phi_r^T}{m_r \omega_r^2} \left[\frac{1}{(1-\bar{\omega}_r^2) + j2\varepsilon_r \bar{\omega}_r} \right] \tag{2}$$

Where ϕ_r, ω_r, m_r, ε_r are modal parameters of rth mode of the rotor system, which can be obtained from modal analysis of rotor system, $\omega_r = \bar{\omega}/\omega_r$.

Having known the FRF and the responses of rotor system, the unbalance forces can be calculated by the following equation

$$T = H^+ D \tag{3}$$

Where H^+ is the generalized inverse matrix of FRF.

3. ESTABLISHMENT OF FRF MATRIX

A multispan rotor system has n supporting bearings and p unbalance foces. i and k denote the ordinal number of bearings and unbalance forces respectively. $i = 1, 2, \cdots\cdots n$; $k = 1, 2, \cdots\cdots p$.

The rotor-bearing system can be divided into three parts of substructures, namely free rotor, oil films and bearings. Taking the modal analysis and test for the three substructures individually, the modal parameters of substructures can then be obtained. The dynamic substructure analysis technique and impedence matching method are used to determine the dynamic characteristics of the substructures and to establish the FRF matrix of rotor – bearing system.

The schematic diagram of rotor system is shown in fig.1. The dynamics model of substurcture is shown in fig.2.

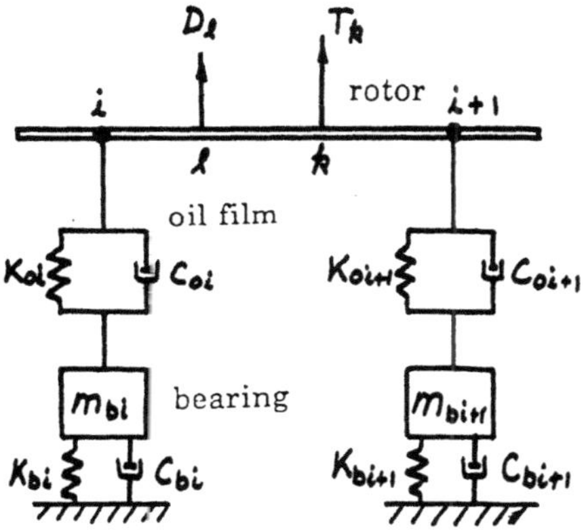

Fig.1 Schematic diagram of rotor system

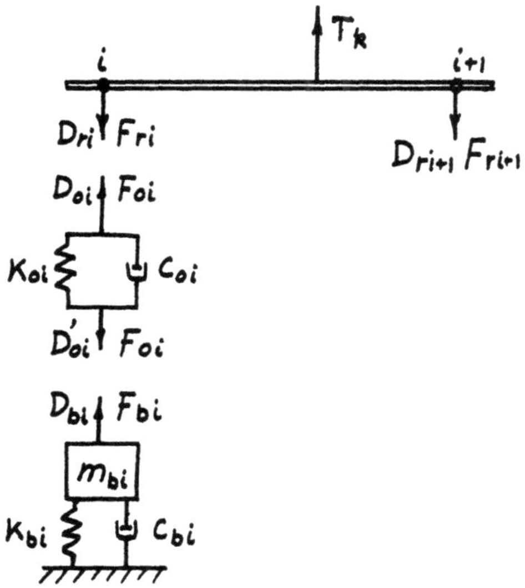

Fig.2 Dynamics model of substructure

Let $D_{r.}$ be the displacement vector of free rotor at the supporting points

$$D_{rj}=[x_{r1},\ y_{r1},\ x_{r2},\ y_{r2},\ \cdots\cdots x_{rn},\ y_{rn}]^T,\quad (2n\times 1) \tag{4}$$

F_{rj} be the vector of bearing forces applied on the rotor.

$$F_{rj}=[f_{r1},\ g_{r1},\ f_{r2},\ g_{r2},\ \cdots\cdots f_{rn},\ g_{rn}]^T,\quad (2n\times 1) \tag{5}$$

Where f,g express the components of force in horizontal and vertical directions,

D_{bj} be the displacement vector of bearings

$$D_{bj}=[x_{b1},\ y_{b1},\ x_{b2},\ y_{b2},\ \cdots\cdots x_{bn},\ y_{bn}]^T,\quad (2n\times 1) \tag{6}$$

F_{bj} be the force vector of bearings

$$F_{bj}=[f_{b1},\ g_{b1},\ f_{b2},\ g_{b2},\ \cdots\cdots f_{bn},\ g_{bn}]^T,\quad (2n\times 1) \tag{7}$$

The displacement vector of rotor at supporting points can be calculated by the following equation

$$D_{rj}=H_{rjj}F_{rj}+H_{rjp}T \tag{8}$$

Where H_{rJ} is the FRF matrix of the rotor between the supporting points, which has the order of $(2n\times 2n)$, H_{rjp} is the FRF matrix of rotor between the supporting points and the applying points of unbalance forces, which has the order of $(2n \times 2p)$.

Let me take the dynamics analysis for ith bearing.

The force vector of oil film of ith bearing is

$$F_{oi} = (K_{oi} + j\omega C_{oi})(D_{oi} - D_{oi}') = H_{oi}^{-1}(D_{ri} - D_{bi}) \tag{9}$$

Where K_{oi}, C_{oi} are the stiffness matrix and damping matrix of oil film of ith bearing

$$K_{oi} = \begin{bmatrix} k_{xx} & k_{xy} \\ k_{yx} & k_{yy} \end{bmatrix}_{oi} \qquad C_{oi} = \begin{bmatrix} C_{xx} & C_{xy} \\ C_{yx} & C_{yy} \end{bmatrix}_{oi} \tag{10}$$

which can be obtained from test or calculation of bearing.

The bering forces are

$$F_{bi} = -(M_{bi}\omega^2 + j\omega C_{bi} + K_{bi})D_{bi} = H_{bi}^{-1}D_{bi} \tag{11}$$

Where M_{bi}, K_{bi}, C_{bi} are the mass matrix, stiffness matrix and damping matrix of ith bearing, H_{bi} is the FRF matrix of ith bearing, which can be obtained from modal test of bearing.

As we know, the system composed of bearing and oil film is a serial system. Its FRF is the sum of the FRF of the subsystems. Therefor, the FRF of ith bearing system will be

$$H_{Si} = H_{oi} + H_{bi} \tag{12}$$

Thus we have

$$D_{ri} = H_{Si}F_{ri} = (H_{oi} + H_{bi})F_{ri} \tag{13}$$

For all supporting points, we have

$$D_{rj} = \begin{bmatrix} H_{S1} & & & \\ & H_{S2} & & \\ & & \ddots & \\ & & & H_{Sn} \end{bmatrix} \begin{Bmatrix} F_{r1} \\ F_{r2} \\ \vdots \\ F_{rn} \end{Bmatrix} = H_{Sj}F_{rj} \tag{14}$$

By substituting D_{rj} into Eq. (8). we obtain

$$(H_{sj} - H_{rjj})F_{rj} = H_{rjp}T \tag{15}$$

This is the equation expressing the relationship between the supporting forces and unbalance forces of the rotor.

Using the consistency conditons between the substructures and considering Eq.(11), we have

$$D_{bj} = H_{bj}[H_{rjj} - H_{Sj}]^{-1}H_{rjp}T = ZT \tag{16}$$

Where Z is the FRF matrix between responses of bearings and unbalance forces of the rotor.

$$Z = H_{bj}[H_{rjj} - H_{Sj}]^{-1}H_{rjp}, \quad (2n \times 2p) \tag{17}$$

The displacement responses of arbitrary points on rotor caused by unbalance forces and bearing forces can be calculated by the following equation

$$D_r = H_{rp}T + H_{rj}F_{rj} = [H_{rp} + S]T = YT \tag{18}$$

$$S = H_{rj}(H_{sj} - H_{rjj})^{-1}H_{rjp} \tag{19}$$

Where D_r is a $(2L \times 1)$ displacement vector of the rotor, L is the number of measuring points on the rotor, H_{rp} is a $(2L\times 2p)$ matrix of FRF between responses points and unbalance forces of the rotor. H_{rj} is a $(2L\times 2n)$ matrix of FRF between the responses points on the rotor and bearing forces.

the generalized displacement vector of the rotor-bearing system can be expressed as follows

$$D=\begin{Bmatrix} D_r \\ D_{bj} \end{Bmatrix}=\begin{bmatrix} Y \\ Z \end{bmatrix}T=HT \tag{20}$$

Where H is the FRF matrix of whole rotor-bearing system

On the basis of knowing the FRF matrix H, we can identify the unbalance forces of rotor from measured date of responses by following equation

$$T=H^{+}D \tag{21}$$

4. EXTENTION OF MEASURING POINTS

For practical rotating machinery, such as trubine, the rotor is covered by casing. We can just measure the vibration displacements of outside part of the rotor. Therefor, the number of measuring point of response of the rotor is usually inadequate.

It is necessary to extend the number of measuring point. According to the theory of modal analysis, the physical coordinats of responses of system can be exprssed by the modal coodinats. thus we have

$$D=\Phi_S Q_S \tag{22}$$

Where Φ_S is the modal matrix of the system, Q_S is the vector of modal coordinats.

Vector D can be separeted into tow parts

$$D=\begin{Bmatrix} D_m \\ D_u \end{Bmatrix}=\begin{bmatrix} \Phi_{mS} \\ \Phi_{uS} \end{bmatrix}Q_S \tag{23}$$

Where D_m is the measurable part and D_u is unmeasurable part of responses. In the same way, the modal matrix can also be divited into tow corresponding parts Φ_{mS} and Φ_{uS} Wheres is the number of interrupted modes.

thus we have

$$D_m=\Phi_{mS}Q_S \tag{24}$$

$$Q_S=\Phi_{mS}^{+}D_m \tag{25}$$

From which we can obtain the following equation

$$D=\Phi_S\cdot\Phi_{mS}^{+}\cdot D_m \tag{26}$$

Thus the number of degree of fredom of responses is to be extended. The equation (21) can then be solved.

For increasing the quantity of informations, the measuring point of response should be as more as possible. For increasing the calculation accuracy and eliminating the noise, the Singular Value Decomposition Algorithm is used to calculate the inverse matrix.

5. SIGNAL EXTRACTION AND ANTI–DISTURBANCE

The factors of disturbance on the vibration signals of the rotor is usually existent in the working case. For extracting the useful informations from measured vibration signals, a kind of generalized Wiener Filter method is used to preprocess the signal so as to increase the signal to noise ratio (SNR).

The measured vibration signal y(t) is the sum of real signal x(t) and noise n(t)

$$y(t)=x(t)+n(t) \tag{27}$$

We assume that the x(t) and n(t) are the stationary radom process, the power spectra of which are known and there are no correlation between each other. The frequency equation of Wiener filter is

$$H_F(f)=\frac{G_{xx}(f)}{G_{xx}(f)+G_{nn}(f)} \tag{28}$$

Where $H_F(f)$ is the FRF of Wiener Filter, $G_{xx}(f)$ and $G_{nn}(f)$ are the power spectra of signals and noise.

From Eq.(28), we can see that the Wiener Filter is a weigted processing to every frequency components of measured signals according to the SNR of system. In the practical case, $G_{xx}(f)$ and $G_{nn}(f)$ are unknown. If the measured signals have the time unvaried power spectra density, the posterior knowledge Wiener Filter method can be used. The estimated values of signal and noise can then be obtained from their generalized average.

Assum that we have N samples. Taking n times average for signal y(t), $\frac{1}{N}\sum_{i=1}^{N} y_i(t)$, the auto power spectrum $\hat{G}_{xx}(f)$ of signal x(t) can then be calculated. Let $\hat{G}_{xx}$ be the estimated value of G_{xx}. To obtain the estimated values of the power spectra of signal and noise, we can take the alternative average for samples and to eliminate the determined composition from the measured signal. For this purpose, we divide N samples into M subset and take the alternative average for each subset, $\frac{1}{L}\sum_{i=1}^{L}(-1)^i y_i^{(m)}(t)$. Where L is the number of samples of each subset, $L=\frac{N}{M}$. $m=1,2,\cdots\cdots M$.
Calculating the power spectrum $\hat{G}_{nn}^{(m)}(f)$ and taking the average of $\hat{G}_{nn}^{(m)}(f)$ for each subset, we can obtain the estimated value of noise signal power spectrum

$$\hat{G}_{nn}(f)=\frac{1}{M}\sum_{m=1}^{M} G_{nn}^{(m)}(f)$$

Substituting the $\hat{G}_{xx}(f)$ and $\hat{G}_{nn}(f)$ into Eq.(29), we can get the weighted function

$$\hat{H}_F(f) = \frac{\hat{G}_{xx}(f)}{\hat{G}_{xx}(f) + \hat{G}_{nn}(f)} \tag{29}$$

Therefcre the final filter estimation of signals in frequency domain is

$$\hat{Y}(f) = Y(f)\hat{H}_F(f) \tag{30}$$

Where Y(f) is the fourier transform of measured signal.

The frequency of vibration signal caused by unbalance forces is equal to that of rotating speed of the rotor. The vibration with frequency of rotating speed can be produced not only by the unbalance forces, but also by the other factors, such as misalingment of rotors. Therefore before identification, it is necessory to extract the vibration signal which is produced only by the unbalauce forces. The expert system and signal analysis techniques can be use.

6. EXPERIMENTAL VERIFICATION

Experimental verification is done with a multiple spans flexible rotor with four ournal bearings and six disks. The simplified diagram of testing set up of the rotor system is shown in fig.3.

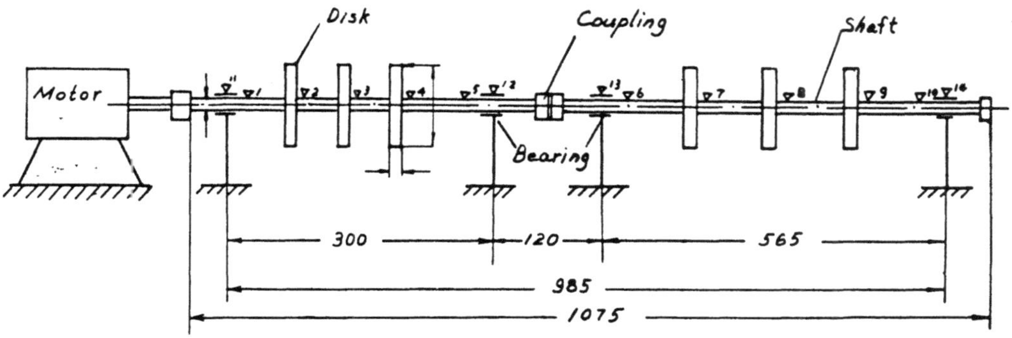

∇——Measuring points

Fig.3, Simplified diagram of testing set up of rotor system

The testing set up can be reassembled in various structural systems of rotor, such as single span with rigid (or flexible) rotor and rigid (or flixible) supporting bearings, multiple spans (1~3) with rigid (or flexible) supporting bearings; The rotor is driven by a frequency-modulated electric motor, the speed of which can be adjusted by adjusting the frequency of input current. The maximum speed of the rotor is 8000rpm. The supporting stiffness of bearings is adjustable. The total length of rotor is 1075mm. The responses of the rotor are measured by contactless transdusers. Data acquisition and processing are performed by using an acquisition system. The unbalance forces are identified by using a software, which is specially programed. The impact excitation technique is used in the modal test of the rotor and bearings. The matrix of FRF is shown in table 1.

Before identification of unbalance forces, the rotor is balanced precisely and to measure the initial responses of the rotor D_0.

For verifying the presented method, we add a certain mass m_k on kth disk and measure the displacements of the rotor and bearings D. The real displacement vector Due to unbalance force is $D-D_0$. Using equation (21), we can identify the unbalance force on kth disk T_k, which should be equal to $m_k r_k \omega^2$.

The identified results of unbalance forces for various kind of rotor systems are given in the tables 2~5. We can see in the tables that the accuracy of identification is satisfied for requirements of engineering.

7. CONCLUSION

It is indicated from research that using the load identification techOniques of the structures to identify unbalance force of the rotor is feasible. Experimental verification indicates that the identified accuracy is satisfied for requirements of engineering.

The following problems should be take care, when this method is used to the rotor of practical rotating machinery

1. Errors result from inadequate measuring point.
2. The influence of non-linearities of oil film of the bearings, Owing to which the principle of superposition and the theoretical basis of modal analysis will be destroyed.
3. Extraction of useful signals and anti-disturbance.

REFERENCES

1. Fu, Zhi-Fang: Vibration Modal Analysis and Parameter Identification, Publishing House of Mechanical industry, 1990.

2. Verhoeven, J.: Excitation Force Identification of Rotating Machines Using Operational Rotor/Stator Amplitude Data and Analytical Synthesized Transfer Functions, J. of Vibration, Acoustics, Stress and Reliability in Design, 1988, Vol.110.

3. Mauvice, L., etc.: Large Unbalance Vibration Analysis of Steam Turbine Generators, Electric Power Research Institute, 1984.

4. Zhu, Ji-Mei, etc.: Modal Parameter Identification Method used to balance of flexible rotors, J. of vibration engineering, vol.1, No.1, 1987.

Table 1. The Matrix of FRF of the Rotor System

Measuring Points	FRF of the Rotor System (μm/N) Locations of Unbalance Plane					
	1#	2#	3#	4#	5#	6#
1	1.1244 + i0.2195	1.1217 + i0.1700	0.8004 + i0.1128	0.6262 − i0.1810	0.9927 − i0.2936	0.7671 − i0.1733
2	1.6793 + i0.3735	1.8101 + i0.4749	1.3210 + i0.3813	0.9133 − i0.0757	1.4574 − i0.1619	1.1416 − i0.0231
3	1.5043 + i0.2466	1.8109 + i0.3063	1.5274 + i0.3272	1.0062 − i0.1713	1.5213 − i0.2913	1.1622 − i0.1631
4	1.0911 + i0.0625	1.4113 + i0.1334	1.3877 + i0.2844	0.9832 − i0.1541	1.2102 − i0.3397	0.9526 − i0.2250
5	0.9004 + i0.0116	1.1891 + i0.0748	1.2642 + i0.2726	0.8098 − i0.1233	1.0898 − i0.3378	0.8316 − i0.2283
6	0.3708 + i0.1343	0.4198 + i0.0822	0.5291 − i0.1764	−1.9308 − i0.0898	−3.3464 − i0.1272	−1.7668 − i0.2968
7	0.8060 + i0.3607	0.9840 + i0.4557	0.8821 − i0.3783	−5.6035 − i0.3359	−8.4868 + i0.7559	−3.2593 + i0.3414
8	1.0771 + i0.4692	1.4800 + i0.5337	1.2799 − i0.4082	−7.4347 − i0.7867	−10.1633 + i0.8874	−7.8667 + i0.8782
9	0.7607 + i0.2799	1.0078 + i0.3265	0.9408 − i0.2864	−5.8209 − i0.4510	−7.6540 + i0.5379	−5.1150 + 0.4826
10	0.2446 + i0.0004	0.1890 + i0.0413	0.3307 − i0.1272	−2.9502 − i0.1833	−3.8120 + i0.0569	−0.490 − i0.2678
11	0.7773 + i0.1058	0.7394 + i0.0368	0.5571 − i0.0484	0.4873 − i0.2607	0.7558 − i0.3955	−0.5636 − i0.2879
12	0.4632 + i0.1658	0.7226 + i0.1516	0.7546 + i0.1535	0.4607 − i0.0196	0.5907 − i0.0076	0.5183 − i0.0220
13	0.2765 + i0.0819	0.3127 + i0.0149	0.4712 − i0.1226	−1.0874 − i0.1627	−2.7034 − i0.3622	−1.8273 − i0.4205
14	0.0684 + i0.0542	0.1585 + i0.0033	0.0988 + i0.0150	−0.8237 − i0.3636	−0.9622 − i0.8358	−0.3781 − i0.4762

Table 2. Results of Unbalance Identification
< single Span, Rigid Rotor, Rigid Bearings>

Location of Unbalance Force	Added Quantity of Unbalance	Identified Quantity of Unbalance	Error	
Disk No.	(g – cm)	(g – cm)	Amplitute (%)	Phase (degree)
4	37.71 < 15°	39.35 < 15.6°	4.3	0.6
5	37.71 < 15°	39.66 < 31.8°	5.2	7.86
5	59.715 < 150°	66.12 < 149.6°	10.7	0.4
6	59.715 < – 60°	57.65 < – 58.6°	3.4	1.4

Table 3. Results of Unbalance Identification
<Single Span, Flexible Rotor>

Location of Unbalance Force	Added Quantity of Unbalance	Identified Quantity of Unbalance	Error	
Disk No.	(g – cm)	(g – cm)	Amplitude (%)	Phase (degree)
4	37.71 < – 165°	38.60 < – 163°	2.3	2.0
5	37.71 < 150°	36.70 < 156°	2.7	6.0
5	27.81 < – 15°	26.80 < – 15.5°	3.6	0.5
6	27.81 < – 15°	28.00 < – 15.4°	0.7	0.4

Table 4 Results of Unbalance Identification
<Double Spans, Flexible Rotor>

Location of Unbalance Force	Added Quantity of Unbalance	Identified Quantity of Unbalance	Error	
			Amplitude	Phase
Disk No.	(g - cm)	(g - cm)	(%)	(degree)
2	27.81 < -210°	28.24 < -241°	1.5	31.0
5	37.715 < -165°	38.89 < -166.8°	3.1	1.8
5	45.765 < 30°	44.87 < 31.3°	1.9	1.3
4	37.71 < -30°	39.23 < -29.3°	4.0	0.7
4	37.71 < -210°	39.90 < -209.3°	5.8	0.7
6	27.315 < 30°	25.37 < 29.7°	7.1	0.3
6	27.315 < -150°	24.37 < -144.6°	10.7	5.4
3	46.765 < 30°	39.55 < 46°	13.4	16.0
3	46.765 < -150°	39.82 < -134°	12.9	16.0

Table 5 Results of Unbalance Identification
<Triple Spans, Flexible Rotor, Flexible Bearings>

Location of Unbalance Force	Added Quantity of Unbalance	Identified Quantity of Unbalance	Error	
			Amplitude	Phase
Disk No.	(g - cm)	(g - cm)	(%)	(degree)
2	88.965 < 15°	74.215 < 12.45°	16.28	2.54
4	129.735 < 330°	118.698 < 337.4°	8.50	0.74
4	109.350 < 330°	103.458 < 331.9°	5.39	1.90
5	88.965 < 15°	95.071 < 12.54°	6.86	2.46

EXPERIMENTAL ESTIMATION OF VIBRATIONAL SYMPTOM LIMIT VALUES FOR LARGE STEAM TURBINES

Z. Orlowski and T. Galka
Polish Association of Diagnostic Engineering, Warsaw, Poland

ABSTRACT

The paper deals with methods of vibrational symptom limit value determination for large steam turbines on the basis of experimental results. Such methods are conceived to provide quantitative data, necessary for turbine operation, maintenance and diagnostics. Various methods are described and discussed in terms of their suitability for this purpose. Two of them, based on vibration trends analysis and statistical data processing, respectively, are described in details and illustrated with examples, taken from the authors' own experience.

1. INTRODUCTION

Shifting from qualitative to quantitative estimation may probably be considered a natural course of development in many forms of human activity. Vibration-based machine condition monitoring is no exception. Providing that we are aware of the fact that vibration, as well as other residual processes (i.e. the ones that accompany the principal process of energy transformation) contains information concerning technical condition of a machine [1], we first must be able to measure it and analyse in time and frequency domains. Having accomplished this, we have to relate certain features of vibration signal to relevant technical condition parameters, which implies construction of a certain model. Next logical step is to determine quantitative relations, which - from the point of view of machine operation - means that certain limit values must be determined or estimated.

Large steam turbines, which now form the backbone of power generating capabilities in most countries, are complex machines, within which many elementary sources of vibration may be distinguished. This means that simple measures, such as e.g. overall vibration levels measured on bearings, are of little, if any, value for technical diagnostics and condition assessment (while, on the other hand, they are indispensable in normal operation). Consequently, one limit value, in the form of maximal acceptable level of vibration velocity or displacement, is adequate only from the point of view of operational safety. For condition monitoring and diagnostics, more elaborate approach is necessary.

There is no generally approved method of vibrational symptom limit value estimation for rotating machines, especialy when broad frequency range is considered. This is one of the reasons for basing our approach on experiments. Such procedure, although somewhat heuristic, yields interesting and consistent results.

2. THE BACKGROUND

Diversity of physical phenomena that give rise to steam turbine vibrations is a well-known fact [1-3]. To put it most briefly, two basic types of such phenomena may be distinguished: those resulting directly from rotational movement of the machine rotor and those connected with interaction between steam flow and the fluid--flow system (rotor blades and guide vanes). Two basic frequency ranges may be specified accordingly: the "harmonic" (low frequency) range and the "blade" (high frequency) range; the meaning of the words "low" and "high" is here, of course, only relative. The former spans from values close to zero to a few hundred Hz (depending on how many harmonics are taken into account); in steam turbines it is convenient to make a further division, sub-harmonics being treated separately, as their origin is different. Lower and upper limits of the latter depend on turbine design (number of blades in individual stages), typical values being some 1 kHz and 10 to 20 kHz, respectively.

Diagnostic model, which may be associated with particular steam turbine, determines specific values of vibration frequency for all elementary sources that have been taken into account. The set of such values must be viewed in the terms of signal analysis in the frequency domain. Narrow-band analysis is precise and allows for high resolution, but levels in individual bands are rather sensitive to operating conditions; furthermore, a lot of data must be processed. It is suitable for purpose-aimed analysis in cases of suspected faults. For technical condition assessment, and consequently limit values estimation, we have employed 23% constant percent bandwidth (CPB) spectra, which provide a reasonable compromise between resolution and amount of data for processing. Each band is referred to by its mid-band frequency.

It should be kept in mind that two turbines, even of the same design, may reveal considerable differences when it comes to quantitative assessment of vibration signal. Even modernization of a turbine may change its vibration pattern significantly [4]. This means that each turbine is, in general, characterized by its own specific set of limit values (i.e. vibration levels in individual frequency bands, for all measuring points).

3. DETERMINATION OF LIMIT VALUES OF VIBRATIONAL SYMPTOMS

Despite remarkable progress in the field of turbine dynamics investigation it is still not possible to determine reliable limit values of vibrational symptoms on the basis of design features; this implies that such determination must be based, at least to certain extent, on experience and experiment.

In general, there are several methods of vibration limit values determination:
- acceptance of existing standards;
- arbitrary determination from reference measurements;
- experiments of "active" type;
- vibration trends analysis;
- statistical methods,

and all of them, except for the first one, are to some degree based on experiments.

Various standards concerning turbine vibration levels are in common use, but their suitability for technical condition monitoring is negligible. Their purpose is to provide alarms and/or trips when exceeded, in order to maintain safe operation, and in most cases they are based on total RMS or peak-to-peak levels of vibration displacement or velocity. Arbitrary determination of limit values is possible only if proper reference measurements may be performed, with an assumption of machine being in strictly determined technical condition; this restricts suitability of such methods for relatively simple machines, certainly not steam turbines. As for "active" experiments, it is rather improbable that any plant manager should allow them to be performed on his turbines, with inevitable generation loss, to say nothing of possible hazards.

The remaining two methods are, in our opinion, the most suitable ones for steam turbines in normal operation.

3.1. Determination of limit values from vibration trends analysis

If certain parameters of vibration signals (e.g. vibration levels in selected frequency bands) have been monitored for some time, and turbine technical condition has been simultaneously traced by means of inspections during overhauls and maintenance works (and by any other means that yield unequivocal results), a sort of empirical relationship may be determined between them; this, in turn, allows for determination of limit values of these parameters. This is, in shortest possible form, the concept of vibration trends analysis method [3].

This method is very useful if suitable database, covering sufficient period of time, is available. It is based on simple measurements and involves virtually no parameters which validity might be controversial. It has, however, certain inherent limitations. Fig.1 shows two plots of vibration velocity trends, both covering a period of one year, recorded on a 55 MW turbine in the 100 Hz band (corresponding to 2nd harmonic). One of them reveals quite unequivocal tendency; such data may provide a

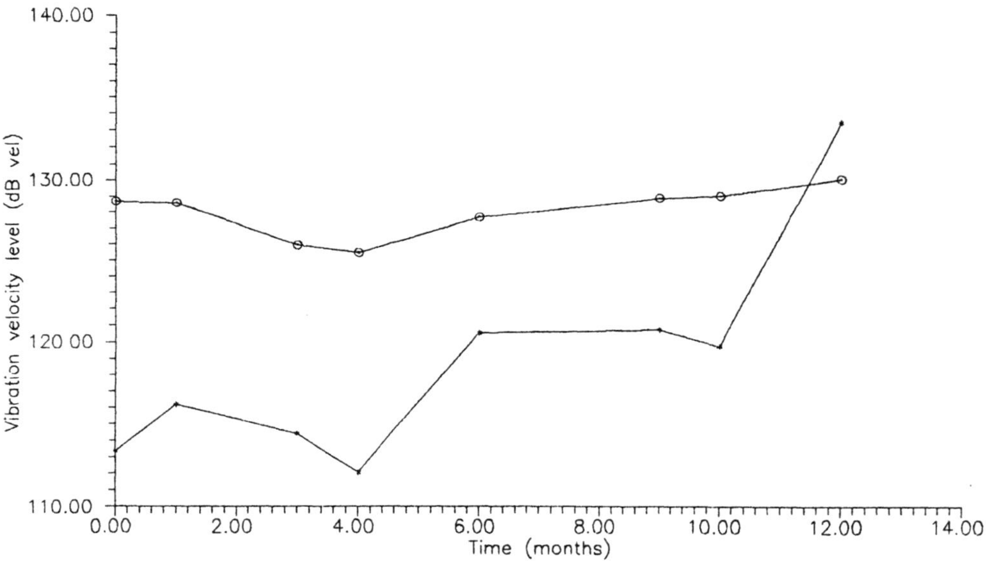

Fig.1. Vibration trends recorded on a 55 MW turbine (100 Hz frequency band) in two points; one trend (asterisks) reveals a clear tendency and allows for limit value estimation, while other (circles) provides no such tendency, as no significant change has occured during investigations.

sound limit value estimation. The other one, however, shows that in fact no significant change has occured during investigations and hence no limit value estimation is possible. These two examples, although taken from data acquired for one machine, illustrate two opposites and, consequently, the above-mentioned limitations.

Such method cannot, in most cases, do without some degree of experience. It does not determine precise values: hence it is a reasonable practice to assume lower limits first, for the sake of safety. These limits may, and should, be verified during subsequent investigations.

3.2. Statistical methods of limit value determination

Such methods have been proposed and analysed by C.Cempel and are based on the symptom reliability concept. We shall not present here the entire theory, which has been described in details elsewhere (see, for example, Refs. 5 and 6). Instead, we shall present conclusion and, later, illustrate them with examples.

Symptom reliability R as a function of symptom value S may be written as

$$R(S) = \int_{S}^{\infty} p(S_o)\, dS_o = \frac{n(S_o > S)}{N} \tag{1}$$

where: So - symptom value readings, $p(So)$ - symptom density histogram, n - number of symptom value readings above given limit, N - total number of symptom value readings.

At the same time, from the Neymann-Pearson rule [6], we have the following general relationship:

$$P \cdot \int_{S^*}^{\infty} p(S)\, dS = A \tag{2}$$

where: S^* - symptom limit value, P - machine availability, A - allowable probability of a needless repair. The S^* value is thus implicitly determined by a simple relation:

$$R(S^*) = A/P \tag{3}$$

Now we have to assume some form of symptom density distribution, in order to obtain $R(S)$ in a functional form. Weibull, Pareto, Frechet and uniform distributions have been proposed for this purpose and, in general, this question still seems open. We have employed Pareto distribution, which results in the following relationship:

$$S^* = (1 - \frac{1}{\gamma}) S_m \sqrt[\gamma]{P/A} \quad (4)$$

where γ is distribution shape factor

$$\gamma = 1 + \sqrt{1 + (S_m/\sigma)^2} \quad (5)$$

and Sm and σ are symptom average value and standard deviation, respectively. It is believed that Weibull distribution gives, for the same case, similar, but slightly higher estimated symptom limit value; we have learnt that in most cases it is really so.

One fundamental problem remains, namely the one of the A/P value. This value may be considered a sort of measure of plant operation policy and allowable risk. Low A/P, which implies low percentage of needless repair and high availability, may result in S^* values being too high. This problem also seems open; reasonable values are probably of the order of 0.01.

4. EXAMPLES

4.1. Vibration trends analysis

As we have already mentioned, the procedure itself is simple, though somewhat time-consuming, and detailed description would have been rather boring. We present, however, some results, in order to show an example of practical application.

We have employed this method for, among others, two 200 MW turbines in a power station in Poland. Our study has covered a period of a few years and estimation has been based on off-line measurements, performed once a month; vibration velocity spectra have been recorded in measuring points located on bearings and LP casing.

Table 1 shows maximal allowable increments of vibration velocity in 23% CPB bands, with respect to average values. What is of interest here is perhaps not the values themselves, but differences between two turbines (which are otherwise the same), as well as between measuring points (HP, IP and LP parts). These differences would have been even greater if absolute limit values had been considered. This points out the need for individual approach to every machine, and also inadequacy of simple limit values, such as provided by standards.

It must be noted here that values listed in Table 1 should not be considered tripping levels; they are rather warning levels and hence, in dubious cases, lower values have been preferred. Subsequent investigations will probably allow for their verification and "tuning".

Table 1: Maximal allowable increments of vibration velocity in 23% frequency bands for two 200 MW turbines

Turbine part	Frequency range	Increments in dB vel	
		Turbine 1	Turbine 2
HP	subharmonic	17	15
	harmonic	17	17
	blade	18	18
IP	subharmonic	23	16
	harmonic	23	17
	blade	22	20
LP	subharmonic	23	20
	harmonic	21	20
	blade	21	22

3.2. Statistical methods

We have applied the above-described approach, based on the Pareto distribution, for a 120 MW turbine. Limit values have been calculated with an assumption of $A/P = 0.01$, which is a comparatively low factor that seems reasonable for important and responsible machines with high demands concerning reliability.

During investigations, the turbine has suffered a damage of the 3rd rotor stage blades and adjacent guide vanes. We have thus had an opportunity to trace the results of this damage in the form of vibration velocity increase in relevant frequency bands, determined from the diagnostic model (see Section 2). Results are shown in Table 2, for two measuring points close to the location where the damage occured. We can find that the damage is most clearly "visible" in bearing vibration levels in the 3150 Hz frequency band (i.e. the one corresponding to the number of blades of this rotor stage): limit values have been exceeded for all directions. Certain excessions may also be observed for other frequency bands that correspond to the interaction between rotor blades and adjacent guide vanes. In general, symptoms of this damage are more evident in bearing vibrations, the signal measured on the LP casing being obviously less sensitive; this is somehow surprising, as absolute bearing vibrations are generally considered to show no relevance to blade frequency vibrations. Similar conclusions may also be drawn from Figs.2 and 3, which show relevant vibration trends in the 3150 Hz band for two measureing points under consideration: noticeable increase of vibration velocity level in all directions, recorded on the bearing, and evident, but much less pronounced increase recorded on the LP casing. Results provided by trends analysis and statistical evaluation are in this case in nearly perfect agreement.

Table 2: Limit values of vibration velocity in relevant frequency bands and actual values, measured with 3rd rotor stage damage (values in dB; 1 mm/s is equivalent to 120 dB)

Frequency band [Hz]	Measuring point	Direction	Vibration level [dB vel] Limit	Measured	Excession yes/no
3150.0	Front LP bearing	vertical	109.86	116.44	yes
		horizontal	106.02	107.80	yes
		axial	104.37	108.02	yes
	LP casing front part	vertical	101.01	105.24	yes
		horizontal	117.37	114.71	no
		axial	105.72	102.76	no
6300.0	Front LP bearing	vertical	110.64	107.10	no
		horizontal	98.01	111.04	yes
		axial	105.97	97.09	no
	LP casing front part	vertical	106.33	107.11	yes
		horizontal	109.47	99.58	no
		axial	105.86	95.39	no
8000.0	Front LP bearing	vertical	108.09	104.45	no
		horizontal	96.67	98.09	yes
		axial	104.19	90.54	no
	LP casing front part	vertical	96.43	103.98	yes
		horizontal	114.22	95.56	no
		axial	107.79	98.12	no

5. CONCLUDING REMARKS

Last years have, for most electric power utilities, been a period of optimization rather than expansion: with little or no new generating capabilities added the stress has been on economy, environment protection and improvement. Technical condition monitoring has also had its share in these processes, as it is indispensable for safe and economical operation of machines.

Introduction of complex systems for on-line vibration monitoring, often with certain diagnostic capabilities, means that some basic "knowledge" must be provided for them, including quantitative reference data. This in turn calls for an adequate and reliable method for estimating such data. The methods dealt with in this paper may be an answer for at least some questions inevitably resulting from these considerations.

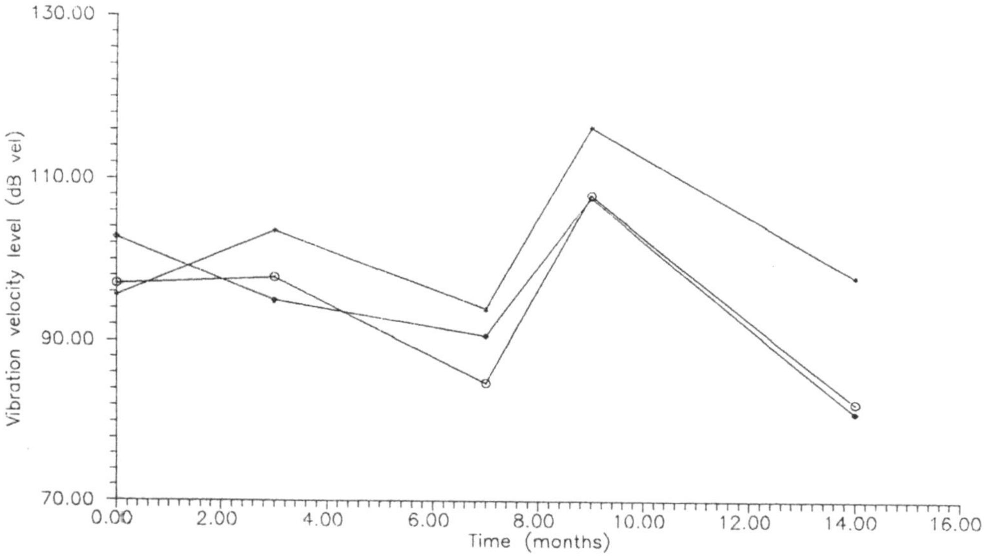

Fig.2. Vibration velocity trends in the 3150 Hz frequency band in vertical (asterisks), horizontal (crosses) and axial (circles) directions, recorded on the front LP bearing casing of a 120 MW turbine; data corresponding to those listed in Table 2.

REFERENCES:

1. Orłowski, Z.: Vibrodiagnostics of Steam Turbines. Proceedings of the Institute of Power Engineering, vol.18, Warsaw, 1989 (in Polish)
2. Morel, J.: Vibrations of Machines and Diagnostics of their Technical Condition, Editions Eyrolles, Paris, 1992 (in French)
3. Orłowski, Z. and Gałka, T.: Diagnostics of Large Turbines on the Basis of Vibration Trends Analysis, in: Proceedings of the 10th School of Diagnostics, Poznań, 1992, 229-234
4. Orłowski, Z. and Gałka, T.: Influence of Turbine Modernization on its Basic Vibration Spectra, in: Proceedings of 5th International Congress on Condition Monitoring and Diagnostic Engineering Management COMADEM'93, Bristol, 1993, 442-447
5. Cempel, C.: Determination of vibration symptom limit value in diagnostic of machinery, Maintenance Management International, 5 (1985), 297-304
6. Cempel, C.: Vibroacoustic Condition Monitoring, Ellis Horwood, London, 1991

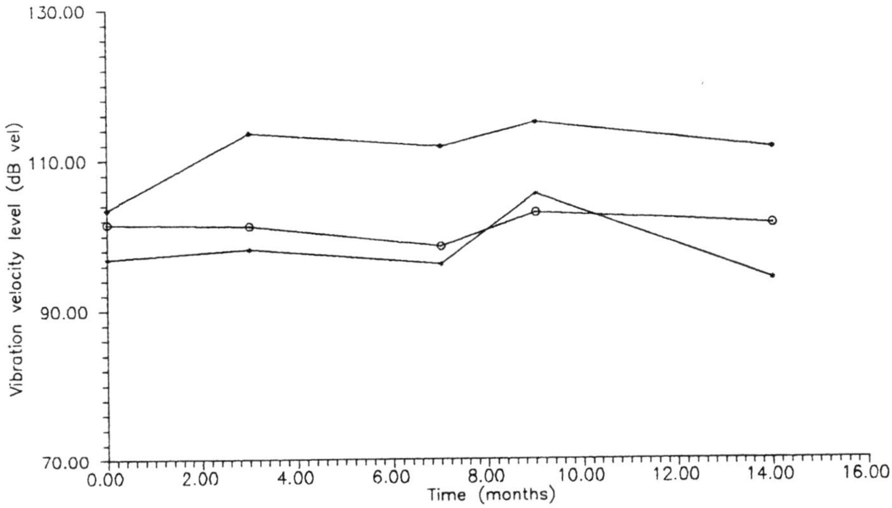

Fig.3. Vibration velocity trends in the 3150 Hz frequency band in vertical (asterisks), horizontal (crosses) and axial (circles) directions, recorded on the LP casing of a 120 MW turbine; data corresponding to those listed in Table 2.

TURBOGENERATOR FOUNDATION VIBRATIONS IN A NEW POWER PLANT

E. Goldfracht and R. Ronen
Israel Electric Corporation Ltd., Haifa, Israel

ABSTRACT

During the first running year of two new 550 MW turbogenerators in Rutenberg Power Station, in Ashkelon, Israel, vibration problems were encountered by the foundation and the generator. A full program for analysis of foundation vibration was prepared, that included displacement measurements, calculations of foundation eigenvalues, modal shapes and dynamic amplitudes by two finite element models, and modal test of the system. The main conclusions from the test program are:

- The natural vibration mode of the turbogenerator foundation at 49.6 Hz has a great influence on the dynamic response of the system at the operating speed of 50 Hz
- The concrete modulus of elasticity, which varies by as much as 40%, is a dominant factor in the determination of foundation eigenfrequencies.

However, the calculated additional stress of the foundation concrete is small, less than 2 kg/cm^2, and no damage it is to be expected at the service frequency.

1. INTRODUCTION

The concrete, low tuned foundation, was designed in 1980 according to the German standard DIN 4024 (Machine foundations), using a finite element model and a computer program. The uppermost portion of the foundation is the deck and it is composed of longitudinal and transverse beams, see figure 1. The deck is supported by columns that are in turn supported by the mat. The mat is a massive concrete plate supported by piles on the soil.

The low tuned turbogenerator foundations were the first ones built in Israel. We had no previous experience with this kind of foundation, and there are almost no cases of low tuned turbogenerator foundation vibration reported in literature.

During the first run up of the turbogenerator high vibration, 25 μm 0-peak (all displacement measurements are 0-peak), was measured on the foundation. Foundation vibration of that level are causing an unpleasant feeling to men, but are not reducing the life expectancy of the structure.

An elaborate plan was prepared and implemented in order to check the influence of foundations vibrations on the system, that comprised:

- Vibration measurements on the foundations.
- Check of the design analytical results.
- Prediction of the dynamic behavior of the foundation.
- Analysis of stiffened turbogenerator foundation proposals.
- Foundation concrete analysis.
- Modal analysis of the system.

2. TURBOGENERATOR FOUNDATION VIBRATION

2.1 Vibration measurements on the foundation

During the turbogenerator first run, the measured vertical foundation displacement was 25 μm. After realignment of the turbogenerator it decreased to 12.5 μm. This value is the "good" limit according to the German guideline VDI 2056 (Criteria for mechanical vibrations of machines).

Foundation vibration was measured on the longitudinal beams and on the transverse beam between the turbine and the generator. The measured vertical vibration and phase angle are presented in figures 4 to 6, for unit No. 1. Turbogenerator shaft and foundation vibrations during run up of the unit are presented in figure 7. Horizontal vibration measured on the foundation deck is small, 4 μm.

The main conclusions from vibration measurements are:

- The maximum measured foundation vibration 25 μm, was a result of machine misalignment.
- The vibration of transverse beam is caused by bending forces, and the vibration of longitudinal beams is characterized by a

combination of bending and torsional forces, according to phase angle diagrams.
- The foundation vibration in the frequency range of 48-52 Hz is increasing by a larger factor than turbogenerator vibration. The natural vibration modes of the foundation have a great influence on its vibration in this frequency range.
- The high vibration amplitude around 39 Hz is dominated by generator and exciter critical speeds.
- According to calculations based on the vibration measurements, the maximum additional stress of the concrete is around 1.5 kg/cm^2.

2.2 Foundation analysis and eigenfrequencies calculations

Turbogenerator foundation model consists of beam elements connected by rigid joints, see figure 3. System analysis was done by two independent consultants, using the same structure model, different mass matrixes and various computer programs. Classical beam elements yields an exact stiffness matrix, depending upon concrete modulus of elasticity, while the mass matrix is only approximated.

The analysis is based on a finite element model of the structure, that has linear characteristics and several degrees of freedom. The mass of foundation is assumed to be distributed at different defined points.

Idealized model-A, original and plain, consist of a plain model of the foundation table supported by springs with elasticity appropriate to the columns stiffness (according to German standard DIN 4024). The beams are represented by concentrated mass. The spatial structure model was obtained by adding the columns mass and stiffness to the beams, and increasing the degrees of freedom of the system. Model-B consists of an enlarged spatial structure, so that each beam element is subdivided into five mass elements. Due to the thickness of the structure compared to the span, the shear deformation effect of the beams was taken into account in this model.

Concrete damping was assumed to be 2% according to the standard DIN 4024. The main conclusions from the foundation analysis are:

- There is good conformity between calculated eigenfrequencies using different computer programs, and the same structure model.
- A major influence on the computed dynamical behavior of the structure is the concrete modulus of elasticity. The dominant frequencies of the structure are considerably shifted if the modulus of elasticity is changing by only 10% and consequently a different response is calculated.
- The computed dynamical behavior of the structure is influenced by the damping ratio. The value of 2% constant for all eigenmodes is a minimum value, thus representing the safe side.
- The eigenfrequencies of the foundation calculated using various models show a considerable difference between them. It is difficult to decide which model is the most suitable.
- Calculations of structural response to dynamic excitation by the turbogenerator around its operational speed may detect coupled resonances.

2.3 Foundation dynamic analysis

Foundation vibration amplitudes were calculated using three dynamical load cases. The dynamic load cases were defined as normal operation by the supplier of turbogenerator. The three load cases are:

- All excitation forces are assumed to act downward simultaneously.
- All excitation forces are assumed to act upward and downward alternating.
- The excitation forces are assumed to act downward on the left part and upward in the right part of the longitudinal beams.

The main conclusions from the dynamic analysis are:

- The maximum response of the foundation is around the frequency of 48-50 Hz, see figure 8, at the middle of cross beam between turbine and generator.
  The maximum response is load dependent, and the second load case is dominant for the response of the structure in this frequency range.
- Additional stress of the concrete calculated using full dynamic analysis is 1.65 kg/cm^2, and concrete stress calculation based on the ratio between dynamic and static displacements is 1.45 kg/cm^2.

2.4 Analysis of stiffened turbogenerator foundation

In order to assess the influence of stiffening the foundation on its vibration, a dynamic analysis was performed. The stiffening included addition of:

- A longitudinal beam between the columns in the area under the generator and the low pressure turbine.
- A column supporting the transverse beam between the turbine and the generator.
- Columns supporting the generator longitudinal beams.
- Mass on the transverse beam between the turbine and the generator.

The main conclusion from the analysis was that all the proposed solutions add some stiffness to the system, however there influence is minimal on the frequencies of the system and on the vibration amplitude. Only the addition of a weight greater than 500 kN on the transverse beam can affect considerably the computed vibration amplitude.

However, it is not feasible to implement any stiffening of the foundation because there is not enough room below the beams due to pipelines and condenser. Any solution will always shift the whole frequency range, and it is nearly impossible to keep the vicinity of the operating speed, 50 Hz, free of computed eigenfrequencies.

2.5 Foundation concrete analysis

Concrete core samples were removed from the foundation, in order to check the concrete strength and its modulus of elasticity. On six cylindrical core samples, a uni-axial compression strength test was performed, see table 2. The conclusions from the concrete test are:

- The minimum measured concrete strength was 41.4 MN/m^2 which is higher than the minimum compression strength required for designed concrete B 25 in German standard DIN 1045, 25 NM/m^2.
- The average measured modulus of elasticity of the concrete samples was more than $30*10^6$ kN/m^2 in accordance to requirements of DIN 1045 and I.S. 466.
- The range of measured concrete modulus of elasticity is large, varying by about 40%.
- The concrete of the foundation is in agreement with the conditions of calculation for strength and modulus of elasticity.

2.6 Modal analysis

Modal analysis of the foundation was performed by a specialist team, using the Impact test techniques, a 4-channel FFT Analyzer and the STARS modal software. In order to obtain accurate results, the entire foundation and generator system was tested.

A model of the system was prepared, see figure 2. The displacements were measured at the model nodes in three directions. A suitable excitation location and power spectrum force input was selected, in order to simulate the actual excitation of the generator. The test was performed when the turbogenerator was shut down.

A polynomial curve fitting techniques was used to calculate the eigenvalues, the corresponding eigenvectors and the mode shape for each frequency of interest. The conclusions from the modal test are:

- The results of the modal analysis indicate that at the frequency of 49.6 Hz, the mode shape can be defined as a system mode.
- The measured foundation vibration levels did not appear to be amplified by any modal resonance.
- Based on the modal analysis results and on the operation deflection shape test it can be concluded that it is not likely that the 49.6 Hz natural frequency of the foundation is contributing a substantial amount to the forced response of the turbogenerator during operation.

3. CONCLUSIONS

3.1 The eigenfrequency of the turbogenerator foundation is excited by the 50 Hz turbogenerator operating speed, and the foundation is near one of its resonant modes, according to:

- Vibration measurements, the foundation resonance is occurring around 52 Hz.
- Finite element analysis using different models, the foundation resonance is occurring around 48 Hz - 50 Hz.
- Modal analysis, the measured resonance of the foundation is at 49.6 Hz.

3.2 The maximum calculated additional stresses in the concrete is small, around 1.5 kg/cm^2.
The foundation dynamic stresses are only 10% of the permitted fatigue stress of the concrete in tension (15 kg/cm^2), and only 1% of the permitted fatigue stress of the concrete in compression (150 kg/cm^2).

3.3 From calculations of the concrete additional stress it may be concluded that even that the foundation is excited near its resonance no damage it is to be expected at the service frequency of 50 Hz.

3.4 It is recommended to require from the producer of the equipment to design the foundation according to his suitable solution.

3.5 The determination of the eigenfrequencies of the foundation is influenced by many factors, including the design model, concrete modulus of elasticity and damping.

3.6 During the design process it is important to consider many modes of vibration and not only the first ten as recommended by the German standard, DIN 4024. It is recommended to calculate the foundation vibration amplitudes for severe dynamic load cases of the turbogenerator.

3.7 The foundation strength and displacement are suitable to the expected requirements. It is not expected that the foundation will have a negative influence on the operation of the turbogenerator.

ACKNOWLEDGEMENT

The authors acknowledge the useful comments of Mr. D. Lowy, Head of Civil Engineering Department, and the help received from Research & Development Laboratory, Rutenberg Power Station Staff, Prof. I. Sheinman, Dr. Rothe, Wetinghouse and MAN Energie.

Table 1 Turbogenerator foundation frequencies calculated by different approaches

Mode Num.	Model A				Model B
	Original	Plain	Spatial 1	Spatial 2	Spatial 2
1			0.742	0.75	0.72
2			0.994	1.01	0.98
3			1.148	1.18	1.12
4			7.929	7.88	7.54
5			8.915	8.92	8.55
6	10.502	10.631	10.684	10.81	10.49
7	11.211	11.399	11.418	11.50	11.15, 11.47, 11.52
8	12.443	12.532	12.604	12.78	12.23, 13.62, 13.80, 13.82, 13.91 13.92, 13.98, 13.99
9	14.071	14.283	14.451	14.50	14.01, 14.11, 14.14, 14.84
10	15.253	15.276	15.365	15.50	15.15
11	15.262	15.311	15.475	15.71	15.90
12			16.606	16.47	16.73
13	17.036	17.073	17.269	17.72	
14			18.481	18.46	18.01, 18.08, 18.21, 18.62, 18.82
15	18.612	18.561	18.655	18.90	18.89, 19.02
16			19.714	19.63	19.12, 19.21, 19.35, 19.39, 19.40 19.58, 19.86, 20.50
17	21.322	21.269	21.386	21.30	21.46
18			21.463	22.00	21.63
19	24.217	24.151	24.232	24.43	24.48
20	25.768	25.521	25.726	25.57	26.06
21			25.931	26.55	26.18
22			26.702	26.73	26.52
23			27.681	27.33	27.24
24	27.788	27.456	27.858	28.07	27.83
25	30.394	29.951	30.449	30.63	30.52, 30.65
26			30.910	31.91	31.16, 32.18, 32.43
27	32.137	31.994	32.286	33.15	32.53
28			33.461	33.48	
29	34.059	33.678	34.111	34.81	34.76, 36.74, 36.87, 37.10, 37.25 37.27, 37.35, 37.36, 37.42, 37.50
30	38.125	37.815	38.176	39.05	39.23
31	39.523	39.255	39.553	39.31	39.43
32			39.745	39.88	39.95
33			40.366		40.46
34			41.636	41.33	41.09
35	41.941	41.628	41.973	41.35	42.90
36	45.867	45.763	45.892	43.75	45.42
37			46.745	46.46	46.55
38			47.292		48.01, 48.63, 48.77
39	48.053	47.903	48.096	46.65	49.88, 50.28, 50.40, 50.89, 51.10 51.24, 51.65
40	49.788	49.702	49.811	46.74	52.90, 53.27
41	49.923	49.674	49.949	51.20	53.77
42			51.900	52.76	58.00
43	54.357	54.241	54.374	53.09	
44			54.757	55.89	
45	58.008	57.925	58.018	58.60	
46			58.204		
47	59.543	59.521	59.587	60.65	
48			59.903	60.96	
49	62.002	61.921	62.040		

The frequency of the mode of vibrating cross beam between turbine and generator is underlined

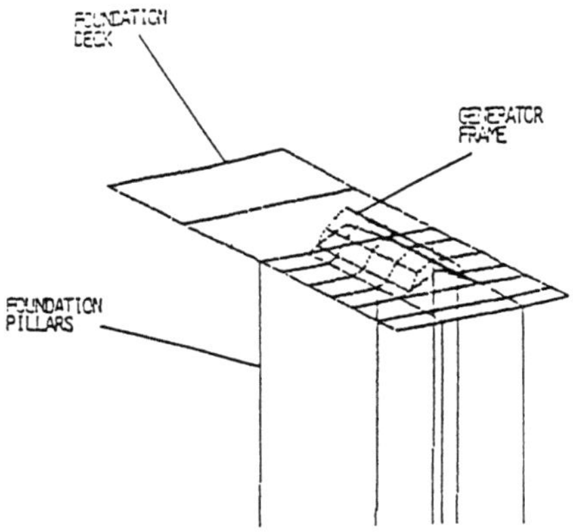

Figure 2 Modal analysis test geometry model.

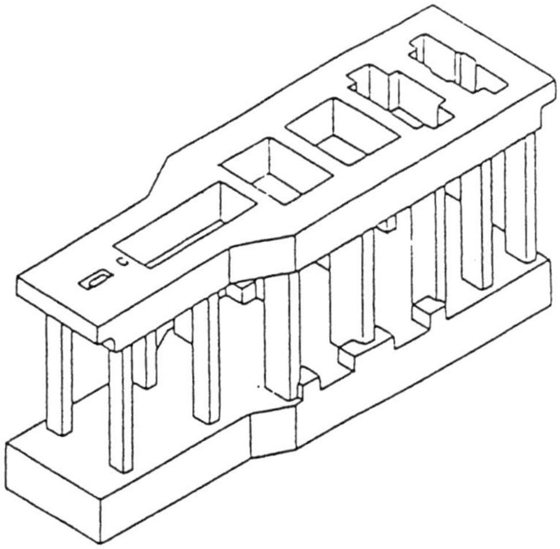

Figure 1 Low tuned turbogenerator concrete foundation.

Table 2 Modulus of elasticity of concrete.

Strength MPa	I.S.466	Rutenberg Turbine Israel	Rutenberg Turbine
15.0	20000		
20.0	25000		
30.0	30000		
40.0	35000		
45.0			
45.0			
46.0			
46.0			
47.7			26328
50.0	37500		
54.0			
56.0			
57.1			26655
58.0		47280	
60.0	40000		
60.5			31570
68.0		32520	

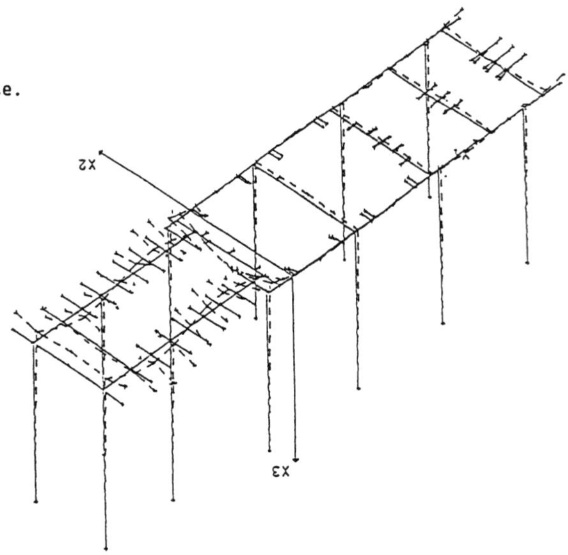

Figure 3 Turbogenerator foundation model.

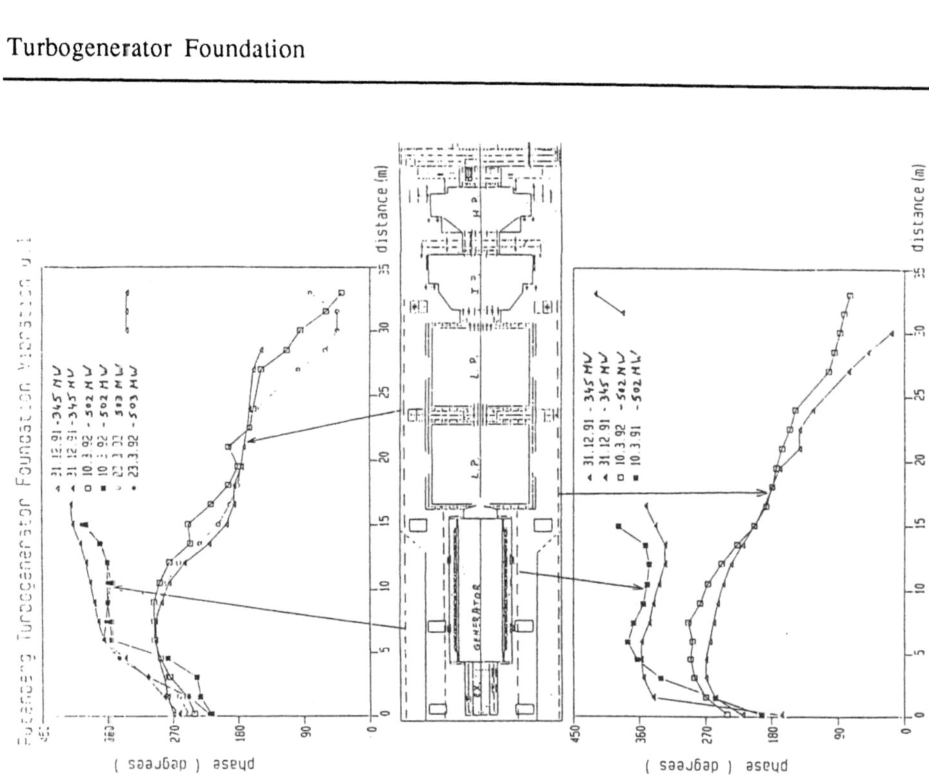

Figure 5 Measurements of turbogenerator foundation phase angle on the longitudinal beams.

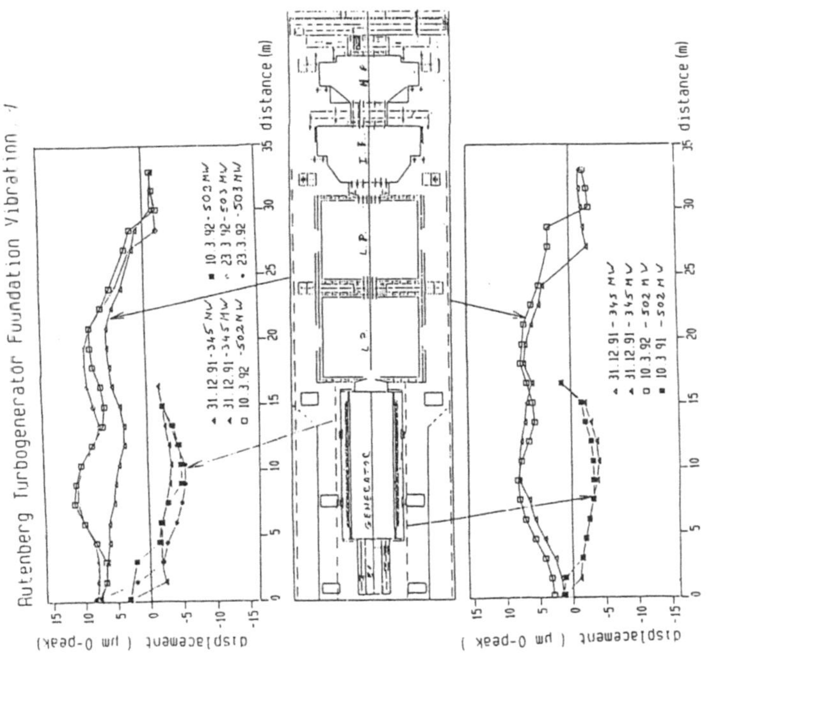

Figure 4 Measurements of turbogenerator foundation vibration on the longitudinal beams.

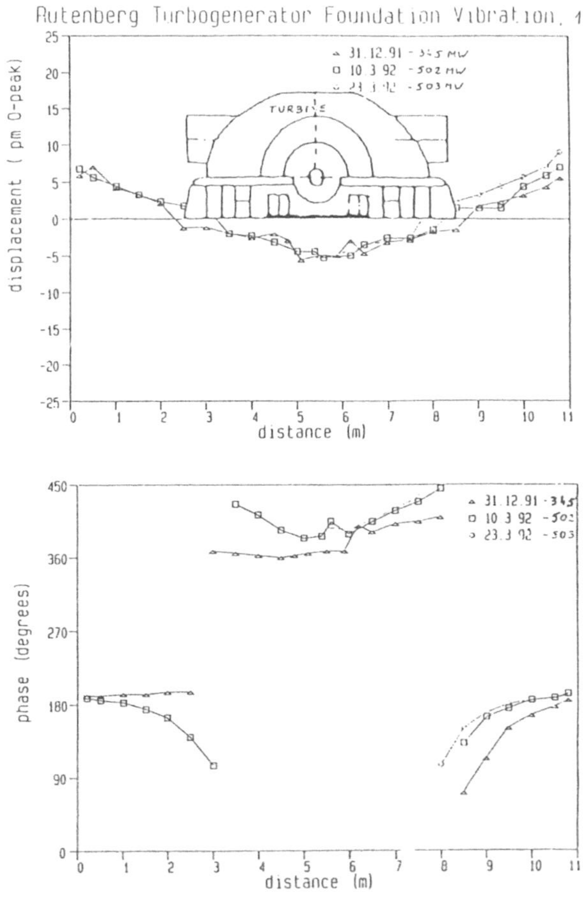

Figure 6 Measurements of turbogenerator foundation vibration and phase angle on the transverse beam between the turbine and the generator.

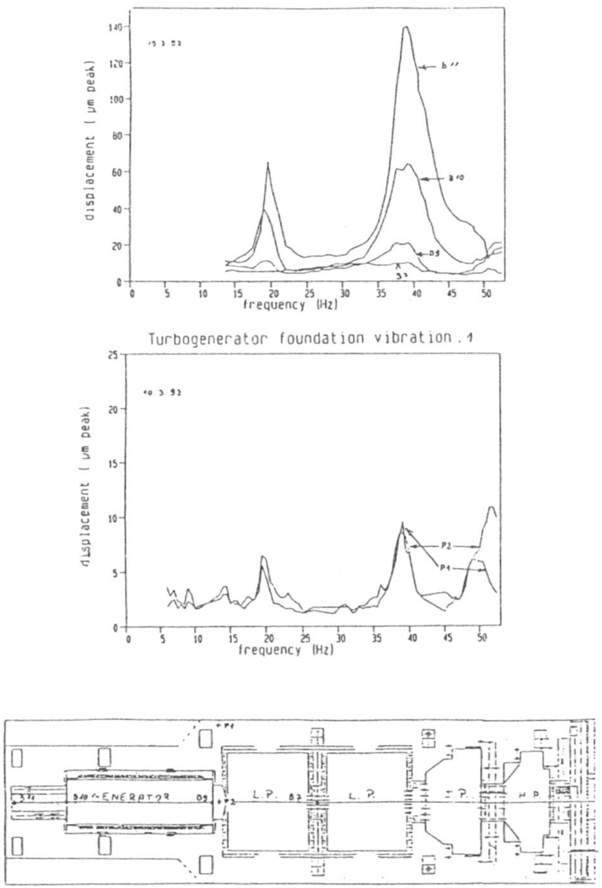

Figure 7 Measurements of turbogenerator shaft and foundation vibrations during run up of the unit.

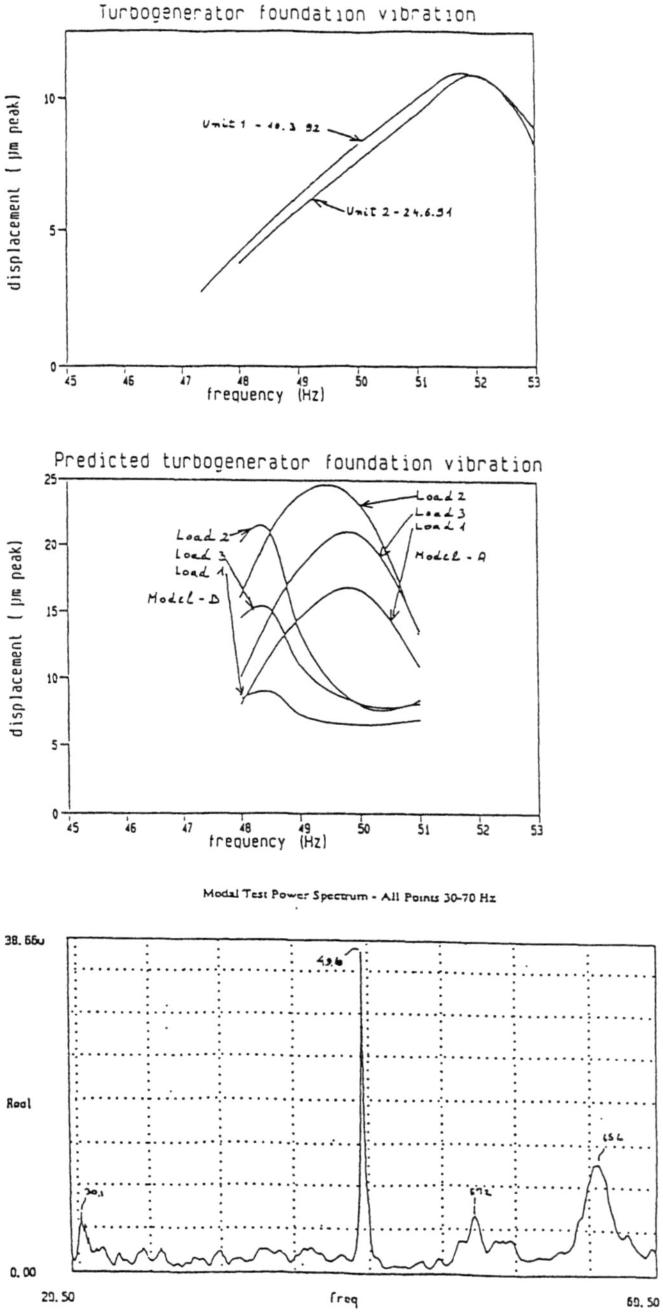

Figure 8 The measured and the predicted foundation displacement in the middle of the transverse beam between the turbine and the generator, and the foundation modal test results.

DIAGNOSTIC TECHNIQUES IN CONDITION MONITORING

G. Diana and A. Vania
Technical University of Milan, Milan, Italy

A. Vallini
ENEL S.p.A. DPT/VDT/STE, Pisa, Italy

G.A. Zanetta
ENEL S.p.A. DSR/CRA, Cologno Monzese, Italy

ABSTRACT

The main purpose of permanent monitoring of rotating machinery is the early detection of fault conditions. This goal can be achieved by identifying abnormal changes in current data in respect with reference values defined on the basis of the data collected during a first monitoring period with the machine in normal state. Furthermore, acceptance regions can be evaluated for each typical operating condition of the plant in order to recognize abnormal changes in the machine vibrations due to fault conditions, deviating from normal changes caused by variations in the process parameters. This monitoring strategy allows to obtain useful information for diagnosing the causes of malfunction conditions.
Some different techniques used to define acceptance regions are shown in this paper. Furthermore, some results obtained by two experiences in monitoring large turbine-generator units are illustrated. The advantages of different diagnostic techniques used to define acceptance regions of monitoring data collected both in steady state and transient conditions are discussed.

1. INTRODUCTION

Nowadays the capabilities of computerized systems ensure the continuous monitoring of rotating machinery vibrations and allow to perform early detection of fault conditions. In addition to that, on-line diagnostic techniques can help the interpretation of malfunction warnings, making it easier to evaluate where the machine should be tripped or operated with particular attention to the vibration trend. As a final result, the diagnostic approach proposed in the paper should make it possible to introduce the predictive maintenance, while improving safety and availability of the plants and, in the end, to extend their life. Furthermore, the process parameters affecting the vibration behaviour can be pointed out and, in case, accordingly controlled. Should these results be attained, a very good costs/benefits ratio would be ensured, when considering introduction of continuous monitoring systems on such large rotating machinery as turbogenerator units.

Advanced monitoring systems analyse both process parameters data and broad-band measurement of support and relative shaft/support vibrations, as well as their harmonic components (i.e. 1X, 2X and 3X vibration vectors) in order to provide more significant information for the diagnostic techniques. Therefore, both scalar and vectorial quantities may be associated to each measurement point. Usually, the early recognition of malfunction conditions is carried out by comparing current monitoring data with reference data previously collected with the machine in normal state. This diagnostic strategy requires to define alert and danger acceptance regions for each quantity associated to each measurement point on each machine. This goal can be achieved by applying different techniques, based on statistical analysis, to the monitoring reference data.

Some of the results obtained by two first experiences in the monitoring of large turbine-generator units in steady state and transient conditions are shown in the paper. Advantages and disadvantages of different diagnostic monitoring techniques are compared by the analysis of some case histories.

2. DIAGNOSTIC STRATEGIES

Fault conditions can be early detected by evaluating abnormal changes in the monitored data with respect to reference values previously defined. This diagnostic strategy requires an initial monitoring period, with the system in learning mode, when the machine can be considered in normal state. If malfunctions are already present in the first monitoring stage it may be sufficient that, during this period, no significant changes in the machine behaviour occur for each typical operating condition of the plant.

If the data collected during the initial monitoring phase are really representative of the machine behaviour for each typical operating condition, they can be considered as reference data. On this basis, it is possible to evaluate mean value and standard deviation of any quantity associated to each measurement point (i.e. overall amplitude and vibration vectors). When the data distribution does not differ too much from a normal (gaussian) distribution, it is possible to define acceptance regions containing pre-established per cent values of the reference observations. Subsequently, during continuous monitoring, the acceptance regions defined for each measurement point can be used to detect alert and danger operating conditions. As a matter of fact, controls based on the acceptance regions can be added to the usual threshold values, based only on the overall vibration level, since they are more precise. Anyhow, standard alarm levels can be considered as maximum limits which should not be exceeded in order to ensure safe operating conditions of the machine. During permanent monitoring, if process parameters data or vibration data are outside their respective acceptance regions warnings of fault conditions can be issued at an early stage. The acceptance regions can be defined both for transient data and for monitoring data collected with the machine in steady state operating condition.

Sometimes, the machine dynamic behaviour is significantly affected by one or more process parameters, usually representative of the plant operating conditions. For example, the shaft vibrations of large turbogenerator units are often affected by load, rotor current, steam temperature and pressure. More effective diagnostic results may be obtained by defining the acceptance regions either over a greater number of subranges of each process parameter or, over a reduced number of subranges, while introducing regression analysis. The greater the number of the subranges, the better the accuracy of the monitoring strategy but, of course, also data processing becomes heavier.

Trend analyses of the monitoring data collected in the same operating conditions before the occurrence of a warning of fault may point out the real cause of incipient changes in the machine behaviour, by the correlation between vibration data and a pre-established subset of process parameters, that showed significant relationships with the machine behaviour during the reference phase of monitoring. Additional process variables can be considered, if their influence was experienced on similar rotating machines.

Trend and correlation analyses allow to estimate the probability factor of the occurrence of some typical faults of the rotating machines. These analyses can be carried out by applying automatic on-line diagnostic techniques. As a matter of fact, the pattern of changes in the harmonic content of the monitored vibrations is compared with the pattern of typical faults. Furthermore, the probability factor of each cause of malfunction can be weighted on the basis of the results of the correlation analyses between vibrations and process variables.

Finally, rotordynamic models, in which the mechanical impedance of the supporting structure and the influence of the dynamic coefficients of the journal bearings are taken into account, can be used to compute the frequency response of the system to pre-established faulty conditions. The results of these off-line analyses can validate the diagnoses of faults carried out by the on-line diagnostic techniques and help to devise and quantify the requested maintenance operations (i.e. balancing or alignment).

3. ACCEPTANCE REGIONS

For scalar quantities, like the overall signal amplitude, the mean value (DC signal component) or the amplitude of an harmonic component, the acceptance region is a segment whose middle point is the mean value of the reference data while the half-amplitude of the segment is proportional to the standard deviation σ of the reference data. In case of normally distributed observations the alert and danger limit values can be settled to two and three times the standard deviation σ respectively, so that the two acceptance regions contain 95.4 % and 99.73 % of the reference data. As for the vectorial quantities, like harmonic signal components, the acceptance regions are represented by an area of the polar plane, whose center is the mean value of the reference vectors. If the real and imaginary components of vector data form a bivariate normal distribution, the acceptance region is an ellipse. By applying statistical analysis algorithms [1], [2] it is possible to evaluate the ellipses

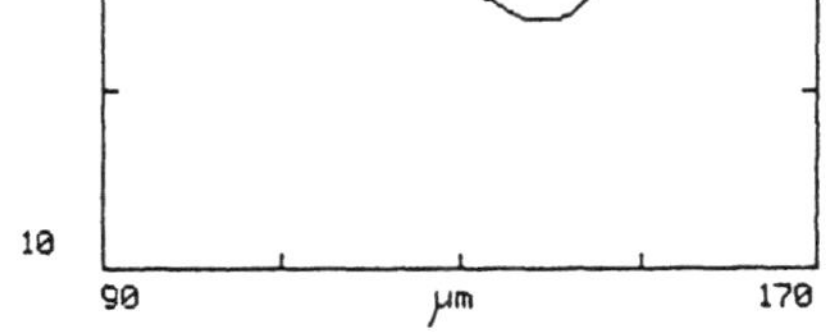

Figure 1 Alert and danger acceptance regions of shaft vibrations (1X) and reference data points.

that contain pre-established per cent values of the reference vector data. In order to differentiate the dimensions of the alert and danger acceptance regions the ellipses that contain 68.3 % and 99.73 % of bivariate normally distributed data can be considered (figure 1). However, also other per cent limit values can be chosen in relation to the machine characteristics or on the basis of the experience.

3.1 Monitoring data correlated to process variables

If the exciting forces applied to a rotating machine do not change their characteristics the vibrational behaviour of the system remains almost unvaried in the time and the standard deviation of the monitoring data distribution associated to each measurement point should be small. That is, the "clouds" of data points should cluster around their mean value. In addition to this the data distributions do not differ too much from a normal distribution (figure 2). However, as the vibrations of turbine-generator units are often affected by one or more process parameters, the machine dynamic behaviour may be significantly influenced by the operating conditions of the plant. If high correlation exists between vibration data and process variables, the data point distribution depends on the number of observations collected in each typical operating condition; furthermore, the differences of the data distribution with respect to a normal (gaussian) distribution may be noticeable.

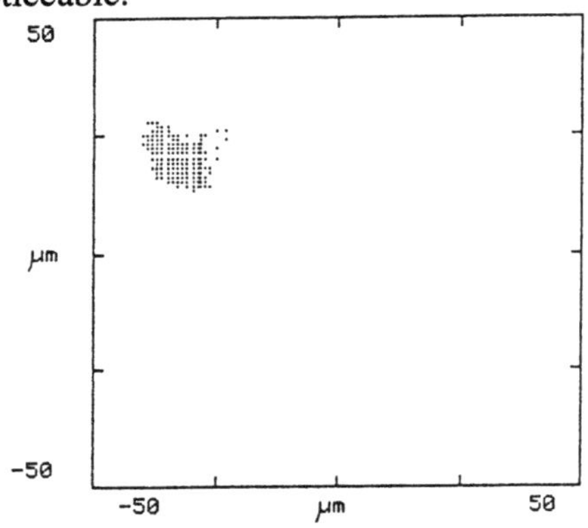

Figure 2 Polar plot of the reference data points of a LP turbine vibrations (1X). Data collected with load ranging from 30 MW to 330 MW.

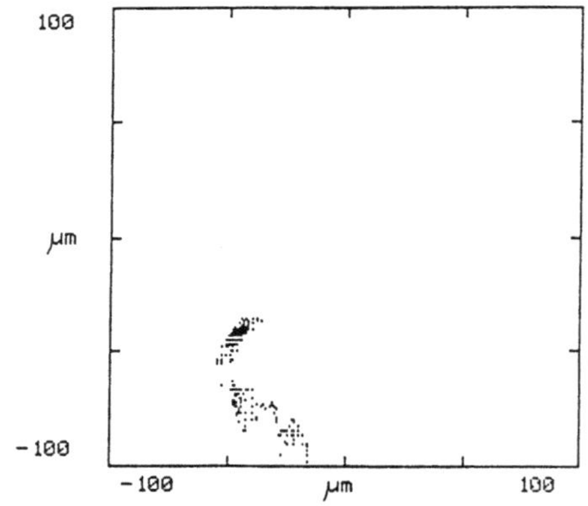

Figure 3 Polar plot of the reference data points of a HP-IP turbine vibrations (1X). Data collected with load ranging from 30 MW to 330 MW.

In figure 3 the 1X vibration vectors of the rotor of a HP-IP turbine are shown. These data were collected with the load ranging from 30 MW to 330 MW. Correlation analyses pointed out a high relationship between the 1X vibration vectors and the load. In consequence of this the data distribution is very different from a bivariate normal distribution. In this case, it would not be reasonable to define acceptance regions over the entire range of the load. Better results can be obtained by selecting data associated to pre-established ranges of variation of the load and defining the acceptance regions

accordingly. This diagnostic strategy allows to define the alert and danger acceptance regions of the monitoring data for each typical operating condition of the machine.
In figure 4 the shaft vibrations (1X) of a generator are shown. The machine dynamic behaviour was affected by the thermal bow due to a rotor windings short circuit. The vibrations were measured with the rotor current ranging from 2400 A to 3600 A.
During a first monitoring phase the 1X vibration data associated to similar values of the rotor current showed only small changes and were assumed as reference data. Due to the high correlation between 1X vibrations and rotor current the data points were not a bivariate normal distribution. Nevertheless, the danger acceptance regions were evaluated (figure 5). The significant changes of the vibration vector phase angles due to the rotor current variations caused a very large confidence region.

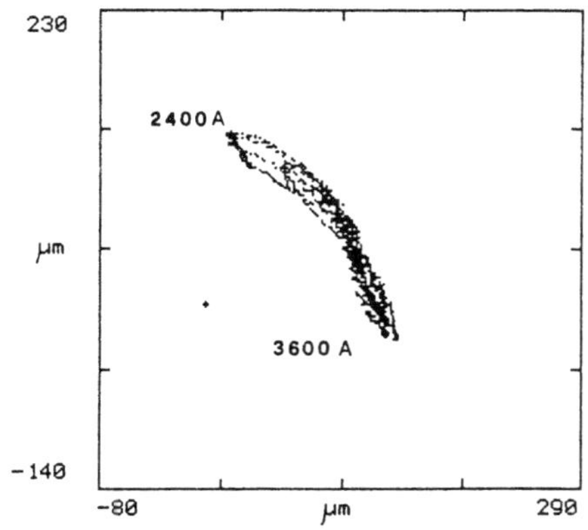

Figure 4 Polar plot of the reference data points of the rotor vibrations (1X) of a generator. Data collected with rotor current ranging from 2400 A to 3600 A.

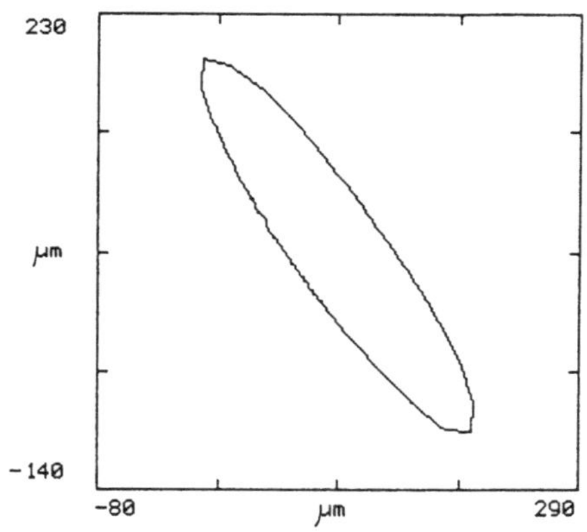

Figure 5 Danger acceptance region of the rotor vibrations (1X) of a generator. Data collected with rotor current ranging from 2400 A to 3600 A.

Therefore, changes in the vibration vector due to a development of a fault could not to exceed the danger confidence region, making it difficult to point out the occurrence of malfunction conditions at an early stage. The rotor current was the process variable more significantly affecting the 1X vibrations of the rotor. The total range of variation of this variable (0 A ÷ 4000 A) was split into ten parts and the danger acceptance regions were evaluated for each data subset.
The regions corresponding to five ranges of variations of the rotor current from 2400 A to 3600 A are shown in figure 6. As a consequence of the good correlation between 1X vibrations and rotor current the acceptance region associated to each operating condition is significantly smaller than the one defined considering the whole set of reference data (figure 6). This make it easier early detection of the faults.
In addition to that, the differences between a normal distribution and the distribution of the data that corresponds to each sub-range of variation of the rotor current are less important. In this case, the danger acceptance region defined for each of the five typical operating conditions were contained inside the danger acceptance region, evaluated using the whole set of reference data (figure 7). In such way it is possible to define more restrictive alarm levels.

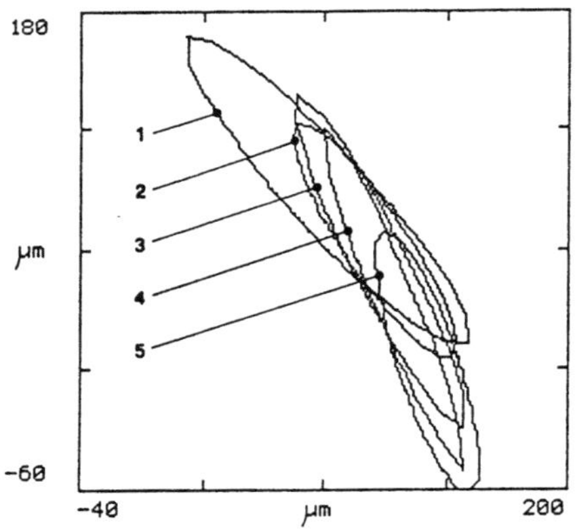

Figure 6 Danger acceptance regions of the rotor vibrations (1X) of a generator. Data collected with rotor current variations inside the following ranges:

1) 2400 ÷ 2700 A 2) 2700 ÷ 3000 A
3) 3000 ÷ 3200 A 4) 3200 ÷ 3400 A
5) 3400 ÷ 3600 A

The two examples emphasises the need for having, in general, a number of different ranges of process parameters for which to define the acceptance regions of the vibration vectors [5].

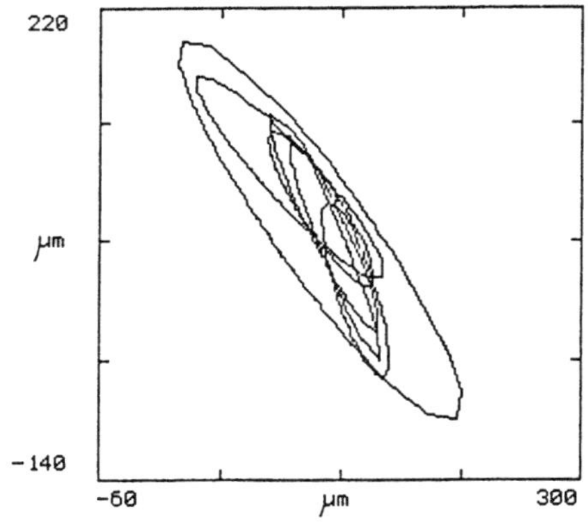

Figure 7 Danger acceptance regions of the rotor vibrations (1X) of a generator. Dotted line: data collected with rotor current ranging from 2400 A to 3600 A.

3.2 Multivariate regression models

Sometimes, due to high correlation with process variables, the distribution of vibration data significantly deviates from the normal distribution, even for comparatively small ranges of the independent variables. Therefore, the accuracy with which the acceptance regions are defined may be not good enough for the diagnostic purposes and, during permanent monitoring, false alarms could be caused. However, just the presence of high correlations between monitoring data and process variables may allow to obtain good results by using multivariate regression analysis techniques. With this diagnostic strategy the reference vibration data associated to each measurement point are fitted by a multivariate regression function, linear or non-linear in the regressor variables [3], [4], [6]. These variables are selected among the process parameters monitored by the system. The difference between each experimental data and the corresponding predicted value obtained with the regression function is called a residual. The residuals associated to a set of reference monitoring data form a data distribution with standard deviation σ and zero mean value. If the accuracy of the regression model is good the residuals are normally distributed. In addition to this, as the residuals of vectorial quantities are vectors, they can generate a bivariate normal distribution.

Therefore, the confidence region of residual data can be defined with the same mathematical techniques used to evaluate the confidence region of raw monitoring data. Then, the alert and danger acceptance regions of raw monitoring data can be replaced by the acceptance regions of the corresponding residual data.

Model identification techniques can be used to select the process parameters included in the model as regressor variables, as well as to determine the relationships between each process parameter and the dependent variable of the regression function [1], [6]. Furthermore, measures of the model adequacy can be carried out in order to validate and optimise the model [1], [6].

3.3 Transient speed data acceptance regions

During run-ups and coast-downs, the vibration amplitudes are speed dependent, because exciting forces depend upon the speed and the frequency response is also variable as a function of the modal parameters of the whole system (rotors and supporting structure). Therefore, the absolute alarm levels of the overall amplitude of vibration signals recommended by the international standards can be exceedingly large over some speed ranges and, in general, not useful for diagnostic purposes. Variable alarm limit values for transient speed data should be defined for each measurement point of a particular machine. Then, the confidence regions of transient data can be evaluated for pre-established values of the shaft angular speed by considering the average frequency response of the machine. The amplitude of the alert and danger acceptance regions can be estimated on the basis of the standard deviation of reference transient speed data collected at each angular speed that has been considered. In order to carry out quick on-line analyses during machine run-ups and coast-downs, the acceptance region of the overall amplitude of vibration only can be considered. However, for more detailed off-line diagnostic analyses, the acceptance regions of vibration harmonic components can be also defined.
In transient condition the vibrational behaviour of rotors and supports may be significantly affected by the machine thermal state, so that, the average frequency response must be estimated by considering transient data collected in similar thermal conditions. Furthermore, the frequency response may be influenced by the angular acceleration of the shaft occurred during the transient speed operating condition. Thus, the monitoring data collected during run-ups and coast-downs should be analysed separately.
In figure 8 the danger acceptance region of the amplitude of the 1X vibrations measured on a bearing of a low pressure turbine (LP) is shown. This region has been evaluated considering the transient data collected during coast-downs with the machine in "warm" thermal state. The dotted line represents the average frequency response (1X) of the rotor, that is, the amplitude and phase reference curves. The half-amplitude of the confidence region was three times the standard deviation of the reference data which correspond to the angular speed values that have been considered. Even if the average frequency response of the rotor shows sharp changes due to the machine resonances, the amplitude of the danger acceptance region is small enough over the whole range of the angular speed.

Once again, the acceptance regions of transient data can fulfil the requirements of the diagnostic techniques making it easier to early detect abnormal changes in the machine vibrations with respect to reference values. However, the acceptance regions must be defined by selecting the transient data with care, otherwise false alarms could be caused.
It is important to outline that the diagnostic information obtained by the analysis of transient speed data are usually more significant in order to detect the causes of the machine malfunction conditions. For example, abnormal changes in the 2X vibration data occurred during a slow coast-down, with the rotor surface getting cold, can clearly point out the presence of a rotor crack.

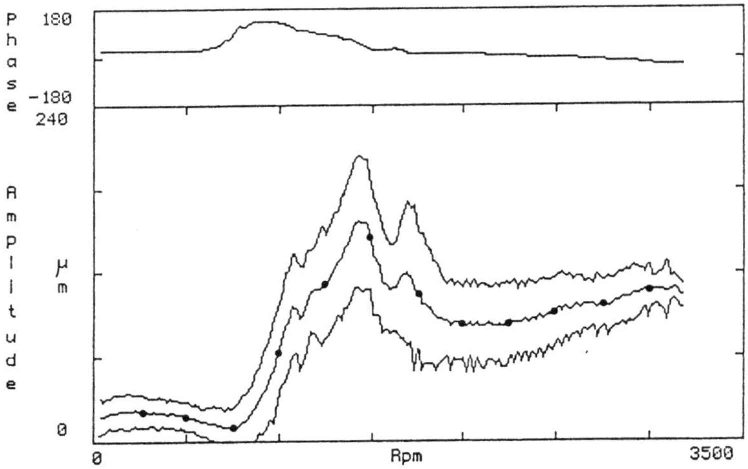

Figure 8 Average frequency response (1X) with the machine in "warm" thermal state (dotted lines). Rotor vibrations (1X) measured on bearing #4 (LP turbine) during coast-downs. Upper and lower control limit curves (solid lines).

4. TREND ANALYSIS

Trend analysis of monitoring data are usually employed in order to point out the development of machine faults, as well as to predict the time interval after which an alarm limit will be exceeded. When the machine behaviour is significantly affected by some process parameters, normal changes in the monitoring data caused by variations in the plant operating condition could be considered themselves as the consequence of the development of a machine fault. Therefore, even for trend analyses, only the monitoring data collected in similar operating conditions should be processed.

For turbogenerator units, many malfunctions can be pointed out by trend analyses carried out using the data collected over a long period of time. Therefore, the monitoring data can be compressed by applying statistical techniques which allow to estimate the mean value and the standard deviation of the data, associated to each typical operating condition, collected during pre-established time intervals (i.e. one day, one week, one month). The figures 9 and 10 show the trend plots of the 1X vibration vector of the rotor of a HP-IP turbine associated to two different ranges of variations of the load. These two quantities were found to be significantly correlated. The comparison between the two trend plots (figures 9 and 10) of the data collected in similar operating conditions during nearly two months confirms that the amplitude of the 1X vibrations were highly influenced by the load, while the vibration changes over the time were less important.

This kind of trend analysis is a very simple but effective tool for evaluating meaningful changes of the reference data over the time. The observations collected during the first monitoring stage in each typical operating condition must be really representative of the machine behaviour in order to be assumed as reference data. This requires that only small changes of data occur during the first monitoring phase. In figure 11 the trend plot of the 1X vibration data of the rotor of a LP turbine collected during a period of nine months is shown. These data were associated to load variations from 300 MW to 330 MW. The trend of the mean values of the 1X vibration amplitude shows that no significant changes in the machine behaviour had occurred during the initial monitoring phase. Therefore, this data can be considered as reference data.

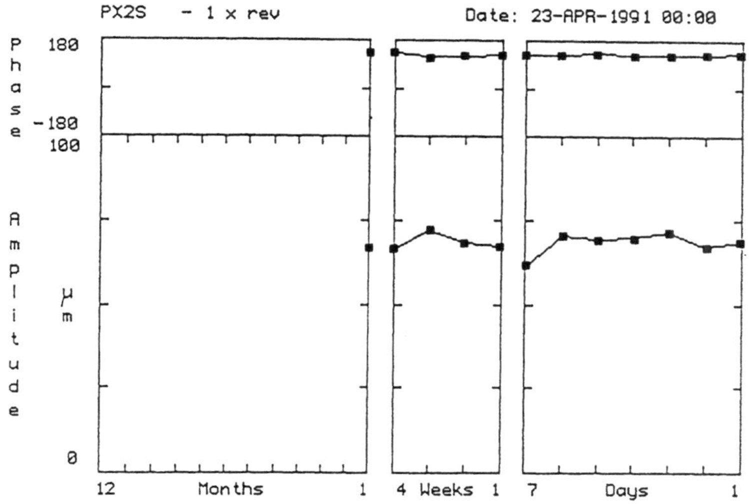

Figure 9 Trend plot of rotor vibrations (1X) measured on bearing #2 (HP-IP turbine) with load ranging from 300 MW to 330 MW (averaged data).

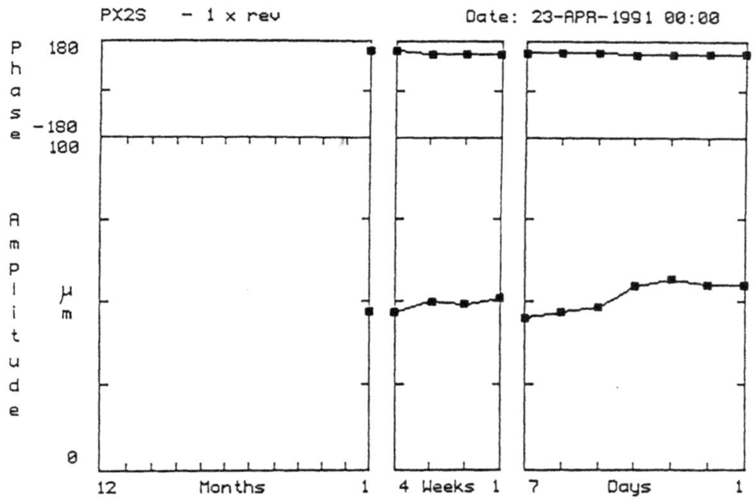

Figure 10 Trend plot of rotor vibrations (1X) measured on bearing #2 (HP-IP turbine) with load ranging from 240 MW to 280 MW (averaged data).

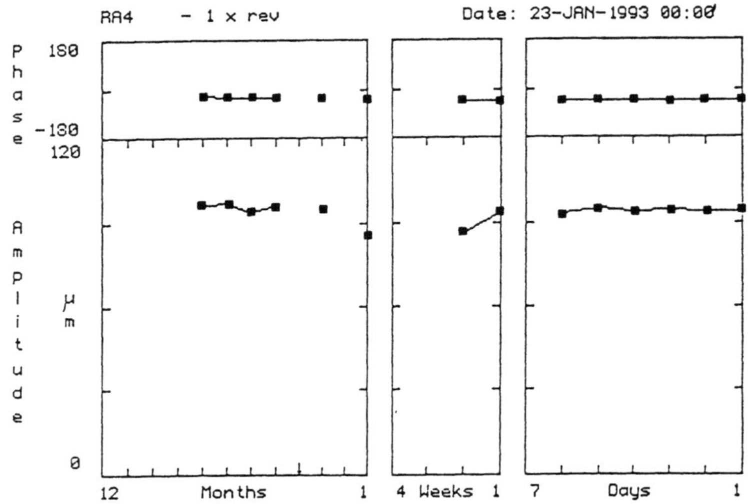

Figure 11 Trend plot of rotor vibrations (1X) measured on bearing #4 (LP turbine) with load ranging from 300 MW to 330 MW (averaged data).

5. CONCLUSIONS

The main purpose of the continuous monitoring of rotating machinery is the early detection of fault conditions. This goal can be achieved by defining alert and danger acceptance regions, for each monitored quantity, evaluated on the basis of reference data collected during a first period of monitoring with the machine in normal state. The acceptance regions of vibration and process parameters data are a characteristic of each machine and represent alarm levels that are more restrictive than the values recommended by international standards. The early detection of machine malfunctions and the identification of the kind of fault make it possible both to carry out predictive maintenance operations and to improve safety and availability of the plants.

When rotor and support vibrations are significantly affected by some process parameters, the acceptance regions depend on the machine operating conditions. This may involve the evaluation of alarm levels on the basis of reference data collected in similar operating conditions. When high correlation exists between vibration data and process parameters, good results can be obtained by multivariate regression analyses. In this case, model identification techniques must be used in order to select the regressor variables among the monitored process parameters, as well as to determine the actual relationship between vibration data and each independent variable included in the model.

The results obtained with the diagnostic techniques proposed in this paper may depend on the characteristics of the reference data distribution, that is on the correlations between vibrations and process parameters. Therefore, a methodology for the automatic evaluation of the acceptance regions may be unsuitable.

Finally, the detection of alarm conditions carried out by verifying if current data are inside their respective acceptance region can provide useful information for diagnosing the causes of the machine faults. Subsequently, these diagnoses can be validated by using rotordynamic models.

Some results obtained by two experiences in monitoring large turbine-generator units are shown in the paper. The advantages of different diagnostic techniques used to define acceptance regions of monitoring data collected both in steady state and transient conditions are illustrated.

ACKNOWLEDGEMENTS

Support of M.U.R.S.T. (Ministero della Università Scientifica e Tecnologica), of C.N.R. (Consiglio Nazionale delle Ricerche) and of ENEL S.p.A. (Italian Electrical Company) is gratefully acknowledged.

REFERENCES

[1] W.W. Hines, D.C. Montgomery, Probability and Statistics in Engineering and Management Science, John Wiley & Sons, 3rd Edition, 1990.

[2] A.M. Mood, F.A. Graybill, D.C. Boes, An Introduction to Statistics, McGraw-Hill, 1988.

[3] D.L. Thomas, "Detecting Changes in Machine Vibrations". VDI Conference on "Schwingungsuberwachung von Maschinen", Hamburg, Germany, 1985. V.D.I. Berichte 568, 31-52.

[4] D.L. Thomas, "Turbine Generator Vibration Monitoring in the CEGB", IFToMM International Conference on Rotordynamics, Tokyo, 1986.

[5] A. Curami, M. Gasparetto, A. Vania, "An On-line Application of a Turbomachinery Supervisory System", IFToMM International Conference on Rotordynamics. Lyon, 1990.

[6] A. Vania, "Identification of Models in Rotormachinery Diagnostics", Politecnico di Milano, Department of Mechanics, Internal Report, 1992.

THE STUDY OF THE ROTOR PUMP VIBRATIONS

J. Wawrzecki and J. Turant
Technical University of Lodz, Lodz, Poland

ABSTRACT

In the paper the results of investigations carried out by the computer simulation method, the aim of which was to determine the effect of the supporting structure elements of the machine on the rotor vibrations have been presented. The dynamic characteristics concatenation receptance method has been employed. The rigid finite element method has been used to determine the receptance of particular elements.

1. INTRODUCTION

In a model of the rotor-machine designated as a system A, the rotor unit as a subsystem B and a supporting structure as a subsystem C have been isolated, Fig. 1. In subsystem all the elements of the bearings, bodies, screws, the isolated, Fig. 2 [1]. Each of these elements is defined by the dynamic characteristics which together constitute the characteristics of the subsystem in the form of the receptance matrix. The dynamic properties of the system are estabilished basing on the dynamic characteristics obtained from the concatenation of characteristics of the subsystems. This method can be used for investigating vibrations of various rotor-machines. The possibility of investigating the effect of particular elements of the

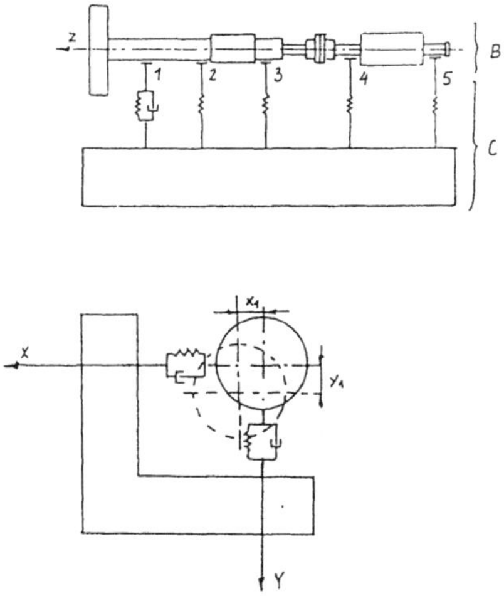

Fig. 1, The model of the system.

supporting structure on the dynamic state of the machine by the computer simulation method in of significance. The possibility of selection of particular elements of the machine structure is important for the constructors of such machines.

2. METHOD

The object of the investigations using the above method is a rotor pump unit of 5AN20 type with an electric motor Sf100L4A, working at a velocity n=1500 rpm^{-1}, which corresponds to the working frequency ω=157 s^{-1} (f=25 Hz). A model of this machine is shown in Fig. 3.

The configuration of the metrix equation has the form:

$$\mathbf{x}_{B1} = \left(\mathbf{I} + \mathbf{B}_{11}\mathbf{C}_{11}^{-1}\right)^{-1} \mathbf{B}_{10}\mathbf{F}_{B} \tag{1}$$

where

$$\left(\mathbf{I} + \mathbf{B}_{11}\mathbf{C}_{11}^{-1}\right)^{-1} \mathbf{B}_{10} = \mathbf{A} \tag{2}$$

is the matrix of unit receptance:
$\boldsymbol{F}_B$ is the matrix of exciting forces.

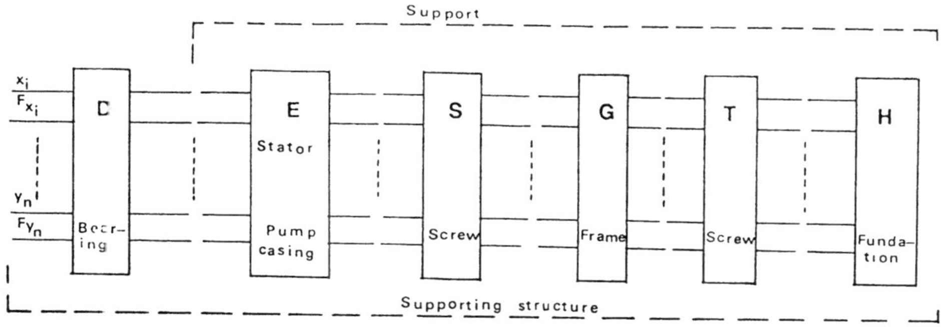

Fig.2, The diagram of the subsystem C.

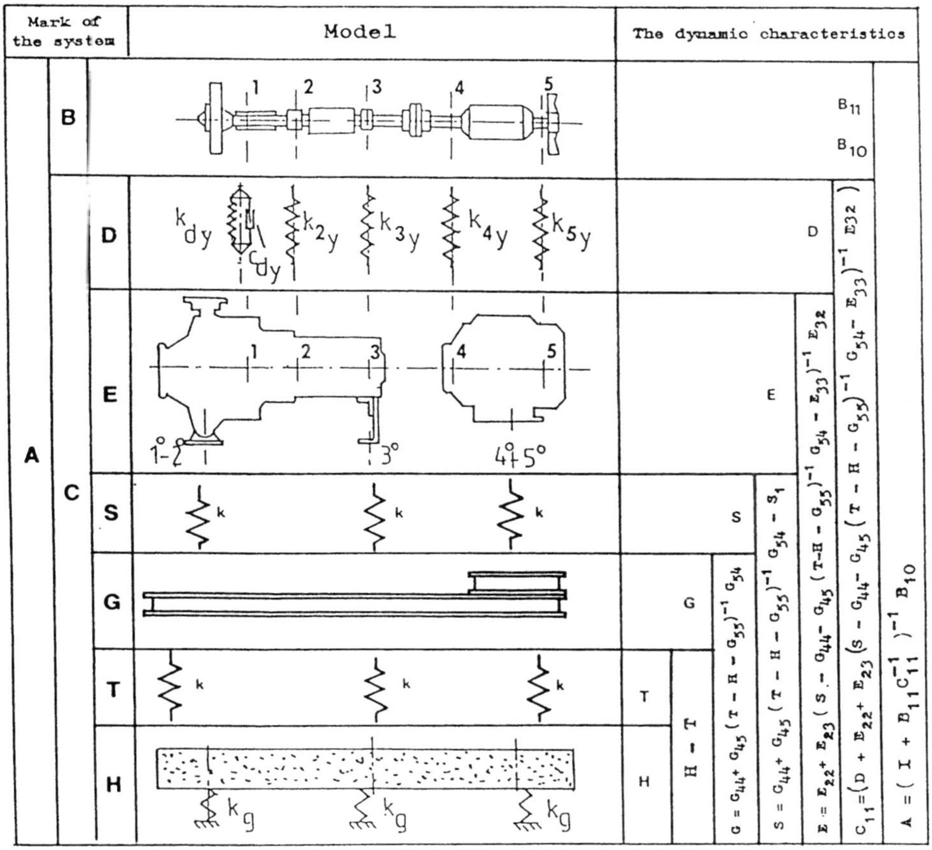

Fig.3, The model of the pump.

The meaning of the remaining symbols is as follows:

$\mathbf{B}_{11}$ a 10x6 receptance matrix of subsystem B,

$\mathbf{B}_{10}$ a 10x10 receptance matrix of subsystem B,

$\mathbf{I}$ elementary matrix

$\mathbf{C}_{11}$ a 10x10 receptance matrix of subsystem C.

The configuration matrix of the supporting structure matrix has the form:

$$\mathbf{C}_{11} = \left(\mathbf{D} + \mathbf{E}_{22} + \mathbf{E}_{23}\left(\mathbf{S} - \mathbf{G}_{44} - \mathbf{G}_{45}\left(\mathbf{T} - \mathbf{H} - \mathbf{G}_{55}\right)^{-1}\mathbf{G}_{54} - \mathbf{E}_{33}\right)^{-1}\mathbf{E}_{32}\right) \tag{3}$$

where:

$\mathbf{D}$ a 10x10 diagonal matrix of the bearings and gland,

$\mathbf{E}_{22}$, $\mathbf{E}_{23}$, $\mathbf{E}_{33}$, $\mathbf{E}_{32}$ matrices of the casings of the measurements: 10x10, 10x14, 14x14, 14x10 which depend on the number of the joint points of the subsystem with the neighbouring subsystems,

$\mathbf{G}_{44}$, $\mathbf{G}_{45}$, $\mathbf{G}_{55}$, $\mathbf{G}_{54}$ 14x14, 14x12, 12x12, 12x14 receptance matrices of the form which shows the effect of the forces on the displacement of the frame mount points,

$\mathbf{H}$ a 12x12 square receptance matrix of the foundation,

$\mathbf{S}$ a 14x14 diagonal receptance matrix of the casings joined to the frame,

$\mathbf{T}$ a 12x12 diagonal receptance matrix of the frame joined to the foundation.

Receptances of all the matrices were determined experimentally or analytically. Receptances of the pump bearings were determined experimentally [3], receptances of the gland were calculated basing on the dumping coefficients c d and the rigidity constants k d of the gland determined experimentally.

Receptances of the rotor [1] and receptances of the frame [4] were determined by the rigid finite element (RFE) method. An actual frame was replaced with a calculation model consisting of 29 rigid elements (RE) and 31 elastic elements (ES). Each part of the frame structure is a channel bar and each RE has 6 degrees of freedom. RFE method is used to determine the matrix of rigidity and the matrix of inertia of the whole frame as an element G of the subsystem, Fig.4. Receptances of the bodies were calculated according to the formulae derived [4]. Receptances of the screw joints were determined basing on the screw joint rigidity. The foundation block as a rigid solid supported elastically was described with differential equations of motion of the mass centre. Receptances were determined in the places of the foundation and frame fastening. The foundation block of the dimensions: width - 0.4 m, height - 0.3 m, length - 0.8 m, was conventionally supported on five springs modeling the ground.

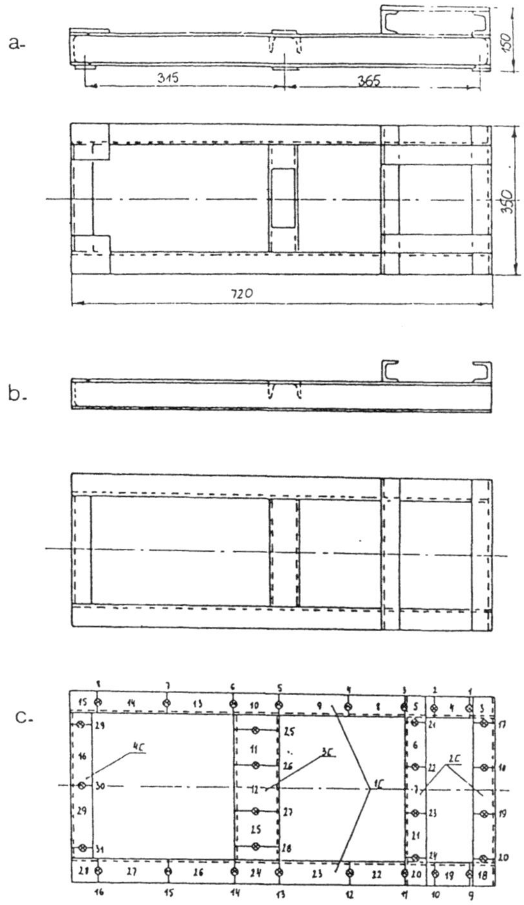

Fig.4, Frame and model of the frame.

3. EXPERIMENTAL INVESTIGATIONS

Investigations of the frame structure were carried out over the frequency of 20 to 1700 rad^{-1}. The system was assumed to be loaded with unbalance forces determined acc. to the VDI findings on the grounds of studies of the Schenck firm. Calculations of the unit rotor vibrations were carried out when the pump and the engine were mounted on the foundation without a frame, and only the screw joint rigidity was taken into consideration. Further investigations referred to calculations of ten different frame structures with an invariable mass of the foundation and an invariable manner of joining the frame to the foundation.

Calculations of the rotor unit vibration amplitudes were per- formed using a POMPROM program in Turbo Pascal 5.0 on an IBM PC microcomputer. The program contained two basic possibilities of calculations of: vibrations of the rotor of the unit mounted on the frame and foundation, vibrations of the rotor of the unit mounted directly on the foundation. The diagrams of vibration amplitudes at points of support of the rotor 1,2,3,4,5 of the system without a frame is shown in Fig. 5 and for four frame structures in Fig. 6.

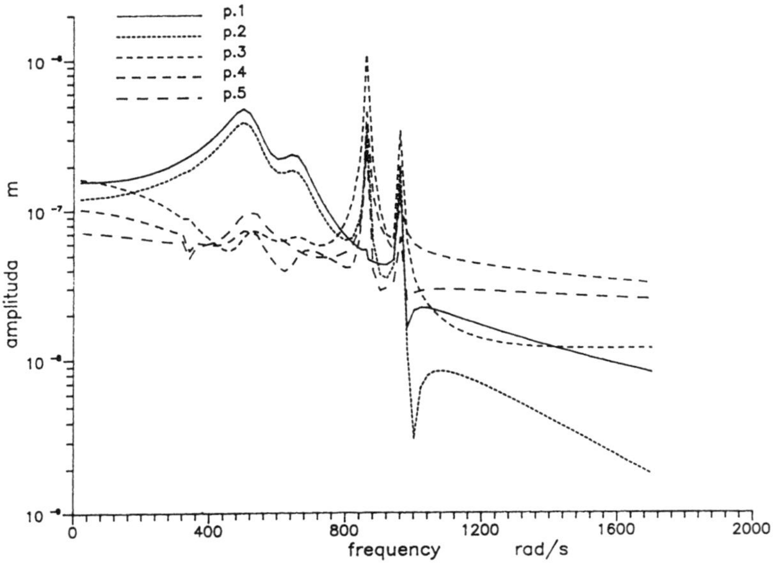

Fig.5, The diagram of the resonance vibrations rotor without a frame.

The frames were made from channel bars. The cross-section dimensions of the side beams of 720 mm were varied. At the working frequency the rotor unit satisfies the dynamic criterion $\omega_r \leq 50\% \; \omega_{kr1}$ and can be considered a rigid rotor.

4. CONCLUSIONS

It can be stated that the use of a frame as a separate structure causes a decrease in the pump rotor vibrations at the working frequency when the unit is mounted directly on the foundation. It results from the calculations that, out of the structures studied, only the frames of beams from channel bars [50 and [80 cause a considerable reduction in vibrations (at the working frequency). A further increase in the cross-sections is of no practical significance and only increases the weight of the structure.

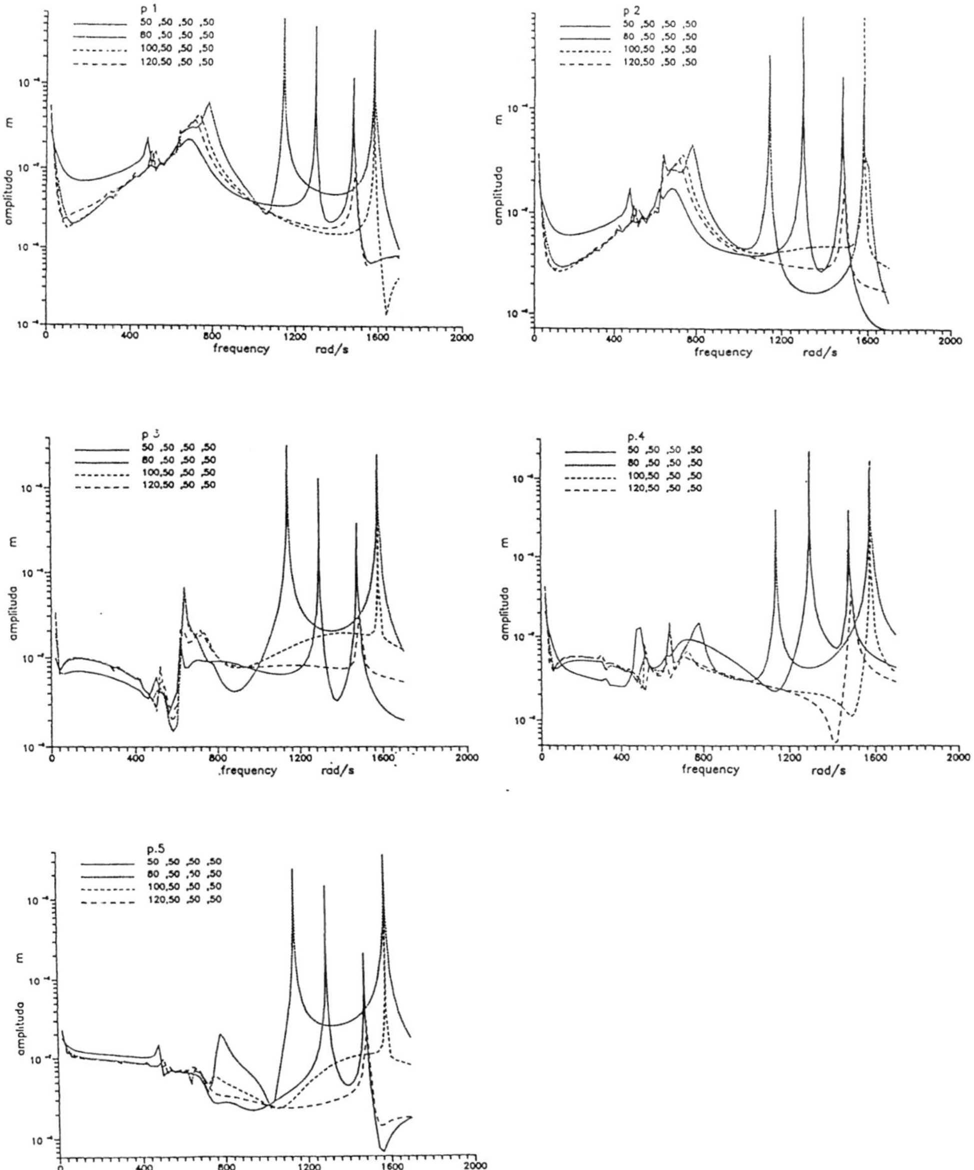

Fig.6,The diagram of the resonance rotor vibrations in the bearings for different construction frame.

REFERENCES:

1. Wawrzecki, J.: Influence of the supporting structure of the pump rotor upon its vibrationns. Proceedings of the Sixth World Congress on the Theory of Machines and Mechanisms. December 15-20 (1983) New Delhi, India Vol.II, 1366-1369.
2. Wawrzecki, J.: Bestimmung der Charakteristik der dynamischen Stütz konstruktion einer Rotorpumpe. 2 Fachtagung mit internationaler Beteiligung "Maschinendynamik - Ferschungsergebnisse zur Unterstützung der Entwicklungsarbeit im Werkzeug und Verarbeitungs maschinenbau". Vom 8-10 Juni 1982 in TH Karl-Marx-Stadt.
3. Wawrzecki, J.: Studies on the Effect of the Support Elements on the Pump Vibrations. Scientific Fascicles of the Technical University of Łódź", No 472 Mechanics 71 (1986), 43-54.
4. Wawrzecki, J. and Turant, J.: The Study of the Influence of the Supporting Construction on Rotor Pump Vibrations. Scientific Fascicles of the Technical University of Łódź", No 635 Mechanics 82 (1993), 70-81.

DIE UNTERSUCHUNG DES EINFLUSSES DER STÜTZUNGKONSTRUKTION AUF DIE SCHWINGUNGEN DER ROTOR PUMPE.

Zusammenfassung

In dieser Arbeit wurden die Ergebnisse der Untersuchungen der Methode der Computers Simulation vorgelegt die zur Feststellung des Einflusses der Rahmenkonstruktion auf die Schwingungen der Rotorpumpe geführt haben. Man benutzte die dynamischen Charakteristiken, deren Bildungsweise in der Bearbeitung geschlidert wurde [1] und [3].

Technische Hochschule Łódź.

AN ON-LINE TECHNIQUE TO MONITOR AND CONTROL THE PERFORMANCE OF PLAIN BEARINGS IN LARGE INSTALLATIONS

S.K. Roy Chowdhury and S. Chandra
Indian Institute of Technology, Kharagpur, India

Abstract

A properly designed hydrodynamic journal bearing should ideally operate smoothly through out its design life. However, they do fail prematurely due to a number of reasons including shaft misalignment, incorrect choice of lubricant and faulty assembly. A need therefore exists for an on-line monitoring and control of plain bearings particularly in larger installations such as a power plant. An alternative to widely used vibration monitoring is to use direct parameters such as film thickness.

The paper describes a feed back control strategy for such bearings using film thickness as the controlled and shaft speed as controlling parameters. An experimental arrangement is developed where the film thickness is continuously monitored while a computer-controlled feed back circuit assesses the parameter and adjusts the shaft speed accordingly. The procedure provides a continuous cheek on the performance of the bearing such that in the event of complete film breakdown pre-emptive measures may be taken to avoid seizure.

An analysis of the stability of the control system is also given. The analysis provides us with the necessary gain of the control module for stability.

1. Introduction

A correctly run-in plain bearing should ideally perform well throughout its designed life and a little wear is expected due to the presence of a relatively thick hydrodynamic oil film under suitable load and speed conditions. They do, however, fail in practice with catastrophic consequences. Such lubrication failures can often be attributed to reasons such as misalignment, incorrect choice of lubricant and materials, incorrect assembly and other design errors. Routine preventive maintenance schedules are useful in ensuring increased plant and component life but are not suitable for predicting or forewarning a failure. A need therefore exists to monitor and control the performance of plain bearings, particularly in large installations.

On-line condition monitoring techniques, hitherto used in predicting bearing failure is largely based on vibration analysis. Tribological parameters such as bulk temperature rise, friction and wear are also useful monitoring parameters but these are either insensitive for developing a control system or are difficult to measure on-line in real situations. An alternative and probably more direct method of monitoring performances of plain bearings based on measurement of oil-film thickness can be contemplated. Although a great deal of work has been carried out to develop different methods of film thickness measurement little attention has been paid to develop techniques of adaptive control using these methods.

Techniques of measuring oil-film thickness based on resistance and capacitance measurement [1,2], optical interferometry [3], UV or laser excited oil fluorescence [4] and R-C oscillation circuit [5] have been developed and Rowe et al and others [6,7] successfully used the resistance method in their study of running-in of plain bearings. The present work deals with the development of a control system using a standard film thickness probe. The object is to assess the film thickness continuously and control a suitable output parameter such as, journal speed to maintain a reasonable film thickness at all times. This implies that a bearing is required to operate between points such as B and C on the well known stribeck curve (Figure 1). Below point B surfaces may not be completely seperated unless EHL conditions are obtained and above point C churning losses may be excessive. Clearly the method is predictive in nature while the methods using bulk temperature, vibration etc. rely on certain amount of damage already taken place in the bearings.

2. Experimental Arrangement

2.1 Bearing test rig

The experimental arrangement uses a standard journal - bearing assembly and essentially consists of a shaft supported at one end by two well spaced deep-groove ball bearings. The shaft is over hung at the other end and supports the test bearing. The test bush is press fitted in a steel housing and the assembly is mounted on the shaft with two thrust bearings at the ends to provide free rotational movement of the bearing while the movement in the axial direction is constrained. The other end of the shaft is coupled to a fractional horse power motor (0.375 KW) with a speed control module. The arrangement is shown schematically in Figure 2. The small vertical movement of the bearing is measured by an inductive type displacement transducer placed at an estimated position of minimum film thickness and this gives a measure of the changes in film thickness. The

B = BOUNDARY LUBRICATION
M = MIXED LUBRICATION
EHL = ELASTOHYDRODYNAMIC LUBRICATION
FH = FULL HYDRODYNAMIC LUBRICATION

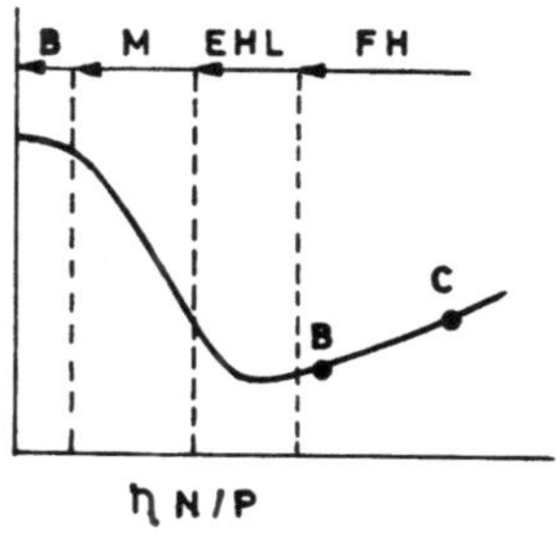

Fig. 1 Typical curves of μ vs. η N/P

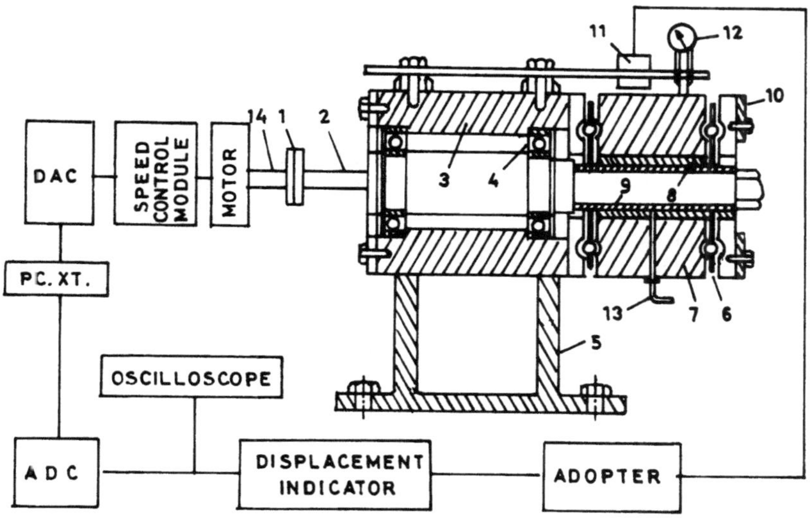

1. COUPLING. 2. MAIN SHAFT 3. TUBE 4. BALL BEARING 5. STAND 6. THRUST BEARING 7. TEST BEARING BLOCK 8. TEST BUSH 9. STUB SHAFT 10. RETAINER PLATES 11. DISPLACEMENT TRANSDUCER 12. DIAL GAUGE 13. OIL INLET 14. MOTOR SHAFT

Fig. 2 Schematic diagram of experimental arrangement

calibration of the transducer is shown in Figure 3. The frictional forces can be measured by arresting the rotational movement of the bearing by a cantilever strip carrying strain gagues (not shown in the figure). However, the choice of the monitoring parameter in the present study is such that the frictional measurement is of limited use.

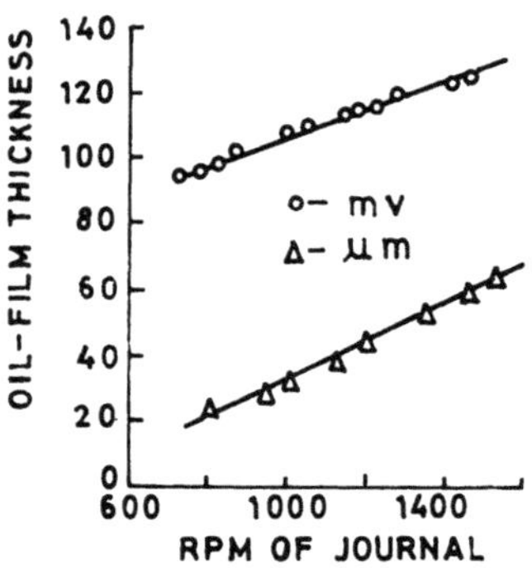

Fig. 3 Experimentally obtained variation of filmthickness with shaft speed

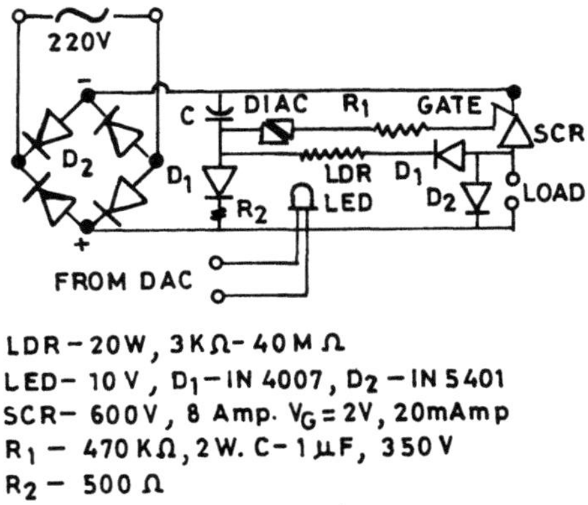

Fig. 4 Circuit diagram of speed control module

2.2 Control circuit layout

Since the objective here is to develop a self-contained bearing unit which corrects itself against any breakdown of fluid-film lubrication the control circuit is essentially a closed-loop one and this is also shown schematically in the Figure 2. The analogue signal from the displacement transducer is fed to the comparator, a PC-XT in this case, via an analogue-to-digital converter (ADC) and the theoretically calculated film thickness is taken as reference input. The difference in the measured and calculated values of film thickness gives the error signal. In the event of a large negative error signal the film thickness may be restored by either increasing the oil viscosity or shaft speed or decreasing the bearing load. Since a moderate speed variation may often be tolerated even in a realistic situation the error signal is fed to a speed control module via a digital-to-analogue converter (DAC) and the journal speed adjusted accordingly.

The standard ADC and DAC used for the present purpose have accuracy of 0.01%, convertion time between 1 μS to 1 mS, analogue input and output voltages of 0-5V and 0-10V respectively. The gains of the ADC and DAC were found from the linear calibration plots to be 819 and 2.44×10^{-3} respectively.

Since the output from the DAC ranges between 0-10V it is necessary to amplify the output to the range of 0-220V to run the drive unit which is a seperately excited DC motor. An amplifier circuit consisting of a LED-LDR unit, SCR and a bridge rectifier was developed to meet the relatively high power rating (Figure 4). The external supply to the SCR is obtained from a bridge rectifier and the SCR is triggered by pulses from the R-C circuit via a diac (32V). The designed values of the components from elementary theory are shown in Figure 4 and the gain of the module from the experimental calibration plot was found to be around 23 volt/volt.

3. System Analysis

Reduction of system error is one of the important objectives of the feed back system and an analysis is required to determine the performance characteristics such as stability, overall gain and sensitivity. Cause and effect relationships of all the sub-system variables are required for formulation and the gain of the forward path may be expressed in terms of the gains of the DAC, speed control module, motor and the plant as

$$G(S) = G_{DAC}(S) \cdot G_C(S) \cdot G_m(S) \cdot G_P(S) \qquad (1)$$

The plant gain G_p may be obtained from the slope of the linear zone of film thickness vs. Sommerfeld's number plot. For finite bearings this is approximately given by $5.7 \times \{\eta LR/W\} \times (R/C)^2$ in m/rpm units where η is the viscosity of the oil, L the bearing length, R the radius, C the radial clearance and W the load. An armature control of the motor is considered here for its low time constant and from the standard analysis the motor gain may be written in usual notations as

$$\frac{\omega(S)}{V_a(S)} = \frac{K_t}{L_a J_m S^2 + (R_a J_m + B_m L_A)S + (B_m R_A + K_b K_t)} \qquad (2)$$

where ω is the angular speed of the motor, V_a the armature voltage, J_m the inertia of the motor, B_m the viscous frictional coefficient, K_b and K_t the proportionality constants given by $V_a/(\phi\ \omega)$ and T/i_a respectively, ϕ being the flux, T the torque and i_a the armature current. The experimental gain of the motor is 7.57 rpm/volt.

Transfer function of the feed back path is given by

$$H(S) = H_T(S) \cdot H_{ADC}(S)$$

where $H_T(S)$ and $H_{ADC}(S)$ refer to the transfer functions of the transducer and ADC respectively. The transducer gain from the Figure 3 is found to be 3.9 x 10^3 volt/m. The closed loop transfer function is G(S)/{1 + G(S).H(S)} and the characteristic equation is

$$1 + G(S) \cdot H(S) = 0 \tag{3}$$

Simplifying equation (2) and substituting gains of DAC, speed control module, plant and the feed back path as 0.0024, K_2, K_4 and 3.2 x 10^6 equation (3) reduces to

$$AS^2 + BS + (D + 9.36 \, K_2K_4K_t) = 0 \tag{4}$$

where $A = L_aJ_m$, $B = R_aJ_m + B_mL$ and $D = B_mR_a + K_bK_t$. From Routh's criterion we may readily show that for stability of the system we have

$$D + 9.36 \, K_2K_4K_t > 0 .$$

Clearly, the stability condition depends on the bearing load along with other system parameters and if, for example, we assume typical values of the parameter as L_a = 8.6 m Henry, J_m = 8 x 10^4 kg m^2, $K_b = K_t$ = 3.3 x 10^{-3}, B_m = 0.025 NS/m^2, R_a = 5.2Ω and the bearing parameters as C = 50 μm, R = 17.8 mm, L = 38 mm and η = 0.025 NS/m^2, we have $K_2 > -4.33 \times 10^{-5} W$. This implies that for all practical values of bearing load the speed control module gain is positive, indicating that the design of the speed control module in suitable.

With the assumed value of the parameters D in equation (4) is relatively small and the characteristic equation of the closed loop transfer function may be written as

$$S^2 + 2\xi\omega_n S + \omega_n^2 = 0$$

where ω_n, the natural frequency of the system and ξ, the damping ratio are given by

$$\omega_n = \pm \left[\frac{4.35 \times 10^8 \, K_2}{W} \right]^{1/2} , \quad \xi = \frac{318}{\omega_n}$$

This implies that for the experimental value of K_2 = 23, ξ lies between 0 and 1 for all reasonable bearing loads upto ~ 70 kN. The system is therefore under damped. In principle, for good performance a damping ratio between 0.5 and 0.8 is suitable and in the present case the ratio with the assumed values of system parameters is probably low. This may however be corrected by decreasing the bearing characteristic number $(\eta LR/W)(R/C)^2$, gain of the speed control module or adjusting the motor parameters [see equation (4)].

4. Application of Control Strategy

A simple programme that uses the I/O port directly for greater flexibility was used to apply the control strategy for experimentation. The reference input h_{ref} is obtained here from the standard plot of non-dimensional load $\bar{W} = WC^2/(\eta UR^2 L)$ vs. ε, the eccentricity ratio since $h_{min} = (1-\varepsilon)c$. The variation is considered to be linear at $\bar{W} <$ 2.5 and logerithimic beyond this value. Since the shaft speed is the controlling parameter a high and low cut-off limits (0.7 - 0.3) of ε was incorporated. At low values of ε metal-to-metal contact is likely while at high values of ε instability may occur. The programme gives a warning in the event of film thickness approaching unacceptable level and also a provision for temporary stoppage till the bearing cools down.

Initial experimental results are encouraging. Constant load test at 40N with the control circuit on shows that the speed initially shoots upto a high value, oscillates and then reaches the steady set value. At a constant load of 80N the speed initially oscillates about the set value and then becomes steady. This probably indicates the stability of the system (Figure 5).

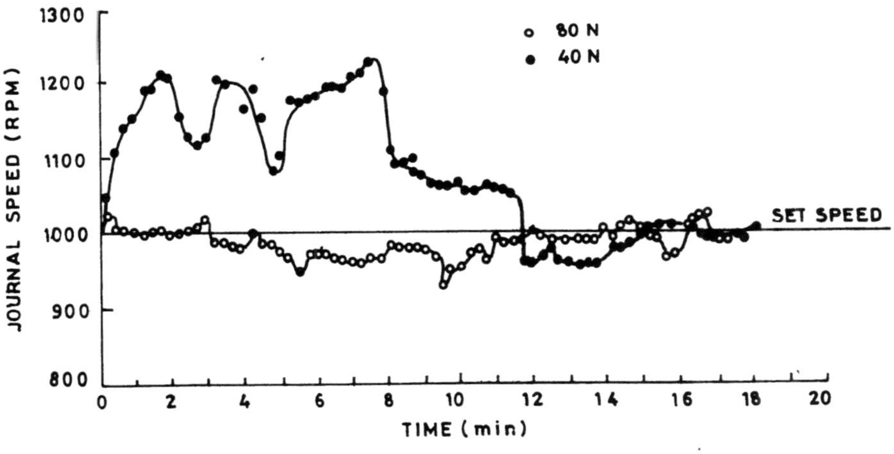

Fig.5 Variation of controlled journal speed with time at a constant load

Another test with step increase in bearing load shows that the journal speed increases with the increase in bearing load in order to maintain an adequate film thickness, implying that the control circuit works well (Figure 6).

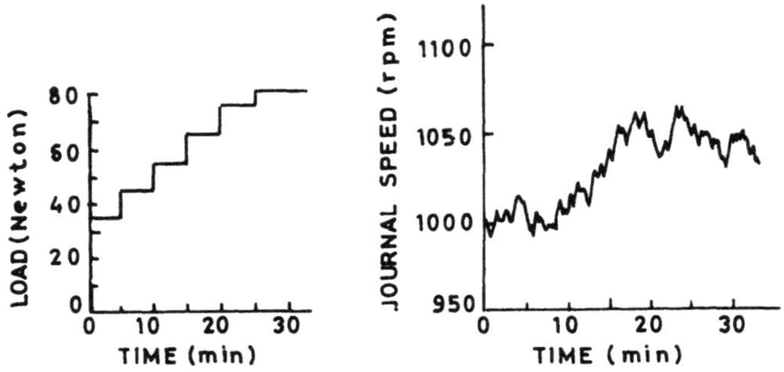

Fig.6. Controlled journal speed vs. time plot for a step-loading bearing

5. Conclusions

The system developed for monitoring and controlling the bearing performance works satisfactorily. The programme developed takes care of all the possible variations in film thickness during operation and provides correction. A dynamic analysis with typical values of the control system-parameters shows that the system is stable and the design of the speed control module is suitable. The shaft speed is used here as controlling parameter purely for demonstation purpose but other alternatives, such as, hybrid type of bearing arrangements may also be considered. In tribological systems such on-line monitoring is rare and the present work is a modest approach to set a trend in this direction. Such systems are considered to be of much use in large installations due to the availability of advanced warning of failures.

References

1. Bowden, F.P.,L. Leben and D. Tabor :Trans. Faraday Soc., 35 (1939), 900.
2. Archard, J.F. and M.T. Kirk : Proc. Roy. Soc. A, 261 (1961), 532-550.
3. Krauter, A.I. : Trans. ASME, J. Lubr. Tech., 104 (1982), 455-459.
4. Ting, L.L. : ASME Trans., 102 (1980), 165-170.
5. Zhang Peng-Shun, Li Shu - Guang and Zhai Wen-Jie, Wear, 148 (1991), 39-46.
6. Rao, B.K.N., H. Kaliszer and G.W. Rowe : Wear, 73 (1982), 157-162.
7. Rowe, G.W., H. Kaliszer and S.K. Roy Chowdhury : Trans. ASLE, 26 (1982), 270-276.

FSC www.fsc.org
MIX
Papier aus verantwortungsvollen Quellen
Paper from responsible sources
FSC® C105338

If you have any concerns about our products,
you can contact us on
ProductSafety@springernature.com

In case Publisher is established outside the EU,
the EU authorized representative is:
Springer Nature Customer Service Center GmbH
Europaplatz 3, 69115 Heidelberg, Germany

Printed by Libri Plureos GmbH
in Hamburg, Germany